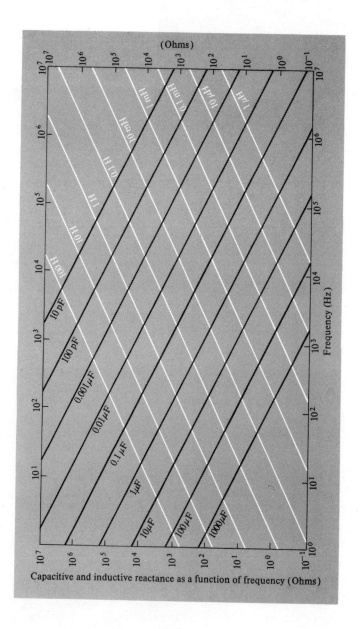

Capacitive and inductive reactance as a function of frequency (Ohms)

Oxford Physics Series

General Editors

E. J. BURGE   D. J. E. INGRAM   J. A. D. MATTHEW

Oxford Physics Series

1. F. N. H. ROBINSON: *Electromagnetism*
2. G. LANCASTER: *D.C. and a.c. circuits. Second edition*
3. D. J. E. INGRAM: *Radiation and quantum physics*
5. B. R. JENNINGS and V. J. MORRIS: *Atoms in contact*
7. R. L. F. BOYD: *Space physics; the study of plasmas in space*
8. J. L. MARTIN: *Basic quantum mechanics*
9. H. M. ROSENBERG: *The solid state. Second edition*
10. J. G. TAYLOR: *Special relativity*
11. M. PRUTTON: *Surface physics*
12. G. A. JONES: *The properties of nuclei*
13. E. J. BURGE: *Atomic nuclei and their particles*
14. W. T. WELFORD: *Optics*
15. M. ROWAN-ROBINSON: *Cosmology. Second edition*
16. D. A. FRASER: *The physics of semiconductor devices. Second edition*
17. L. MACKINNON: *Mechanics and motion*

GORDON LANCASTER
SENIOR LECTURER IN PHYSICS, UNIVERSITY OF KEELE

dc and ac circuits

SECOND EDITION

Clarendon Press · Oxford · 1980

Oxford University Press, Walton Street, Oxford OX2 6DP
OXFORD LONDON GLASGOW
NEW YORK TORONTO MELBOURNE WELLINGTON
KUALA LUMPUR SINGAPORE JAKARTA HONG KONG TOKYO
DELHI BOMBAY CALCUTTA MADRAS KARACHI
NAIROBI DAR ES SALAAM CAPE TOWN

© OXFORD UNIVERSITY PRESS 1973, 1980
First edition published 1973
Second edition published 1980

Published in the United States by
Oxford University Press, New York

All rights reserved. No part of this publication may be reproduced, stored in a retrieval system, or transmitted, in any form or by any means, electronic, mechanical, photocopying, recording, or otherwise, without the prior permission of Oxford University Press

British Library Cataloguing in Publication Data

Lancaster, Gordon
    Dc and ac circuits. — 2nd ed. — (Oxford
    physics series).
    1. Electric circuits
    I. Title   II. Series
    621.319'2   TK454   80-40060

ISBN 0-19-851849-8 Pbk
ISBN 0-19-851848-X

Printed in Great Britain at the Alden Press
Oxford London and Northampton

Editors' Foreword

THE study of electricity and magnetism must be one of the central topics in any physics course and for this reason two of the 'core' texts of the Oxford Physics Series are devoted to this area. This text on *d.c. and a.c. circuits* is designed to take the school leaver on from ideas that he should already have met, and develop them into a consistent and coherent theory that will both link with other areas of physics, and provide the basis for Robinson's book on *Electromagnetism*, which moves into a second year of undergraduate study.

In order to ensure that all readers will be able to follow the text from the beginning, the first chapter is presented as a quick revision exercise, dealing with basic ideas of electromagnetism presented in a simple and straightforward way, but which most readers will, one suspects, find well worth studying carefully. The subject of electromagnetism can become very theoretical and abstract, and, being well aware of this, Dr. Lancaster takes every opportunity to point out the practical implications of any concepts that are being considered, and quotes actual values of parameters for various materials, whenever possible. In this way the text has more of a ring of experimental reality about it than many others on the subject.

This chapter sequences follow through from a consideration of the theorems associated with d.c. circuits, including the use of mesh and nodal analysis, to the treatment of a.c. circuits where a graphical representation is first discussed, and then the method of complex-number representation explained. The mathematical steps are all presented carefully and logically so that even those with a weaker mathematical background should have no difficulty.

Specific chapters are then devoted to resonance and pulsed circuits, and the more general theorems on four-terminal networks are introduced systematically in the penultimate chapter. The correlation of the different types of circuit parameters together with a discussion of the advantages and disadvantages of each set is particularly helpful in this section.

The final chapter is devoted to a.c. measuring techniques and again

Editors' Foreward

serves to underline the balance between theory and experiment that is maintained in this book, and which should make both it and its subject more understandable and enjoyable to the interested reader.

Note on the second edition

The considerable interest shown in the first edition of Dr. Lancaster's books has encouraged us to go forward with this second edition in which the opportunity has been taken to expand and amplify many of the original features.

Although the general layout of the book has been preserved three chapters have been added, namely 'Electromagnetics', 'Transmission Lines', and 'Transducers', and other chapters have significant extensions. Students should find no difficulty in following the whole text through in a logical and carefully reasoned manner, and the inclusion of additional problems at the end of each chapter should also be of considerable assistance in enabling a careful check to be made on the understanding of the various points which have been covered.

We believe that the text will be appreciated even more than the first edition as covering some of the most important and fundamental aspects of the whole of Physics.

D.J.E.I.

Preface

THIS revised and extended edition of '*d.c. and a.c. circuits*' is based almost entirely on first- and second-year undergraduate lecture courses which I have given in recent years in the Physics and Electronics Degree courses at the University of Keele. The basic properties of electrical circuits are treated ranging from d.c. circuits through lumped parameter a.c. networks to distributed parameter transmission lines. As such the text should be suitable for students who are attending courses in physics, applied physics, physics-with-electronics, and electronics at the level of first- and second-year courses in English universities. Such courses ought to provide a broad and firm foundation for the specialized studies which students undertake in the third year of a degree course and so the simplest examples have been chosen to illustrate the important general features of circuit theory. By this means the amount of mathematical analysis in which this subject is so often enshrouded has been reduced. Of course it is impossible to avoid completely the use of mathematical techniques but in every context the intention has been to emphasise those features which are of practical importance; in particular great emphasis has been laid on the importance of making sensible approximations which will curtail long-winded analyses or enable abstruse analyses to be by-passed.

No attempt has been made, deliberately, to designate individual chapters as 'first-year level' or 'second-year level'; no doubt there would be a large measure of agreement as to the designations of many chapters but for the remainder there will be variations from institution to institution.

Keele   G.L.
November 1979

Contents

1. **FUNDAMENTAL ELECTROMAGNETISM**   1

   Electrostatics. Current electricity. Magnetic effects of electric currents. Electromagnetic induction.

2. **DIRECT-CURRENT CIRCUITS**   22

   Circuit elements. Thévenin's theorem. Kirchhoff's laws. Mesh analysis. Nodal analysis. The superposition and reciprocity theorems. Equivalent circuits. Direct-current measurements.

3. **ALTERNATING CURRENTS**   42

   Transient currents in series $R-C$ and $R-L$ circuits. The graphical and mathematical representation of alternating currents and voltages. Impedance. Admittance. Power in a.c. circuits. Real capacitors and inductors. Transformers. a.c. networks. a.c. bridges.

4. **RESONANCE**   85

   Free, damped oscillations. Forced, damped oscillations. Some practical aspects of resonant circuits. Coupled resonant circuits.

5. **ELECTROMAGNETICS**   99

   Static electric fields in solids. Alternating electric fields and dielectric losses. Types of dielectric. Alternating electromagnetic fields in conductors. Magnetic fields in solids.

6. **PULSES AND TRANSIENTS**   133

   Introduction. Amplitude and phase spectra. Pulse trains. Fourier integrals and transformations. The Laplace transform. Practical details.

7. **NETWORK ANALYSIS AND SYNTHESIS**   169

   Introduction. T and π networks. The interconnection of two-port networks. Equivalent circuits for active devices. Basic features of linear amplifiers. Excitations, system functions, and responses. General aspects of filters. The synthesis of driving point immitances.

8. **TRANSMISSION LINES AND WAVEGUIDES**   230

   Introduction. Transmission line equations. Classes of transmission line. Examples of calculations on transmission lines. Waveguides considered as transmission lines.

9. TRANSDUCERS 272

General features of transducers. Resistive transducers. Capacitive transducers. Piezo-electric transducers. Inductive transducers. Photo-detectors. Magnetic fields.

10. INSTRUMENTATION 294

Voltage and current measurements. Power. Impedance measurements. Frequency measurements. Interference suppression, noise reduction, and signal enhancement.

APPENDIX 1. COLOUR CODES FOR THE VALUES OF RESISTORS AND CAPACITORS 311

APPENDIX 2. PARTIAL FRACTIONS – GENERAL RULES 313

APPENDIX 3. MAXWELL'S EQUATIONS 315

ANSWERS TO PROBLEMS 317

INDEX 323

1. Fundamental electromagnetism

THIS introductory chapter is concerned with the definitions of, and relationships between, those physical concepts which form the essential basis of a study of electric current networks. Thus, for many readers the study of this chapter will be a revision exercise largely but nonetheless important for that particularly as, for some of the topics, the emphasis of the treatment is different to that found in many elementary texts.

1.1. Electrostatics

Up to date it is believed that the smallest electric charge it is possible to isolate has a magnitude equal to that of the ('negative') charge associated with an electron, namely 1.602×10^{-19} coulomb (C). Also the results of experiments indicate that the magnitudes of the positive and negative charges carried by protons and electrons, respectively, are equal to at least 1 part in 10^{19}. It is further believed that in any isolated physical system the algebraic sum of the electric charges remains constant; this is the principle of conservation of electric charge.

If two bodies have charges q_1, q_2 and if the separation between the bodies is much greater than the dimensions of either of them (so that they can be regarded as 'point-sized' bodies separated by distance x, say) then the force of interaction, F, between them is proportional to $q_1 q_2 / x^2$ (Coulomb's law). In SI units with F measured in newtons (N), q_1 and q_2 in coulombs, and x in metres (m), the constant of proportionality is taken to be $(4\pi\epsilon_0)^{-1}$ where ϵ_0, the so-called 'permittivity of free space', has the value 8.854×10^{-12} farad per metre (F m^{-1}) and so we have

$$F = \frac{q_1 q_2}{4\pi\epsilon_0 x^2}. \tag{1.1}$$

1.1.1. Force, field and potential

The experimental fact that an electric charge q_1 experiences a force **F** due to another charge q_2 located some distance away is represented

2 Fundamental electromagnetism

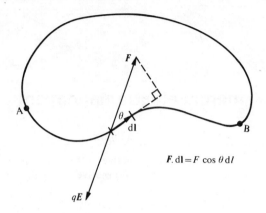

FIG. 1.1. The force required to move a charge in an electric field.

by introducing the concept of an electric 'field' E in the surrounding space. The field E at the site of q_1 due to q_2 is such that the force F on q_1 is given by

$$\mathbf{F} = q_1 \mathbf{E}. \tag{1.2}$$

This relationship defines field as force per unit charge and the units are newtons per coulomb; obviously the field is a vector quantity in the same direction as the force acting on the charge.

If a charge q is moved from point A to point B in an electric field E (see Fig.1.1), then mechanical work is done by the external agency. If there are no dissipative forces (e.g. friction) or acceleration, then the externally applied force F just balances the force $q\mathbf{E}$ and the work done, which equals the change of potential energy of the charge ΔU_p, is given by

$$\Delta U_\text{P} = \int_A^B F \cos\theta \, dl. \tag{1.3}$$

Using eqn (1.2),

$$\Delta U_\text{P} = -q \int_A^B E \cos\theta \, dl. \tag{1.4}$$

The electric potential difference (p.d.) ΔV between A and B is defined through $\Delta V = \Delta U_\text{P}/q$ and the units are joules per coulomb which are called volts (V). The point B is said to be at a higher potential than A if work is done against electrostatic forces in moving a positive charge from A to B; this energy is recoverable if the constraining force is removed and the charge returns to A.

Fundamental electromagnetism 3

If the path from A to B were linear, in the 'x-direction' say, then eqn (1.4) could be written as

$$\Delta U_P = -q \int_A^B E_x \, dx \quad \text{or} \quad \Delta V = -\int_A^B E_x \, dx \qquad (1.5)$$

where $E_x = E \cos \theta$ is the component of E in the x-direction. It can be seen that eqn (1.5) is satisfied by

$$E_x = -d(\Delta V)/dx. \qquad (1.6)$$

This equation and eqn (1.5) are alternative expressions relating electric field and p.d. So it can be seen that alternative units for electric field are volts per metre ($V\,m^{-1}$).

A consequence of the earlier statement about the recoverability of work done against purely electrostatic forces is that $\oint E \cos \theta \, dl = 0$ where $\oint$ indicates a closed path of integration; an electrostatic field is said to be a *conservative field*.

In the analysis of electric circuits it is usual to refer all potential differences to a common reference potential ('earth' or 'common'). Therefore, in the remainder of this text the term potential (symbol V) will be used where, strictly speaking, a difference of potential relative to some named potential is implied.

A law which is of crucial importance is Gauss's law. This states that the integral of the flux of electric field E over a closed surface S (either a real or hypothetical surface) is proportional to the total electric charge in the volume $\mathcal{V}$ which is contained by the surface; symbolically

$$\int_S \mathbf{E} \cdot d\mathbf{S} = \frac{1}{\epsilon_0} \int_{\mathcal{V}} \rho \, d\mathcal{V}. \qquad (1.7)$$

Here ρ is the density of charge (not necessarily constant) within the volume and the scalar product[†] $\mathbf{E} \cdot d\mathbf{S}$ gives the element of outward flux over an elementary area of surface $d\mathbf{S}$[‡] (see Fig.1.2); the element of flux is a scalar quantity which is equal in magnitude to the normal component of E multiplied by the area of the element of surface.

1.2.2. Capacitance

Consider the spherical body of radius x, which has a net charge q_1

[†] The 'dot' product of two vectors A and B is defined to be a scalar quantity of magnitude $AB \cos \theta$ where A, B are the magnitudes of the vectors A, B respectively and θ is the angle between them.
[‡] It is conventional to represent an element of a surface by a vector drawn in the direction of the outward normal to the surface at the site of the element; the magnitude of the vector is made equal to the magnitude of the area of the element, e.g. dS in the example illustrated in Fig.1.2.

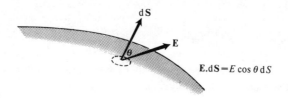

FIG. 1.2 An element of electric flux E.dS is defined as shown.

distributed over its surface. In order to bring a charge q_2, of the same sign as q_1, from an infinite distance and add it to the charge on the body, work must be done against the electrostatic repulsive force. From eqns (1.1), (1.2) and (1.5) the increase in the mutual potential energy of q_1 and q_2 can be calculated:

$$\Delta U_P = -q_2 \int_{\infty}^{x} \frac{q_1}{4\pi\epsilon_0 x^2} \, dx = \frac{q_1 q_2}{4\pi\epsilon_0 x}. \quad (1.8)$$

Since the mutual potential energy of q_1 and q_2 is zero when their separation is infinite, the potential energy of q_2 when it reaches the surface of the charged body is $U_P = q_1 q_2 / 4\pi\epsilon_0 x$ and its potential $V = U_P/q_2$. This must be the potential of all the charges comprising the total charge q_1 since otherwise they would move† so as to adjust their positions on the body until there were no potential gradients (and hence no resultant forces) remaining.

The quantity $4\pi\epsilon_0 x (= q_1/V)$ which is a characteristic property of the spherical body, independent of its charge and potential, is called the capacitance C of the body. The equivalent calculation for a non-spherical body is not so easy but nevertheless the capacitance C of an isolated conductor is defined by

$$C = \frac{q}{V} \quad (1.9)$$

where q is the charge on the conductor, in coulombs, and V is its electric potential relative to an assumed zero of potential at infinity. The unit of capacitance is the farad (F).

In practice, charged bodies can never be completely isolated but are

---

† The material of the body doesn't have to be that of a conventional good conductor, such as copper, necessarily, in order for this to occur. For instance glass when in a normal condition of uncleanliness and dampness is a sufficiently good conductor in this context.

always influenced by other bodies to some degree. For example, if one conductor A has charge q and would have potential V if completely isolated, then in a situation where it will be influenced by other conductors having charges of the same sign as q the potential of A will be raised and its capacitance lowered. Hence, to increase the capacity of the conductor A the proximity of another conductor, or conductors, having charges of the opposite sign to q is required. Suppose A is influenced by one other conductor B only: all the lines of force, or electric field lines, from A terminate on B and the charge on B is equal and opposite to that on A (Gauss's law). Such an arrangement of conductors is called a capacitor. The capacitance of the capacitor is defined by $C = q/(V_A - V_B)$. For an idealized arrangement of two concentric spherical conducting shells of radii a and b ($a < b$),

$$V_A - V_B = \frac{q}{4\pi\epsilon_0 a} - \frac{q}{4\pi\epsilon_0 b}$$

and $C = 4\pi\epsilon_0 ab/(b-a)$.

An ideal capacitor geometry is one in which one conductor is surrounded completely by the other so that their charges are equal and opposite; such an arrangement is impossible in practice because of the resultant inaccessibility of the inner conductor. The common practical forms are the 'parallel plate' and the 'concentric cylinder' (both either air- or dielectric-filled).

Consider a parallel plate capacitor as shown in Fig. 1.3. The total charge q on one plate is equal in magnitude to that on the other plate and opposite in sign. If the lateral extent of the plates is much greater than their separation, then the entire charge on each plate can be assumed to be distributed over that part of its surface which faces the other plate and hence the electric field is zero except in the space between the plates. Furthermore, the electric field will be uniform in the space between the plates with a value E, say.

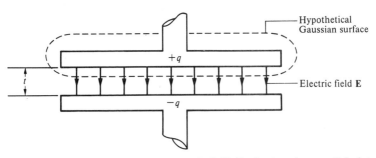

FIG. 1.3. The electric charge and electric field distributions in a parallel plate capacitor.

6 Fundamental electromagnetism

If the area of the part of the surface of the plates over which the charge is distributed is A then, using Gauss's law

$$EA = q/\epsilon_0. \qquad (1.10)$$

The uniformity of E implies that the total charge $\pm q$ is distributed uniformly over a plate with a uniform charge density σ, where $\sigma = q/A$. Thus

$$E = \sigma/\epsilon_0. \qquad (1.11)$$

The approximation which was made regarding the lateral extent of the plates being much greater than their separation means that this result gives the electric field outside a plane conductor which is infinite in extent. However the usefulness of the expression is greatly enhanced by the fact that, to an approximation, it gives the electric field at a point 'close to' a non-planar conductor. A good analogy is that of a person sitting in a small boat on a perfectly calm sea; the surface of the sea appears flat and (almost!) infinite in extent.

Since the electric field between the plates is uniform, the potential difference V between the plates is equal to Et and

$$V = qt/\epsilon_0 A.$$

Thus, the capacitance C, which is equal to q/V, is given by

$$C = \frac{\epsilon_0 A}{t}. \qquad (1.12)$$

The dielectric constant, or relative permittivity, ϵ, of a medium is defined through $\epsilon \equiv C/C_0$ where C, C_0 are the capacitances of a capacitor when the space between its places is 'filled' with the medium and vacuum respectively. For a parallel plate capacitor, $C = \epsilon_0 \epsilon A/t$ where A is the area of the plates and t is their separation, and, for a concentric cylinder, $C = 2\pi\epsilon_0 \epsilon l/\ln(b/a)$ where l is the length and a, b are the radii of the inner and outer conductors respectively (see later in this Section).

Commonly used dielectric media are paper, mica, and polythene with dielectric constants of 3.7, 6.0, 2.3, respectively. The values of capacitance obtainable in capacitors (usually in the concentric cylinder form) using such dielectric media range from about 10 picofarad (pF) (1 pF = 10^{-12} F) to the order of $10\,\mu$F. The physical origins of the dielectric constant are discussed in Chapter 5.

It is possible to obtain a large value of capacitance in a physically comparatively small capacitor by utilizing the electrolysis of certain salts to form a very thin insulating film on aluminium foil. In this way values of capacitance ranging up to $1000\,\mu$F can be obtained. Two

Fundamental electromagnetism

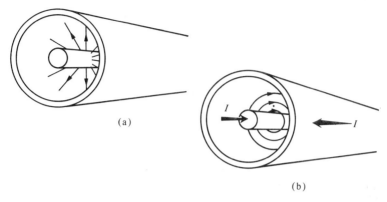

FIG. 1.4. (a) The electric field pattern and (b) the magnetic field pattern between the inner and outer conductors of a coaxial line.

disadvantages of such 'electrolytic' capacitors are their low working voltage (~ 10 V) and the necessity of ensuring that the capacitor is always electrically polarized in a particular sense (usually indicated by $+$ and $-$ signs on the body of the capacitor, or some other code).

A system of practical importance in which a knowledge of the capacitance is of interest is the coaxial line or cable. This consists of a cylindrical outer conductor of radius b, say, and a coaxial cylindrical inner conductor of radius a, say (see Fig.1.4(a)). Consider a hypothetical cylindrical Gaussian surface of radius r which is concentric with the inner and outer conductors ($a < r < b$). From the symmetry of the situation the electric field in the space between the conductors must be radially directed and so the contribution to $\int \mathbf{E} \cdot \mathrm{d}\mathbf{S}$ from the end surfaces of the hypothetical cylinder is zero. So, noting that the electric field varies with r,

$$\int E(r)\,\mathrm{d}S = E(r) \cdot 2\pi r \quad \text{per unit length of line.}$$

If the charge per unit length of conductor is q then on applying Gauss's law to the hypothetical closed cylindrical surface

$$E(r) \cdot 2\pi r = q/\epsilon_0$$

or

$$E(r) = \frac{2q}{4\pi\epsilon_0 r}. \tag{1.13}$$

The potential difference V between the inner and outer conductors is given by

Fundamental electromagnetism

$$V = \int_{r=a}^{b} E(r)\,dr$$

$$= \frac{2q}{4\pi\epsilon_0} \int_a^b \frac{dr}{r}$$

$$V = \frac{2q}{4\pi\epsilon_0} \ln\left(\frac{b}{a}\right).$$

Hence, the capacitance per unit length, which is equal to q/V, is given by

$$C = \frac{4\pi\epsilon_0}{2\ln(b/a)}. \tag{1.14}$$

Another physical system which is of great practical importance is the twin line consisting of two parallel cylindrical conductors of radius a, say, which are spaced a distance b apart (see Fig.1.5). The calculation of the capacitance per unit length of this system involves the use of a mathematical technique which is much more sophisticated than those which have been used so far in this text. You may find it appropriate, at a first reading at least, to bypass the following analysis and just take note of the result given in eqn (1.20).

The essence of the argument lies in recognizing that the surface of a

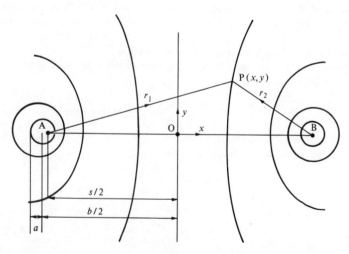

FIG. 1.5. The geometry of the equipotentials used in the calculation of the capacitance per unit length of a twin line consisting of two parallel, cylindrical conductors.

conductor (strictly speaking an ideal conductor of infinite conductivity) is an equipotential surface since the electric field in an ideal conductor is zero. Now the solutions of electrostatic field calculations are given in the form of potential functions ($V(x, y, z)$ in cartesian coordinates) and equipotential surfaces are specified by $V(x, y, z) = $ constant. Consequently, a conductor whose surface fits the form of a particular equipotential surface, and which is at the appropriate potential, can be placed in the field without disturbing in any way the form of the field over the remainder of the space. As we shall see shortly, the equipotential surfaces of the field due to two parallel lines of charge are cylindrical and so we can place two parallel cylindrical conductors so that their cross-sections fit exactly to a pair of equipotential surfaces. We can then determine easily the potential difference between the two conductors in terms of their charge per unit length.

Imagine that two lines of charge, having charge $\pm q$ per unit length respectively, are spaced distance s apart as shown in Fig.1.5. First we need to find the potential V_P at a general point P: We will assume for convenience, that the potential is zero over the plane which is equidistant from both lines of charge; in particular the potential is zero at $O(x=0; y=0)$.

By using the expression for electric field, eqn. (1.13), in eqn (1.5) the potential difference between A and O is

$$V_A - V_O = \frac{q}{2\pi\epsilon_0} \int \frac{dr}{r} = \frac{q}{2\pi\epsilon_0} |\ln r|_{r=s/2}.$$

Similarly,

$$V_P - V_A = -\frac{q}{2\pi\epsilon_0} |\ln r|_{r=r_1}.$$

Thus $V_P - V_O = (V_P - V_A) + (V_A - V_O)$ and on putting $V_O = 0$, as we assumed earlier, the potential at P due to the positive charge on line A is

$$V_P = \frac{q}{2\pi\epsilon_0} \ln(s/2r_1).$$

Similarly, the potential at P due to the negative charge on line B is

$$V_P = -\frac{q}{2\pi\epsilon_0} \ln(s/2r_2)$$

and so the total potential at P is

$$V_P = \frac{q}{2\pi\epsilon_0} \ln\left(\frac{r_2}{r_1}\right). \tag{1.15}$$

An equipotential line in the xy-plane is defined by

$$r_2/r_1 = K \quad \text{(a constant)}. \tag{1.16}$$

Now $r_1^2 = \left\{\left(\frac{s}{2}+x\right)^2 + y^2\right\}$ and $r_2^2 = \left\{\left(\frac{s}{2}-x\right)^2 + y^2\right\}$ and on substituting in eqn (1.16) we find that the equipotential line defined by a particular value of K is given by

$$x^2 - sx\frac{(1+K^2)}{(1-K^2)} + y^2 = -\frac{s^2}{4}.$$

By adding the term $\left\{\frac{s(1+K^2)}{2(1-K^2)}\right\}^2$ to both sides of this equation we obtain

$$\left\{x - \frac{s(1+K^2)}{2(1-K^2)}\right\}^2 + y^2 = \left\{\frac{sK}{(1-K^2)}\right\}^2. \tag{1.17}$$

Now the equation of a circle can be written in the general form

$$x^2 + y^2 + 2gx + 2fy = -c$$

or

$$(x+g)^2 + (y+f)^2 = g^2 + f^2 - c.$$

The centre of this circle is at $x = -g$; $y = -f$ and the radius is equal to $(g^2 + f^2 - c)^{1/2}$. By comparison we can see that eqn (1.17) represents a family of circles with K as parameter; for a particular value of K the circle is centred at

$$x = \frac{s(1+K^2)}{2(1-K^2)}; \quad y = 0$$

and has radius $sK/(1-K^2)$.

Now let us consider P to lie on the surface of the left-hand member of a pair of conductors of radius a and separation between their axes equal to b (see Fig.1.5); for the equipotential defined by the left-hand conductor we have that

$$a = \frac{sK}{(1-K^2)} \qquad b = \frac{s(1+K^2)}{(1-K^2)}.$$

On elminating s we find

$$\frac{b}{a} = K + \frac{1}{K}. \qquad (1.18)$$

From eqns (1.15) and (1.16), $K = e^{2\pi\epsilon_0 V_P/2}$ and on substituting in eqn (1.18)

$$\frac{b}{a} = e^{2\pi\epsilon_0 V_P/2} + e^{-2\pi\epsilon_0 V_P/2} \quad \text{or} \quad \frac{b}{a} = 2\cosh\frac{2\pi\epsilon_0 V_P}{q}. \qquad (1.19)$$

Since the p.d. between the two conductors is $2V_P$ it follows that

$$C = \frac{q}{2V_P} = \frac{4\pi\epsilon_0}{4\cosh^{-1}\left(\dfrac{b}{2a}\right)}. \qquad (1.20)$$

Since $\cosh^{-1}(b/2a) = \ln\{b/2a + \sqrt{(b^2/4a^2 - 1)}\}$, this means that $C = \pi\epsilon_0/(\ln b/a)$ to an accuracy better than 0.5 per cent for $b/a > 10$; for $b/a = 5$ then $C = 17$ pF m^{-1}.

You should notice that the net line charge on a conductor is not distributed uniformly over its cross-section due to the influence of the charge on the other conductor; this feature is known as the proximity effect and is exemplified by the fact that $b \neq s$.

The capacitance per unit length between a long cylindrical conductor of radius a and an infinite conducting plane can be obtained easily from the result of the foregoing analysis. The physical situation of a line charge q per unit length at a distance $s/2$ above a conducting plane of infinite extent can be represented in terms of a hypothetical charge, or 'image' charge, $-q$ per unit length, situated at a distance $s/2$ below the position of the plane, the plane itself having been imagined to have been removed. The equipotential surfaces for the line charge and its image are identical to those illustrated in Fig.1.5; the conducting plane is an equipotential surface and its position would coincide with the plane equipotential surface containing the line $x = 0$ and lying perpendicular to the plane of the diagram.

From eqn (1.19)

$$2\pi\epsilon_0 V_P/q = \cosh^{-1}(b/2a).$$

Now in the present case the potential difference between the two conductors is $V_P/2$ in terms of the definition given in the foregoing analysis and so

$$C = \frac{q}{V_P/2} = \frac{4\pi\epsilon_0}{\cosh^{-1}(b/2a)}$$

12 Fundamental electromagnetism

or, if we put $b/2 = h$, being the distance of the conductor from the plane, then

$$C = \frac{4\pi\epsilon_0}{\cosh^{-1}(h/a)}. \qquad (1.21)$$

For $h/a = 2$ then $C \approx 80\,\text{pF}\,\text{m}^{-1}$ which could have a very significant effect on the properties of a circuit at a high operating frequency.

If any of the electric field lines emanating from one of a pair of charged conductors do not terminate on the other conductor of the pair, then there will be 'stray' capacitance associated with the other conductors on which the stray lines do terminate. Since the reactance of a capacitor is inversely proportional to frequency the effects of stray capacitance can be very significant in high frequency circuits by providing undesirable low impedance paths (see the Frontispiece and Section 3.3 for example).

1.2. Current electricity

1.2.1. Electromotive force

Changes may be moved by agencies other than electrostatic forces; a good example for illustrative purposes is the Van de Graaff generator (see Fig.1.6). A moving belt made from a good electrical insulator transports charges in the direction shown producing separated assemblies of positive and negative charges which in turn produce an electric field **E**. An external agency must exert a force **F**, say, on the charges, and do work, in order to move them against the field **E**. There will be,

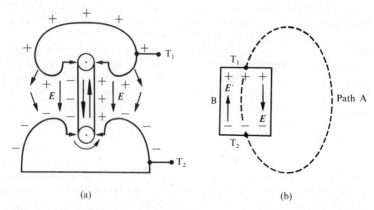

FIG. 1.6.   (a) A Van de Graaff generator: the separation of charges and the directions of the electric field E are shown schematically. (b) A schematic representation of a general source of e.m.f.

obviously, an electric potential difference between the terminals T_1 and T_2.

Now consider a more general physical system B (see Fig.1.6(b)) which could be a Van de Graaff generator or might be a battery or thermocouple for instance. From definition (1.2) a non-electrostatic field E' can be defined through $E' = F/q$ and, in contrast to the electrostatic field, for a closed path $\oint E' \cdot dl \neq 0$ in general i.e. this field is not conservative.

Imagine that such a field E' exists inside B: charges of opposite sign will be separated until the resultant electrostatic field is equal and opposite to E'; there is now a zero net field and there is no further separation of charges. Imagine further, for simplicity, that path A lies in vacuum: Then in this equilibrium situation

$$\oint_A E' \cdot dl = \int_{T_2}^{T_1} E' \cdot dl$$

since $E' = 0$ outside B. The right-hand side of the above equation is called the electromotive force (e.m.f.) of the system, or source. Since there is no external conducting path between the 'terminals' T_1, T_2 the source is said to be on 'open circuit' and in this situation it follows that, since $E' = -E$ inside the source and $E' = 0$ outside the source

$$E = \int_{T_2}^{T_1} -E \cdot dl = V_{T_1} - V_{T_2} \qquad (1.22)$$

where eqn (1.6) has been used. Thus the e.m.f. of a source is equal to the p.d. between its terminals when it is on open circuit.

If T_1, T_2 are connected by an external conducting path then a current will flow, the charges at T_1, T_2 being replenished by F. From the definition of E it follows that $qE = \int_{T_2}^{T_1} F \cdot dl$ is the work done by a source in supplying a charge q and, since current is equal to rate of flow of charge in a circuit, the rate of working, or power supplied in maintaining a current I is EI. In SI units E and I are measured in volts and amperes, respectively, work in joules, and power in joules per second or watts (W).

The e.m.f. of a Van de Graff generator may be millions of volts but that of a chemical cell is only of the order of a volt since the changes in energy of an atom or molecule involved in the chemical reaction which transports charges are of the order of one electron volt.

A source of e.m.f. is reversible in the sense that a current can be driven through it by a greater source of e.m.f., as when a lead accumulator is 'charged'. An ideal source of e.m.f. would be perfectly reversible but in practice there are always some internal dissipative effects, which can be represented by an internal resistance, and no source is perfectly reversible.

14 Fundamental electromagnetism

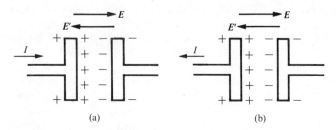

FIG. 1.7. Currents and fields in a capacitor (a) during charging and (b) during discharging.

An example of a transient source of e.m.f. is charged capacitor. As the capacitor is being charged work must be done in producing the separated positive and negative charge distributions, the e.m.f. opposes the current, and the potential energy acquired by the charges is conceived as being stored in the electric field which is set up between the capacitor plates (see Fig.1.7). On discharge the e.m.f. is in the direction of the current to which it supplies energy. It is left as an exercise to show, by using a parallel plate capacitor as a convenient example, that the energy density associated with an electric field is $\epsilon\epsilon_0 E^2/2$ J m^{-3}.

1.2.2. Ohm's law

Assume that an average electric field E exists throughout the length of a metal wire; the mobile charge carriers (conduction electrons) are accelerated in this field, gaining momentum and kinetic energy. However a state of equilibrium is reached quickly, in $\sim 10^{-13}$ s after the field is applied, in which the average rate of gain of momentum and kinetic energy by the conduction electrons is equal to their average rate of loss of these two quantities to the crystal lattice of the metal through random interactions (collisions) with the lattice and with impurities and defects. In this steady state the conduction electrons have an average drift velocity in the direction of the field superimposed on their otherwise random motion due to the collisions: This net drift of conduction electrons constitutes the current in the wire. The p.d. V between the ends of a wire of length l is equal to El and, if the current is I, then a property of the wire called its resistance R is defined through $R \equiv V/I$. For many conductors (metals and alloys) the resistance is independent of current, for most practical purposes, providing that their temperatures are kept constant: this is Ohm's law.

In many situations it is correlated changes of current δI and voltage δV which are of interest and then a so-called dynamic resistance is defined through $R \equiv \delta V/\delta I$.

Materials whose resistances are a function of current are said to be 'non-ohmic'. Some important solid state electronic devices rely for their operation on the non-ohmic properties of certain materials such as silicon and gallium arsenide for instance.

We reserve the term 'resistor' for the material object which has resistance as its most important physical property in this context.

1.2.3. Dissipation of energy

In the steady state current flow described in the preceding Section there is no net gain of kinetic energy by the conduction electrons as they traverse the length of wire and so all of the potential energy qV lost by a charge q in 'falling' through a p.d. V is given to the conductor where it manifests itself as heat. Hence the rate of production of heat, or power dissipation, is $V \mathrm{d}q/\mathrm{d}t = VI = I^2R$ watts.

The most elementary form of electric circuit consists of an active component (an ideal source of e.m.f.), the two terminals of which are connected to the two terminals of a passive component (the 'load') by connecting wire whose resistance is negligible compared with that of the load. The source of e.m.f. provides energy at a rate EI to create a p.d. which causes the charge carriers to move through the passive parts of the circuit in which, if the total resistance is R ohms (Ω), energy is dissipated at a rate $I^2R = EI$.

1.3. Magnetic effects of electric currents

A fundamental description of the nature and origin of what are called magnetic field lies outside the scope of this text and it will have to suffice to say that if a moving electric charge, either in vacuum or in a metal wire, say, experiences a *velocity-dependent* force then the charge is moving through a region in which a magnetic field is said to exist. Experiments have shown that a magnetic field exists in conjunction with an electric current and the results are compatible with an expression for $\mathrm{d}B$, the element of magnetic field due to an element of a circuit carrying a current I, of the form

$$\mathrm{d}B = \frac{\mu_0}{4\pi} \frac{I \sin \theta \, \mathrm{d}l}{r^2} \tag{1.23}$$

where $\mu_0/4\pi$ (equal to 10^{-7} henry per metre) is the appropriate constant of proportionality in SI units. The respective directions of $\mathrm{d}B$ and $\mathrm{d}l$ are shown in Fig.1.8(a). The total magnetic field B at P due to a complete circuit is given by

$$B = \frac{\mu_0 I}{4\pi} \oint \frac{\sin \theta \, \mathrm{d}l}{r^2} \tag{1.24}$$

where $\oint$ indicates an integral taken round the complete circuit.

16 Fundamental electromagnetism

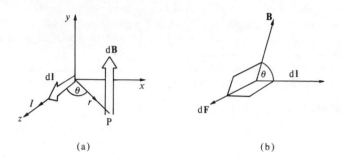

FIG. 1.8. (a) The magnetic field dB due to a current element. (b) The force dF on a 'current element' situated in a magnetic field (the 'left-hand rule').

It is left as an exercise to show that the magnetic field at a point distant r from an infinitely long, straight wire carrying a current I is $B = \frac{\mu_0}{4\pi} \cdot \frac{2I}{r}$ and that at a point on the axis of a circular loop of wire of of radius a carrying current I, the magnetic field is axially directed and given by $B = \frac{\mu_0}{4\pi} \cdot 2\pi I a^2 (a^2 + x^2)^{-3/2}$ where x is the distance of the point in question from the plane of the loop.

The following expression (eqn (1.25)) for the force dF acting on an element of circuit dl carrying current I and situated in a magnetic field B has been found to be in accordance with experimental measurements which, of course, had to be made using complete circuits:

$$dF = I\, dl \sin \theta \cdot B. \tag{1.25}$$

The relative directions of B, dl, and dF are shown in Fig.1.8(b). Notice that the force is proportional to the component of the element of circuit $dl \sin \theta$ which is perpendicular to B.

Eqns (1.23) and (1.25) can be combined to yield a general expression for the force F between two circuits but a particular result which is of interest is that which applies to two long straight parallel wires separated by a distance d and carrying currents I_1 and I_2, namely $F = \frac{\mu_0}{4\pi} \cdot \frac{2I_1 I_2}{d}$ N m^{-1}. The ampere (A) is so defined that for $I_1 = I_2 = 1$ A and $d = 1$ m then $F = 2 \times 10^{-7}$ N m^{-1} which fixes the value of μ_0 as $4\pi \times 10^{-7}$ henry per metre (H m^{-1}).

The equivalent expression to eqn (1.25) for a single particle of charge q moving at speed v is

$$F = qv \sin \theta \cdot B \tag{1.26}$$

Fundamental electromagnetism 17

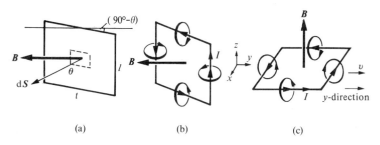

(a)   (b)   (c)

FIG. 1.9. (a) The magnetic flux Φ linked with the loop of wire is given by $\Phi = \int \mathbf{B} \cdot \mathrm{d}\mathbf{S} = Blt \cos\theta$. (b) Electromagnetic induction when $\mathbf{B}$ is spatially uniform but decreasing with time. (c) Electromagnetic induction when $\mathbf{B}$ is time-independent but decreases in magnitude in the y-direction and the loop is moving in the y-direction. In both (b) and (c) the induced flux is in the same direction as $\mathbf{B}$, i.e. the induced flux is in such a sense as to try to sustain $\mathbf{B}$.

The equivalence may be seen through writing $I = \mathrm{d}q/\mathrm{d}t$ and $v = \mathrm{d}l/\mathrm{d}t$.

1.4. Electromagnetic induction

The magnetic flux Φ linked with a closed circuit (see Fig.1.9) may change due either to motion of the circuit in a time-independent magnetic field $\mathbf{B}$ or to a time-dependent field linked with the circuit, or to a combination of these two circumstances. In any case it is observed that a current is induced in such a sense as to oppose the relevant change in physical circumstances which is producing the induced e.m.f. which is driving it (see Fig.1.9): The magnitude of the induced e.m.f. is given by

$$E = \mathrm{d}\Phi/\mathrm{d}t. \qquad (1.27)$$

If the change in flux is associated with only part of a circuit then the e.m.f. is localized in that part of the circuit; otherwise the e.m.f. is non-localized. A change in flux linked with a circuit may be a change in self-flux due to its own current being time-dependent or it may be a change due to a time-dependent current in a neighbouring circuit. For these two cases a coefficient of self-inductance L and of mutual inductance M, respectively, are defined through

$$L \text{ or } M = \frac{E}{\mathrm{d}I/\mathrm{d}t} \qquad \text{(unit: 'henry' H)} \qquad (1.28)$$

where $\mathrm{d}I/\mathrm{d}t$ is the appropriate rate of change of current.

Using eqn (1.27) for E then eqn (1.28) can be written also as

$$L \text{ or } M = \frac{\mathrm{d}\Phi}{\mathrm{d}I}. \qquad (1.29)$$

18 Fundamental electromagnetism

Every component of a circuit has a coefficient of self-inductance, however small, but a component designed to have a specific value (an 'inductor') is, except for operation at very high frequencies (~ 100 MHz and higher), usually in the form of a coil consisting of many turn of wire.

In the calculation of the self-inductance and mutual inductance of circuit elements it is necessary to know the magnetic field produced by the currents flowing in the elements and a law of electromagnetism which is useful very often in this context is Ampere's Circuital law: this states that the line integral of the magnetic field **B** round a closed path is equal to the total current I linked with that path multiplied by the fundamental magnetic constant μ_0

$$\oint \mathbf{B} \cdot d\mathbf{l} = \mu_0 I. \qquad (1.30)$$

As an example of the use of this law let us calculate the field inside a 'long' cylindrical solenoid, i.e. a solenoid whose length is much greater than its diameter (see Fig.1.10). If the solenoid has n turns per metre and the current in each turn is I, then for the closed path ABCD which is situated far from either end of the solenoid

$$\oint \mathbf{B} \cdot d\mathbf{l} = \int_A^B \mathbf{B} \cdot d\mathbf{l} + \int_B^C \mathbf{B} \cdot d\mathbf{l} + \int_C^D \mathbf{B} \cdot d\mathbf{l} + \int_D^A \mathbf{B} \cdot d\mathbf{l} = Bz.$$

Here it has been assumed that the magnetic field is uniform inside this part of the solenoid and is negligibly small outside the solenoid in the region containing the element of path CD. Also for those portions of BC and DA which lie within the solenoid the magnetic field is perpendicular to the elements of path and so the contributions to $\oint \mathbf{B} \cdot d\mathbf{l}$ are zero. Now the total current linked with the closed path ABCD is nzI and so using Ampere's Circuital law we have

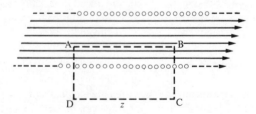

FIG. 1.10. The path of integration ABCD in the calculation of the magnetic field inside a 'long' cylindrical solenoid.

or
$$Bz = \mu_0 nzI$$

$$B = \frac{\mu_0}{4\pi} \cdot 4\pi nI. \quad (1.31)$$

The magnetic flux Φ linked with the solenoid due to the current in its own winding is given by $\Phi = BAnl$ if end effects are neglected, where l is the length of the solenoid and A is its area of cross-section. Thus

$$\Phi = \frac{\mu_0}{4\pi} \cdot 4\pi A l n^2 I$$

and, using eqn (1.29), the coefficient of self-inductance L_0, is given by

$$L_0 = \frac{\mu_0}{4\pi} \cdot 4\pi A l n^2. \quad (1.32)$$

If a long cylindrical solenoid is filled with a material medium, then the relative permeability μ of the medium is defined through $\mu \equiv L/L_0$ where L, L_0 are the self-inductances of the solenoid with and without the medium respectively. For ferromagnetic materials μ may be as large as several thousand (see Section 5.5.3).

A transformer is an example of a mutual inductor; two coils which are electrically insulated from each other are wound on a common ferromagnetic core and, as a result, a changing current in the 'primary' coil induces an e.m.f. in the 'secondary' coil. Since the commonly used ferromagnetic materials are also electrical conductors currents will be induced in the core also. If the primary current has a time dependence of the form $I = \hat{I} \sin \omega t$, then the resultant magnetic flux will have a time dependence $\sin \omega t$ and the amplitude of the induced e.m.f., and hence of the induced currents ('eddy' currents) will be proportional to ω since it is proportional to the time derivative of $\sin \omega t$. Thus there will be power losses proportional to ω^2 in the core; in order to reduce these eddy current losses transformer cores are made up from suitably oriented insulated laminations thus reducing the effective electrical conductivity. Even so the upper limit for the use of such transformers is about 30 kHz; at higher frequencies cores made from ferrites (materials with fairly high values of μ but low electrical conductivities) are used.

Consider a circuit which can be represented by a resistor R and an inductor L in series with a source of e.m.f. E. Remembering that the induced e.m.f. in the inductor opposes any change in the current through it and remembering also that the total e.m.f. in the closed circuit must be equal to the total p.d. we have

$$E - L\frac{dI}{dt} = RI.$$

Fundamental electromagnetism

The source of e.m.f. is supplying power EI and so

$$EI = LI\frac{dI}{dt} + RI^2. \tag{1.33}$$

For I increasing with time the induced e.m.f. is in the opposite direction to I and energy can be considered as being stored as potential energy in the magnetic field associated with the inductor. If I is decreasing with time the induced e.m.f. is in such a sense as to try to maintain the magnitude of the current.

The energy W stored in the inductor is given by

$$W = L \int_I I \frac{dI}{dt} dt$$

or

$$= L \int_{I=0}^{I} I \, dt$$

$$W = \frac{LI^2}{2}. \tag{1.34}$$

It is left as an exercise to show that the energy density associated with a magnetic field $\mathbf{B}$ is $B^2/2\mu_0\mu$ J m^{-3}.

PROBLEMS

1.1. A 12-V battery supplies 5 A continuously for 2 hours. How much energy does the battery deliver? To what voltage would a 1000 μF capacitor have to be charged in order to store an amount of energy equal to that supplied by the battery?

1.2. Show, by considering a parallel plate capacitor, that the energy density associated with an electric field of strength E is $\epsilon\epsilon_0 E^2/2$, where ϵ is the relative permittivity of the medium between the capacitor plates.

1.3. What is the resistance of a 12 V d.c., 24 W light bulb whilst it is in operation? How much charge passes through the element of the bulb per second?

1.4. Show that (a) the magnetic field B at a point distant r from an infinitely long straight wire carrying current I is $B = \mu_0 2I/(4\pi r)$ and (b) that the magnetic field at a point on the axis of a circular loop of wire of radius a is axially directed and given by $B = \mu_0 2\pi I a^2/\{4\pi(a^2 + x^2)^{3/2}\}$ where x is the distance of the point in question from the plane of the loop.

1.5. Show, by considering a current flowing in a long cylindrical solenoid, that the energy density associated with a magnetic field B is $B^2/(2\mu\mu_0)$, where μ is the relative permeability of the medium filling the solenoid.

1.6. Derive an expression for the coefficient of mutual inductance between a long straight wire and another wire which is bent into the shape of a closed rectangle of length a and breadth b. The long wire is coplanar with the rectangle, is parallel to the sides of length a, and is distance d from the nearer such side.

2. Direct-current circuits

2.1. Circuit elements

A simple closed circuit consisting of an ideal 'active' element (a source of e.m.f. E) and a 'passive' element (a resistor R) connected by wires of negligible resistance was described in Section 1.2.3: all the energy supplied by the source is dissipated in the resistor and appears as heat and other forms of energy. There are other physical systems, namely inductors and capacitors, for which the energy supplied to them via an electric current is partly stored in the system and partly dissipated as a result of physical processes in the system. Actually, for time-dependent currents there is an additional source of loss of energy from the circuit, namely by the radiation of electromagnetic waves, but this process has practical significance only at very high frequencies or in circuit elements designed specifically for the purpose such as aerials.

In analysing direct-current (d.c.) and alternating-current (a.c.) circuits an idealization will be made in which the actual physical components will be represented by circuit elements such that each element has only the property conveyed by its name i.e. a resistor has resistance only, an inductor has inductance only, and a capacitor has capacitance only. The wire which constitutes the coil of an inductor has resistance and so a real inductor can be represented by an inductance L in series with a resistance R. Similarly, a small current can leak between the plates of a real capacitor and this situation can be represented by a capacitance C with a resistance R in parallel with it which represents the high resistance leakage path. Real resistors have small values of capacitance and inductance associated with them. A non-inductive resistor can be made by using 'doubled' wire so that adjacent equal and opposite currents produce cancelling magnetic fields. Also practical sources of e.m.f. always have internal resistance, inductance, and capacitance although in many practical situations one or more of these properties may be ignored. In this chapter only d.c. situations are treated and so the only circuit elements which are of interest are pure resistors and active elements which have internal resistance only.

Let us consider a power source of e.m.f. E_0 which has an internal

Direct-current circuits   23

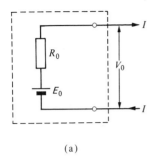 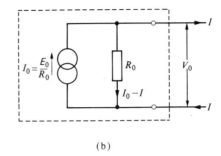

(a)   (b)

FIG. 2.1. Representations of a source of e.m.f. E_0 of internal resistance R_0 in terms of (a) a constant voltage generator E_0 and (b) a constant current generator $I_0 (= E_0/R_0)$.

resistance R_0 (see Fig.2.1); if a current I is drawn then the p.d. V_0 between its terminals falls and is given by

$$V_0 = E_0 - IR_0 \qquad (2.1)$$

(note that for the open circuit situation $I = 0$ and $V_0 = E_0$ as discussed in Section 1.2). Nominal values for E_0 and values for R_0 for some commonly used sources are shown in Table 2.1.

Table 2.1
Characteristics of some sources of e.m.f.

Source	E_0 (volts)	R_0 (ohms)
Weston–Cadmium cell	1.01864	500
Lead acid accumulator	1.95	0.01
Nickel–Iron alkaline accumulator ('Nife' cell)	1.2	0.1
'Dry' cell	1.5	30

In the situation represented in Fig.2.1(a) the source of e.m.f. is regarded as an ideal source, with zero internal resistance, in series with a pure resistance R_0 which represents the dissipative processes present in the real source. Since an ideal source has $R_0 = 0$, so that V_0 is independent of I, it is described as a 'constant voltage generator'.

A real source of e.m.f. can be represented usefully in another way: imagine a source of e.m.f. which supplies a constant current I_0 no matter what value of load resistor is connected between its terminals (if you think about it this means that this imaginary source must have an infinite internal resistance). If current I is drawn from the real source, then the p.d. between its terminals is (see Fig.2.1(b))

$$V_0 = (I_0 - I)R_0 \qquad (2.2)$$

since I_0 divides between the external load and R_0. Also, for the open circuit condition ($I = 0$) we have $V_0 = I_0 R_0 = E_0$ (i.e. $I_0 = E_0/R_0$) and it follows that eqn (2.2) can be written as

$$V_0 = E_0 - IR_0 \tag{2.3}$$

which is identical to eqn (2.1).

Real sources of e.m.f. lie closer to the ideal constant voltage generator than to the ideal constant current generator. The latter is not a very good representation of a battery, say, since it is constantly dissipating power within itself to maintain its terminal voltage even on open circuit. Nevertheless, as shall be seen, the concept of the constant current generator has great utility in the analysis of current networks.

Imagine that a load resistor R_L is connected between the terminals of the source of e.m.f. E_0 in Fig.2.1; from Ohm's law

$$I(R_0 + R_L) = E_0. \tag{2.4}$$

The power dissipated in R_L is

$$P = I^2 R_L = E_0^2 R_L/(R_0 + R_L)^2. \tag{2.5}$$

From this equation $dP/dR_L = 0$ for $R_L = R_0$ and, furthermore, since $d^2P/dR_L^2 < 0$, for $R_L = R_0$, this is the condition for maximum power dissipation in the load.

It has been shown already that E_0 is equal to the open circuit voltage between the accessible terminals of the source. If these terminals are short-circuited then

$$I_{sc} R_0 = E_0 \quad \text{or} \quad R_0 = E_0/I_{sc}, \tag{2.6}$$

i.e. $R_0 = $ (open-circuit voltage)/(short-circuit current).

2.2. Thévenin's theorem

An important feature of networks containing resistors and sources of e.m.f. which are much more complicated than the network of Fig.2.1 is expressed in Thévenin's theorem: 'Any linear[†] circuit having two terminals can be represented by a source of e.m.f. and a resistance in series'.

This means that a complicated network which may be, or may imagined to be, contained within a 'black box' behaves, as far as a circuit connected to its two accessible terminals is concerned, just like the simple battery of Fig.2.1.. The values of E_0 and R_0 for the simple 'equivalent generator' may be calculated from a knowledge of the

---

[†] The significance, in this context, of 'linear' will be explained in Section 2.6.

network by methods to be described in the succeeding sections, or may be determined experimentally by measurements on open and short circuit as explained in the preceding section. In, practice, of course, it may not be wise to short-circuit the terminals of a network or source in which case it is useful to notice that, if R_L is varied, then $R_L = R_0$ when the voltage across the load resistor is equal to $E_0/2$ since $V_L = E_0 R_L/(R_0 + R_L)$. Alternatively, from eqn (2.4) $1/I = (R_0/E_0) + (R_L/E_0)$ and so from a graph of $1/I$ against R_L, R_0/E_0 and E_0 may be found by extrapolation to the intercept at $R_L = 0$ and from the slope respectively.

An illustration, which is almost trivial, of Thévenin's theorem is provided by the network shown in Fig.2.2(a) which consists of series

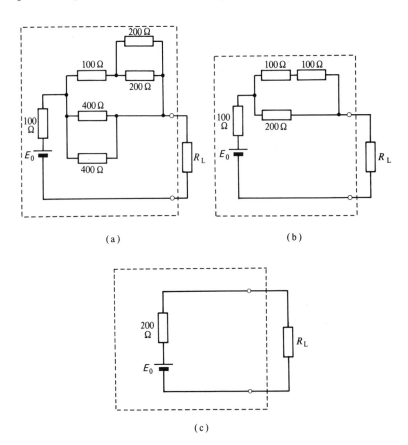

FIG. 2.2. An illustration of Thévenin's theorem for a simple network contained within a 'black box'.

26 Direct-current circuits

and parallel branches only, together with a single source of e.m.f. This network can be simplified, using

and
$$R = R_1 + R_2 + R_3 \ldots = \sum R_i$$
$$1/R = 1/R_1 + 1/R_2 + 1/R_3 \ldots = \sum 1/R_i$$

to a source in series with a single equivalent internal resistance, and the load resistor, as shown.

However, even a simple Wheatstone bridge circuit cannot be reduced in this way and a different approach, based on Kirchhoff's two laws must be adopted.

2.3. Kirchhoff's laws

(i) *The 'voltage' law.* The sum of the potential differences across the circuit elements, taken in order, around a loop or mesh of a network is zero.

(ii) *The 'current' law.* The algebraic sum of the currents into a node of a network is zero.

Consider the circuit shown in Fig. 2.3 which, as for the Wheatstone bridge, does not consist of simple series and parallel branches; (it should be noticed that if the branch containing R_3 is open-circuited ($R_3 = \infty$) then the circuit becomes a 'potentiometer' in which E_2 may be compared with E_1 ($E_2 < E_1$) by adjusting the ratio R_5/R_6 until no current flows through the branch containing R_4 (which would represent the resistance of the indicating meter).

Let the branch currents be $i_1, i_2, \ldots, i_6$ as shown and the application of Kirchhoff's voltage law to mesh ABDA gives

$$i_1 R_1 + i_5 R_5 + i_6 R_6 = E_1 \qquad (2.7.\text{a})$$

and to the other meshes

$$i_3 R_3 - i_4 R_4 - i_5 R_5 = 0 \qquad \text{(ACBA)} \qquad (2.7.\text{b})$$
$$i_2 R_2 - i_4 R_4 + i_6 R_6 = E_2 \qquad \text{(CBDC)} \qquad (2.7.\text{c})$$
$$i_1 R_1 + i_5 R_5 + i_4 R_4 - i_2 R_2 = E_1 - E_2 \qquad \text{(ABCDA)} \qquad (2.7.\text{d})$$
$$i_1 R_1 + i_3 R_3 - i_4 R_4 + i_6 R_6 = E_1 \qquad \text{(ACBDA)} \qquad (2.7.\text{e})$$
$$i_1 R_1 + i_3 R_3 - i_2 R_2 = E_1 - E_2 \qquad \text{(ACDA).} \qquad (2.7.\text{f})$$

Application of the current law at nodes A, B, C, and D respectively gives

$$i_1 - i_3 - i_5 = 0 \qquad (2.8.\text{a})$$
$$i_5 - i_4 - i_6 = 0 \qquad (2.8.\text{b})$$

Direct-current circuits 27

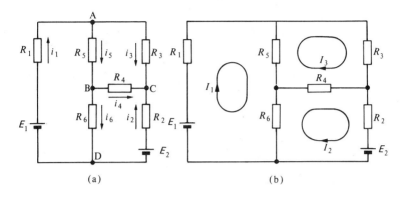

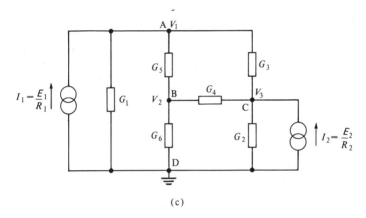

FIG. 2.3. A general network illustrating (a) branch currents, (b) mesh currents, (c) node voltages. (N.B. I_1, I_2 are not the same as I_1, I_2 of (b).)

$$i_2 + i_3 + i_4 = 0 \qquad (2.8.c)$$

$$-i_1 - i_2 + i_6 = 0. \qquad (2.8.d)$$

On examination of these ten equations it becomes apparent that they are not all independent of each other, e.g. for eqns (2.7), (a) + (b) − (c) = (f) and for eqns (2.8), (a) + (b) + (c) = (d).

In fact it is a general feature of network analysis that the number of independent mesh equations is equal to the number of independent meshes N_m and that the number of independent node equations is equal to $(N_j - 1)$, where N_j is the number of nodes. Hence the total number of independent network equations is equal to $(N_m + N_j - 1)$: for the network under consideration $N_m = 3$, $N_j = 4$, and hence six independent equations can be obtained. Since it can be shown that, in general, the number of branches of a network is equal to $(N_m + N_j - 1)$, the number of independent equations is equal to the number of branch currents, which can therefore be determined: Thence the various potential differences can be calculated from $V = RI$ (henceforth the colloquial term 'voltage' will be used for potential difference).

2.4. Mesh analysis

Consider a set of mesh currents I_1, I_2, I_3 which traverse complete loops of the circuit of Fig. 2.3(b). Since every mesh current passes right through any junction in its path, Kirchhoff's current law is satisfied, automatically, and the number of independent equations obtained is equal to N_m. Now $N_m < N_m + N_j - 1$ and so the labour involved in the mathematical solution of the simultaneous equations is much reduced by using mesh currents as opposed to branch currents, even for the comparatively simple circuit under consideration at present where $N_m = 3$ and $N_m + N_j - 1 = 6$.

Applying the voltage law to the circuit of Fig.2.3(b),

$$I_1 R_1 + (I_1 - I_3)R_5 + (I_1 - I_2)R_6 = E_1$$

$$I_2 R_2 + (I_2 - I_1)R_6 + (I_2 - I_3)R_4 = -E_2$$

$$I_3 R_3 + (I_3 - I_2)R_4 + (I_3 - I_1)R_5 = 0$$

or

$$I_1(R_1 + R_5 + R_6) - I_2 R_6 - I_3 R_5 = E_1$$

$$-I_1 R_6 + I_2(R_2 + R_4 + R_6) - I_3 R_4 = -E_2 \qquad (2.9)$$

$$-I_1 R_5 - I_2 R_4 + I_3(R_3 + R_4 + R_5) = 0.$$

The total resistance traversed by I_1 in mesh 'one', $(R_1 + R_5 + R_6)$ is called the 'self-resistance' R_{11} of mesh 'one' and the mutual resistances between mesh 'one' and meshes 'two' and 'three' are R_6 and R_5 which will be denoted by R_{12} and R_{13}, respectively (if the two mesh currents flow through a mutual resistance in opposite directions, then the sign of the mutual resistance is taken as negative). Following this convention, eqns 2.9 take the form

$$I_1 R_{11} + I_2 R_{12} + I_3 R_{13} = E_1$$

$$I_1 R_{21} + I_2 R_{22} + I_3 R_{23} = -E_2 \qquad (2.10)$$

$$I_1 R_{31} + I_2 R_{32} + I_3 R_{33} = 0.$$

For a general network containing n meshes, say, the set of simultaneous equations analogous to eqns (2.10) is

Direct-current circuits 29

$$I_1 R_{11} + I_2 R_{12} + \ldots + I_n R_{1n} = E_1$$
$$I_1 R_{21} + I_2 R_{22} + \ldots + I_n R_{2n} = E_2 \qquad (2.11)$$
$$\vdots \qquad \vdots \qquad \vdots \qquad \vdots$$
$$I_1 R_{n1} + I_2 R_{n2} + \ldots + I_n R_{nn} = E_n.$$

These equations can be written in matrix form as

$$\begin{bmatrix} I_1 \\ I_2 \\ \vdots \\ I_n \end{bmatrix} \begin{bmatrix} R_{11} & R_{12} & R_{13} \ldots R_{1n} \\ R_{21} & R_{22} & R_{23} \ldots R_{2n} \\ \vdots & \vdots & \vdots \quad \vdots \\ R_{n1} & R_{n2} & R_{n3} \ldots R_{nn} \end{bmatrix} = \begin{bmatrix} E_1 \\ E_2 \\ \vdots \\ E_n \end{bmatrix}. \qquad (2.11)$$

For $n > 3$ an analysis of the simultaneous equations through successive eliminations of variables becomes exceedingly tedious (and prone to error on that count alone). A mathematically more elegant, and more systematic, method of analysis makes use of the theory of determinants. It will suffice here merely to state some of the salient properties of determinants and to give some simple examples of their use in network analysis.

The determinant D of the coefficients of the currents in eqns (2.11) is the array

$$D = \begin{vmatrix} R_{11} & R_{12} \ldots R_{1n} \\ R_{21} & R_{22} \ldots R_{2n} \\ \vdots & \vdots \quad \vdots \\ R_{n1} & R_{n2} \ldots R_{nn} \end{vmatrix}$$

and the so-called co-factor D_{ij} of element R_{ij} is the same determinant, but with the ith row and jth column removed, and multiplied by $(-1)^{i+j}$. The useful formal result, known as Cramer's rule is

$$I_j = \frac{1}{D}(E_1 D_{1j} + E_2 D_{2j} + \ldots E_n D_{nj}) = \frac{1}{D} \sum_{i=1}^{n} E_i D_{ij} \quad (2.12)$$

but the following simple examples of the use of this technique will be more illuminating. Consider the circuit of Fig. 2.4(a) where the independent mesh equations are

$$I_1 R_{11} + I_2 R_{12} = E_1$$
$$\qquad\qquad\qquad\qquad\qquad (2.13)$$
$$I_1 R_{21} + I_2 R_{22} = E_2$$

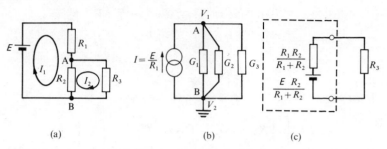

(a) (b) (c)

FIG. 2.4. A potential divider circuit: (a) mesh analysis; (b) nodal analysis; (c) a 'black box' equivalent circuit.

where $R_{11} = (R_1 + R_2)$, $R_{12} = -R_2$, $R_{21} = -R_2$, $R_{22} = (R_2 + R_3)$, $E_1 = E$, and $E_2 = 0$. Hence, using eqn (2.12), $I_1 = E_1 D_{11}/D$ where

$$D = \begin{vmatrix} R_{11} & R_{12} \\ R_{21} & R_{22} \end{vmatrix} = (R_{11}R_{22} - R_{21}R_{12}),$$

$$D_{11} = R_{22}, \quad \text{and} \quad E_1 = E.$$

Also $I_2 = E_1 D_{12}/D$ where $D_{12} = -R_{21} = R_2$. On evaluating these expressions for I_1 and I_2†,

$$I_1 = E \left\{ R_1 + \frac{R_2 R_3}{(R_2 + R_3)} \right\}^{-1} \quad \text{and} \quad I_2 = \frac{ER_2}{(R_1 R_2 + R_1 R_3 + R_2 R_3)}. \tag{2.14}$$

If R_3 is called the load resistor and is denoted by R_L, then the voltage V_L across it is given by

$$V_L = \frac{ER_2}{(R_1 R_2/R_L) + (R_1 + R_2)}. \tag{2.15}$$

If $R_L \gg R_1 R_2/(R_1 + R_2)$ then $V_L = ER_2/(R_1 + R_2)$ and the circuit becomes what is known as a 'potential divider' network.

†A useful test of the validity of the solution to a problem which can be applied in many situations is to let one or more of the variables or parameters tend to extreme values (e.g. zero or infinity) or to other values which would greatly simplify the problem.

In the problem of which eqns (2.14) are solutions, if $R_3 = \infty$ then $I_1 = E/(R_1 + R_2)$, $I_2 = 0$ which is as it should be. Additionally, if $R_3 = 0$ then $I_1 = I_2 = E/R_1$ which again is what is to be expected. N.B. This test is a necessary, but not sufficient, condition for the analysis to be correct.

An alternative but equivalent recipe for finding a current I_j, say, from eqns (2.11) which may be preferred in practice is

$$I_j = \frac{\begin{vmatrix} R_{11} & R_{12} & \ldots & E_1 & \ldots & R_{1n} \\ R_{21} & R_{22} & \ldots & E_2 & \ldots & R_{2n} \\ \vdots & \vdots & & \vdots & & \vdots \\ R_{n1} & R_{n2} & \ldots & E_n & \ldots & R_{nn} \end{vmatrix}}{D} \quad (2.16)$$

i.e. the determinant in the numerator is the same determinant D but with the jth column replaced by the elements of the right-hand side of eqns (2.11). It is worth verifying that this gives the same results for I_1 and I_2 as in eqns (2.14).

On referring back to the circuit of Fig.2.3(b), $DI_2 = E_1 D_{12} - E_2 D_{22}$ and $DI_3 = E_1 D_{13} - E_2 D_{23}$ where

$$D = \begin{vmatrix} (R_1 + R_5 + R_6) & -R_6 & -R_5 \\ -R_6 & (R_2 + R_4 + R_6) & -R_4 \\ -R_5 & -R_4 & (R_3 + R_4 + R_5) \end{vmatrix}.$$

Consider the condition under which $(I_2 - I_3) = 0$. From above,

$$D(I_2 - I_3) = E_1(R_3 R_6 - R_2 R_5)$$
$$- E_2(R_1 R_3 + R_1 R_5 + R_3 R_5 + R_3 R_6).$$

If R_3 is very large (corresponding to the resistance of some insulating material, say), then the circuit becomes that of a potentiometer with $(R_5 + R_6)$ corresponding to the 'slide wire' and R_4 the resistance of the indicating meter. Further if it is assumed, for simplicity, that $R_1 = 0$ so that all of the e.m.f. E_1, appears across $(R_5 + R_6)$, then the balance condition $I_2 - I_3 = 0$ becomes

$$E_1(R_3 R_6 - R_2 R_5) - E_2(R_3 R_6 + R_3 R_5) = 0. \quad (2.17)$$

Now since $R_3 \gg R_2, R_5, R_6$, this condition becomes

$$E_1 R_3 R_6 = E_2 R_3 (R_5 + R_6) \quad \text{or} \quad \frac{E_2}{E_1} = \frac{R_6}{(R_5 + R_6)} \quad (2.18)$$

which is the familiar balanced potentiometer condition.

If, in the circuit of Fig.2.3, the source of e.m.f. E_2 is removed (leaving resistance R_2 in the branch), the circuit becomes a Wheatstone bridge and the balance condition of eqn (2.17) ($I_2 - I_3 = 0$ with $E_2 = 0$, remember) takes the familiar form $R_3 R_6 = R_2 R_5$, i.e. the

32 Direct-current circuits

products of the resistances in opposite arms of the bridge are equal. If R_2, R_3, R_6 are calibrated variable resistors then the value of R_5 may be determined experimentally by adjusting R_3 and R_2 (the 'ratio' arms) to a convenient value and then adjusting R_6 to obtain the balanced condition.

2.5. Nodal analysis

In Section 2.4, in applying Kirchhoff's voltage law, the equation through which the resistance of a resistor was defined was used in the form $V = RI$, with I the independent variable. This equation can be transposed and written as $I = GV$ where $G (\equiv 1/R)$ is the conductance, in siemens (S), of the resistor. The relevance of this is that current networks can be analysed with voltages as unknowns, resistances transposed to conductances, and with constant current generators in place of constant voltage generators. As a simple example the circuit of Fig.2.4(a) can be redrawn as in Fig.2.4(b); as far as the loop containing R_2 and R_3 is concerned, A and B are the accessible terminals of a source of e.m.f. E having an internal resistance R_1. Since only differences in potential have physical significance it is possible to make a convenient choice of junction B as a reference node ('common' or 'earth') and refer all voltages to it and, of course, Kirchhoff's voltage law is satisfied thereby. From the current law, $(N_j - 1)$ equations are obtained; in this example $N_j = 2$ and, hence, there is only one equation to be solved, namely,

$$(V_1 - V_2)(G_1 + G_2 + G_3) = I \qquad (2.19)$$

where the left-hand side represents the sum of the currents leaving the node A and I is the current entering it. Putting $V_2 = 0$ for convenience,

$$\begin{aligned} V_1 &= IR_1R_2R_3/(R_1R_2 + R_1R_3 + R_2R_3) \\ &= ER_2R_3/(R_1R_2 + R_1R_3 + R_2R_3) \end{aligned} \qquad (2.20)$$

which, on putting $V_1 = V_L$, $R_3 = R_L$ is identical to eqn (2.15).

The situations in which it is advantageous to use nodal analysis are firstly, where voltages are of more direct interest than currents (for instance in transistor circuits) and, secondly, in networks of a parallel nature where the number of independent node equations may be chosen to be significantly smaller than the number of mesh equations because of common connections. This second point is very important since, for complex networks, the evaluation of the determinants may be the most time-consuming part of the analysis, and this labour is obviously related to the dimensions of the determinants involved. The preceding example, although very simple, does illustrate this general point since there is only one node equation compared with two mesh equations.

As another example of the use of nodal analysis consider again the network of Fig.2.3; note that the voltage sources E_1 (in series with R_1) and E_2 (in series with R_2) have been replaced, as outlined in Section 2.1, by equivalent current generators I_1 (in parallel with $G_1 = 1/R_1$) and I_2 (in parallel with $G_2 = 1/R_2$). On applying Kirchhoff's current law to nodes A, B, and C respectively the following equations are obtained (assuming $V_D = 0$):

$$V_1 G_1 + (V_1 - V_2)G_5 + (V_1 - V_3)G_3 = I_1$$
$$(V_2 - V_3)G_4 + (V_2 - V_1)G_5 + V_2 G_6 = 0$$
$$V_3 G_2 + (V_3 - V_1)G_3 + (V_3 - V_2)G_4 = I_2$$

or (2.21)

$$V_1(G_1 + G_3 + G_5) - V_2 G_5 - V_3 G_3 = I_1$$
$$-V_1 G_5 + V_2(G_4 + G_5 + G_6) - V_3 G_4 = 0$$
$$-V_1 G_3 - V_2 G_4 + V_3(G_2 + G_3 + G_4) = I_2.$$

The formal similarity between these equations and eqns (2.11) should be noticed; the methods for solving them are the same.

The procedure for obtaining nodal equations can be extended obviously to more complex networks but the general rules have emerged already:

(a) Choose a convenient reference node (usually common or earth) and assume a voltage for every other node.

(b) In the equation for a particular node the coefficient of the voltage of that node is the sum of the conductances connected to it; the coefficients of the other voltages are the conductances joining them to the particular node, always taken negative.

For a circuit containing n nodes

$$V_j = \frac{1}{D} \sum_{i=1}^{n} I_i D_{ij} \qquad (2.22)$$

(compare eqn (2.12) or use the analogue of eqn (2.16)). It is left as an exercise to obtain the balance condition for the network of Fig.2.3 using nodal analysis.

2.6. The superposition and reciprocity theorems

In the analyses leading to eqns (2.12) and (2.22), an implicit assumption was that the resistors were linear, i.e. $V \propto I$ or V/I independent of the current. A non-linear relation between V and I can be written in the form

$$V = AI + BI^2 + CI^3 + \ldots \quad \text{or} \quad V/I = A + BI + CI^2 + \ldots$$

where V/I is not independent of I. A circuit element with such a characteristic is non-linear (and non-ohmic).

A consequence of the assumed linearity of the circuit elements is that the algebraic equations in the mesh and nodal analyses are linear and the practical significance of this is manifested in the superposition theorem: the current flowing in a particular mesh of a network is the sum of the currents which would flow if each of the sources of e.m.f. were, in turn, present alone (the internal resistances of all sources must be left in place of course). This statement is exemplified by eqn (2.12) and the equivalent statement for voltages is eqn (2.22). Networks containing two or more sources can often be solved in inspection by using this theorem.

Also, if in a network there is only one source of e.m.f. E and a current I is measured in a particular branch then the source and ammeter (an ammeter of zero resistance whose existence is implied by the operation of measuring a current) may be interchanged and the measured current will remain unchanged. This is a statement of the reciprocity theorem and an example which you will have met is the interchangeability of the source of e.m.f. and detector in a Wheatstone bridge.

2.7. Equivalent circuits

Frequently the current and voltage in only one branch of a network are of immediate interest; for instance, a simple example is the circuit of Fig.2.4(a) where the voltage across the load resistor R_3 is of particular interest and R_1, R_2, and the source of e.m.f. could be contained within a 'black box'. The expression (eqn (2.14)) for the current through the load resistor R_3 can be written as

$$I_2\left(\frac{R_1 R_2}{(R_1 + R_2)} + R_3\right) = \frac{ER_2}{(R_1 + R_2)} \tag{2.23}$$

and by comparison with eqn (2.4) the internal resistance 'R_0' and the e.m.f. 'E_0' of the Thévenin equivalent source are $R_1 R_2/(R_1 + R_2)$ and $ER_2/(R_1 + R_2)$ respectively.

Since $I_2 R_3 (= $ 'V_0' say) is the voltage at the accessible terminals of the imagined black box when current I_2 is drawn by the load R_3, we can arrange eqn (2.23) in the form

$$V_0 = \frac{ER_2}{(R_1 + R_2)} - I_2 \frac{R_1 R_2}{(R_1 + R_2)}. \qquad (2.24)$$

It can be seen that eqn (2.24) is in the same form as eqn (2.3) with 'E_0' = $ER_2/(R_1 + R_2)$ and 'R_0' = $R_1 R_2/(R_1 + R_2)$.

So if the Thévenin equivalent of a network has to be found it is useful to aim at expressing the relation between the terminal voltage and the load current in the form of either eqn (2.3) or (2.4). The *Norton equivalent network* follows immediately from these considerations i.e. it consists of an ideal current source 'I_0' = 'E_0'/'R_0' shunted by a conductance 'G_0' = $1/$'R_0'.

In Section 2.4 it was pointed out that the circuit of Fig.2.3 became a conventional potentiometer circuit if $R_3 \to \infty$ and it can be redrawn as shown in Fig.2.5 where, for simplicity, R_1 has been taken to be zero, and $R_4 = R_M$, the resistance of the indicating meter. The balance condition ($I_2 = 0$) has already been shown to be $E_2/E_1 = R_6/(R_5 + R_6)$ but another crucial practical consideration is that of the magnitude of I_2 for a given degree of unbalance of the circuit, since this will determine the precision with which the balance can be found. From the arguments above it can be seen quite simply, by inspection of the equivalent circuit, that

$$I_2 = \left(\frac{E_1 R_6}{(R_5 + R_6)} - E_2\right) \Big/ \left(\frac{R_5 R_6}{(R_5 + R_6)} + R_M + R_2\right). \qquad (2.25)$$

If $E_1 R_6/(R_5 + R_6)$ differs from E_2 by x per cent, say, then

$$I_2 = \frac{xE_2}{100} \Big/ \left(\frac{E_2 R_5}{E_1}\left(1 + \frac{x}{100}\right) + R_M + R_2\right) \qquad (2.26)$$

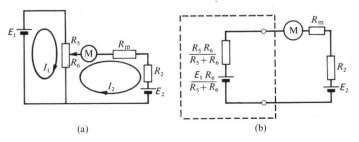

FIG. 2.5. The potentiometer: (a) a mesh representation; (b) a representation as a 'black box' connected to the secondary source of e.m.f. and the indicating meter.

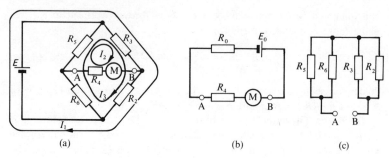

FIG. 2.6. (a) A Wheatstone bridge and (b) its equivalent circuit as 'seen' from the indicating meter. The method for calculating R_0 is illustrated by (c).

and for $x = 1$, $E_1 = 2$ V, $E_2 = 1.5$ V, $R_M = 450\,\Omega$, $R_2 = 50\,\Omega$, $R_5 = 660\,\Omega$, then $I_2 \approx 15\,\mu A$ which can be detected with commonly available galvanometers.

Another illustration of the usefulness of equivalent circuits is the problem of obtaining an expression for the out-of-balance current in a Wheatstone bridge. In an approach using mesh analysis the currents are chosen so that only one such current traverses the branch containing the detector, since the detector current is the quantity of particular interest (see Fig.2.6). It is a useful, but somewhat laborious, exercise in setting up mesh equations and solving determinants (try it) to show that $I_2(R_0 + R_4) = E_0$ where

$$R_0 = \frac{R_5 R_6}{(R_5 + R_6)} + \frac{R_2 R_3}{(R_2 + R_3)} \qquad E_0 = \frac{E(R_3 R_6 - R_2 R_5)}{(R_5 + R_6)(R_2 + R_3)}.$$

(2.27)

For the simple source of Fig.2.1, R_0 is equal to the resistance between the terminals with the ideal source of e.m.f. E_0 removed. This statement is also true for more complex networks; to calculate the equivalent resistance R_0 remove all sources of e.m.f. and replace them by short circuits, or by their internal resistances if these are not negligible.

For the Wheatstone bridge of Fig.2.6(a) the network as 'seen' from terminals A and B is as shown in Fig.2.6(c), if the source is assumed to have negligible internal resistance, i.e. R_5 is in parallel with R_6 and R_3 with R_2. Hence,

$$R_0 = R_5 R_6/(R_5 + R_6) + R_2 R_3/(R_2 + R_3)$$

Furthermore, the open-circuit voltage between A and B is just

$$ER_3/(R_2 + R_3) - ER_5/(R_5 + R_6) = E_0$$

These expressions for R_0 and E_0 are identical with eqns (2.27). Thence, as with the potentiometer, the smallest detectable out-of-balance, and the precision with which the 'unknown' resistance R_5 can be measured, can be established.

To conclude this section the analogue in nodal analysis to Thévenin's theorem will be stated: 'Any linear network having two terminals can be represented by a constant current generator in parallel with a conductance.' This is Norton's Theorem.

2.8. Direct-current measurements

The most commonly used current measurement instrument is the d'Arsonval meter, which is of the moving-coil type. The deflection of the coil and pointer is proportional to the magnetic field in which the coil resides, the coil area and number of turns, the current through the coil, and inversely proportional to the torque constant of the restraining springs. Meters with full-scale deflection (f.s.d.) sensitivities ranging from 10 mA down to about $50\mu A$ are commonly available. The range of a meter may be extended to larger currents by placing a shunt of suitable magnitude in parallel with the meter, thus diverting a calculable proportion of the current to be measured around the meter. The resistance of a current-measuring meter should be as small as possible, for obvious reasons, but in situations where it is not negligible in comparison with other resistances in the circuit then it can be allowed for; the resistances of 10-mA and 100-μA meters are, typically, of the order of $1\,\Omega$ and $10^3\,\Omega$ respectively.

As an example of calculating the value of a shunt resistance consider a 100-μA meter having a resistance of $1\,k\Omega$ and suppose that it is wished to use this meter to measure currents in the range $100\mu A$ to 100 mA. If the resistance of the shunt and the meter are denoted by R_s and R_m respectively and if the current through the meter, the current through the shunt, and the total current are denoted by I_m, I_s, and I respectively, then $I = I_m + I_s$ and $I_m R_m = I_s R_s$ and for $I = nI_m$, where n is an integer, then $R_s = R_m/(n-1)$. For $n = 10^1$, 10^2, 10^3 then $R_s = 111.1\,\Omega$, $10.1\,\Omega$, and $1.001\,\Omega$ respectively. If a tapped shunt is connected across a 100-μA meter, as indicated in Fig.2.7(a) then the range of the meter can be extended simply by the factors 10, 10^2, 10^3; this is the principle of the 'universal' shunt.

An ammeter can also be used as a voltmeter since a deflection of the coil can be considered as being due to a voltage IR_m across the meter resistance R_m; obviously the range of measurable voltages can be extended by including a suitable resistance in series with the meter. The resistance of a voltmeter should be high enough so that it drains an insignificant current between the two points of the circuit between

38 Direct-current circuits

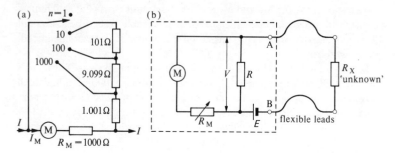

FIG. 2.7. (a) A universal shunt to increase the current range of a 1000-Ω meter by factors $n = 10, 100,$ and 1000. (b) The use of a sensitive moving-coil meter to give direct measurements of resistance.

which it is connected (otherwise the voltmeter is said to 'load' the circuit to which it is connected).

Suppose in the circuit of Fig.2.4(a) that R_3 represents the resistance of a voltmeter (i.e. meter resistance plus series resistance) which is to be used to measure the voltage across R_2. Further, suppose that a condition of the problem is that $I_2 \leqslant I_1/100$; then from eqns (2.14), $R_3 \geqslant 99R_2$. If $E = 1.5$ V, $I_1 = 10$ mA, then $R_1 + R_2 = 150$ Ω, very closely, and if $R_2 = 100$ Ω, say, then $R_3 \geqslant 9900$ Ω. An implication is that the meter employed must give a significant deflection for $I_2 = 100 \mu$A.

A moving-coil meter, if incorporated into a suitable circuit, can be used to measure, directly, the resistance R_x of a resistor, or indeed of a two-terminal network. A simple circuit is shown in Fig.2.7(b). Assuming that the meter mesh draws a negligible current, then the voltage across R (i.e. the voltage measured by the voltmeter, M) is

$$V = \frac{ER}{(R + R_x)} \quad \text{or} \quad R_x = R\left(\frac{E}{V} - 1\right). \quad (2.28)$$

If the terminals A, B are now shorted together then the voltage across R is E; in this situation R_m is adjusted to give f.s.d. deflection on the meter scale, which point is marked as 'zero ohms'. For $V = E/2$, $R_x = R$ and so the midpoint of the meter scale can be marked as 'R ohms'. Similarly for $V = E/4$, $R_x = 3R$ and a zero scale reading corresponds to $R_x = \infty$ or 'open-circuit'; the ohmmeter scale is non-linear, clearly, but it is direct reading.

Since the mid-scale reading depends on the value of R it is obvious that it is possible to encompass different ranges of resistance by switching suitable values of R into the circuit.

In commercial 'multimeters' there is a mode switch to convert the instrument into an ammeter or voltmeter (d.c. or a.c.) or into

Direct-current circuits 39

an ohmmeter, together with range switches for each mode. A slight complication, compared with the simple ohmmeter circuit just described is that the meter current may not be negligible when R is large, and this factor has to be allowed for.

A frequently-used figure of merit for voltmeters is the 'ohm per volt'; for instance a meter of resistance $1\,\text{k}\Omega$ for which the f.s.d. current is 1 mA has a figure of merit of $1\,\text{volt}/10^{-3}$ Ampere = 10^3 ohm per volt. For a 10^4 ohm per volt voltmeter operating on a 10-volt scale the meter's resistance is $10^4 \times 10 = 10^5\,\Omega$ and the f.s.d. current sensitivity is $10\,\text{volt}/10^5$ ohm = $100\,\mu\text{A}$.

It is of some interest to calculate the out-of-balance current for Wheatstone bridge circuits which might be used in practice and hence determine the type of detector which will be most suitable. Assume in the circuit of Fig.2.6 that $R_2 = R_3 = R_6 = R$; $R_5 = R_x$ (the 'unknown' resistance), and $R_4 = R_m$ (the resistance of the detector). It follows that the expressions for the out-of-balance current I_m through the detector and the out-of-balance e.m.f. E_0 (eqns (2.27)) take the forms

$$I_m = \frac{E(R - R_x)}{2R_m(R + R_x) + R(R + 3R_x)} \quad \text{and} \quad E_0 = \frac{E(R - R_x)}{2(R + R_x)}.$$

(2.29)

Near to 'balance' $(R - R_x) \ll R$ and these expressions reduce to

$$I_m \approx \frac{E(R - R_x)}{4R(R + R_m)} \quad \text{and} \quad E_0 \approx \frac{E(R - R_x)}{4R},$$

i.e. I_m and E_0 are linearly dependent on $(R - R_x)$. Assuming $(R - R_x)/R = 10^{-3}$ then consider two extreme practical situations namely (a) where $R = 10\,\Omega$ and $R_m = 100\,\Omega$ and (b) where $R = 100\,\text{k}\Omega$ and $R_m = 1\,\text{M}\Omega$. In situation (a) if $E = 2\,\text{V}$ then $I_m \approx 5\,\mu\text{A}$ and in (b) if $E = 200\,\text{V}$ then $E_0 \approx 50\,\text{mV}$. Hence in (a) a low-resistance current-measuring meter would be used whereas in (b) a high-resistance voltmeter would be used.

PROBLEMS

2.1. A power supply has a terminal voltage of 999 V and 990 V when the currents drawn from it are 10 mA and 100 mA respectively. If the load resistances in the two situations are $99.9\,\text{k}\Omega$ and $9.9\,\text{k}\Omega$ respectively, calculate the e.m.f. and internal resistance of the power supply.

2.2. In the circuit of Fig.2.4(a) for what range of values of R_3 is the voltage across R_3 equal to $ER_2/(R_1 + R_2)$ to at least 1 per cent if $R_1 = 6\,\text{k}\Omega$, $R_2 = 4\,\text{k}\Omega$?

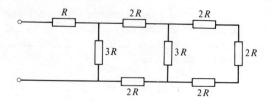

FIG. 2.8.

2.3. Show that the resistance between the input terminals of the network shown in Fig.2.8 is equal to $4R$.

2.4. Obtain eqn (2.27) for the out-of-balance current in the Wheatstone bridge circuit of Fig.2.6 using mesh analysis.

2.5. Find the Norton equivalent of the network shown in Fig.2.9.

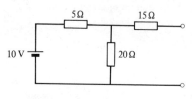

FIG. 2.9.

2.6. For a two-terminal 'black-box' which contains resistors and current or voltage sources only the voltage V and current I at the terminals are related by $I = (-V/4 + 3)$A. Find (a) the Thévenin equivalent circuit and (b) the Norton equivalent circuit, of the 'black box'.

2.7. Find (a) the Thévenin and (b) the Norton equivalent network for the network shown in Fig.2.10.

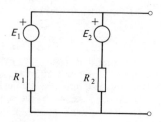

FIG. 2.10.

Direct-current circuits   41

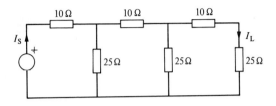

FIG. 2.11.

2.8. Calculate the value of the ratio I_L/I_S for the circuit shown in Fig.2.11 using the methods of both mesh and nodal analysis.

2.9. Find the Thévenin and Norton equivalents of the network shown in Fig.2.12.

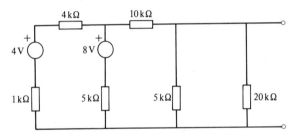

FIG. 2.12.

2.10. Two batteries A and B act in the same circuit as shown in Fig. 2.13. What power does each battery deliver?

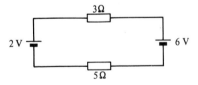

FIG. 2.13.

3. Alternating currents

3.1. Transient currents in series $R-C$ and $R-L$ circuits

THE applicability of Kirchhoff's laws to circuits containing inductors and capacitors as well as resistors, and in which the currents, voltages, and e.m.f.s are time-dependent, will now be examined.

The idea of a current flowing continuously in a closed circuit, with no local accumulations of charge, is retained, i.e. for any element of the circuit the rate at which charge is carried into the element is equal to the rate at which charge leaves it. In a capacitor, charges accumulate on the plates but since these two accumulations are of opposite sign the statement of the preceding sentence remains valid. So it will be assumed that Kirchhoff's voltage law can be applied to the instantaneous values of currents.

Kirchhoff's current law is related to the principle of conservation of energy: the rate of working of the source of e.m.f. is equal to the rate at which energy is being dissipated in the circuit. For the case of a circuit containing resistors only the only drainage of energy from the source is to the moving electrons which form the current, as outlined in Section 1.2.3. However, in a complete circuit containing inductors and capacitors, there is potential energy associated with the magnetic and electric fields in these circuit elements. Hence the rate of working of the source is equal to the rate of dissipation of energy as heat in resistors plus the rate of increase of energy stored in inductors and capacitors. This statement does not apply necessarily, of course, to a loop which is part of a larger complete circuit. Nevertheless a single charge carrier making a complete traversal of a particular closed loop can have no net gain or loss of energy. This energy balance is exemplified by: e.m.f. = sum of all the potential differences around the loop (due to (a) current flowing through resistors, (b) induced e.m.f.s in inductors, and (c) charge accumulations on capacitors). If, in a particular closed loop, there is one resistor R, one inductor L, and one capacitor C, connected in series then

$$E = RI + L\frac{dI}{dt} + \frac{q}{C}. \tag{3.1}$$

Alternating currents

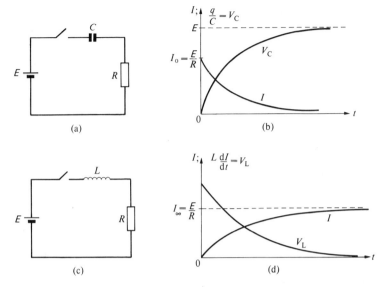

FIG. 3.1. The time-dependence of the voltage V_C across the capacitor C, and the current I, on closing the switch in the circuit (a) are illustrated in (b). The time-dependence of the voltage across the inductor L and the current I in the circuit (c), are illustrated in (d).

For a series $R-C$ circuit (Fig.3.1(a)) eqn (3.1) becomes

$$E = RI + q/C \quad \text{or} \quad E = R\frac{dq}{dt} + \frac{q}{C} \quad (3.2)$$

since for a capacitor $I = dq/dt$ or, alternatively, $q = \int_0^t I\,dt$. This equation can be rewritten as

$$EC - q = CR\, dq/dt \quad \text{or} \quad \frac{1}{CR}\int dt = \int \frac{dq}{(EC-q)}.$$

These indefinite integrals give

$$t/CR = -\ln(EC-q) + \text{'constant'}$$

or

$$(EC-q) = \text{(constant)} \times e^{-t/CR}.$$

Since $q = 0$ at $t = 0$ (the moment when the switch is closed and charging of the capacitor begins), this equation becomes

Alternating currents

$$(EC - q) = ECe^{-t/CR} \quad \text{or} \quad q = EC(1 - e^{-t/CR}). \quad (3.3)$$

Also,

$$I = dq/dt = \frac{E}{R} e^{-t/CR} = I_0 e^{-t/CR} \quad (3.4)$$

where I_0, the current at $t = 0$, is the steady current which would flow if the capacitor were short-circuited. The quantity CR, which has the dimensions of time, is called the time constant of the circuit i.e. the time taken for the current to decrease to a value I_0/e.

For the series $R-L$ circuit of Fig.3.1(c), eqn (3.1) becomes

$$E = RI + L \, dI/dt \quad \text{or} \quad E/R - I = (L/R) \, dI/dt \quad (3.5)$$

and thence

$$\frac{R}{L} \int dt = \int \frac{dI}{(E/R - I)}$$

which gives $Rt/L = -\ln(E/R - I) + \text{'constant'}$ or $E/R - I = \text{'constant'} \times e^{-Rt/L}$. Since $I = 0$ at the moment when the switch is closed ($t = 0$) then the constant must equal E/R and so

$$I = \frac{E}{R} (1 - e^{-Rt/L}) = I_\infty (1 - e^{-Rt/L}) \quad (3.6)$$

where $I_\infty (= E/R)$ is the steady current which would be established after an infinite time. The quantity L/R is called the time constant of this circuit i.e. the time taken for the value of $(I_\infty - I)$ to fall to I_∞/e.

A useful fact to remember from the practical point of view is that in an $R-C$ circuit $I/I_0 = 0.007$ for $t = 5CR$ i.e. the current has fallen to what would be considered to be a negligible magnitude for most purposes. A similar argument applies to $R-L$ circuits, i.e. the current rises to within 0.7 per cent of I_∞ in a time equal to $5 L/R$.

It is left as an exercise to show that if, in the circuit of Fig.3.1(a), the source of e.m.f. is disconnected after the capacitor has been fully charged, and the switch is then closed so that the capacitor discharges through the series resistance R, then the charge q on the capacitor is given by

$$q = q_0 e^{-t/CR}$$

where $q_0 = CE$ is its initial charge. A potentially hazardous situation may arise in circuits where large-value capacitors are charged to a high voltage, e.g. in a high-voltage power supply. If, after such a power supply is switched off, the series resistances through which the capacitors are able to discharge have large values (possibly the input resistance of the device to which the power supply was connected), then the voltages on the capacitors will remain high for a long time. For instance, in an extreme case, the resistance may be that of the leakage path between the terminals of the capacitor (leakage via a film of dirt and/or moisture on the surface of the insulators) which may be $10^{10}\,\Omega$ or higher. In this case if $C = 10\,\mu\text{F}$ then $CR = 10^5$ s and so, if $E = 1\,\text{kV}$, then 28 hours (approximately) after switching off the capacitor would be charged still to a voltage $1000/e \approx 370\,\text{V}$. So provision should be made always for the safe discharge of large-value capacitors which have been charged to a high voltage.

If a series $R-L$ circuit is broken, then the induced e.m.f. in the inductor tries to maintain the current and will cause a spark across the opening contacts or, if the inductor consists of a tightly wound coil of insulated wire, the large induced e.m.f. may cause a breakdown of the insulation. For instance if a current of 10 A in an electromagnet of inductance 10 H were reduced to zero in 1 ms then the induced e.m.f. would be $10 \times 10/10^{-3} = 100\,\text{kV}!$. Thus power supplies for large electromagnets are designed so that the current cannot be altered rapidly and, additionally, it is common practice to connect a rectifier (such as a silicon diode) in parallel with the coil of the magnet. This rectifier is connected so that it presents a high resistance to the normal current from the magnet power supply but a low resistance to induced currents which would flow if the main magnet current decreased rapidly for any reason.

3.2. The graphical and mathematical representation of alternating currents and voltages

3.2.1. Graphical representation

Consider two identical plane rectangular coils each consisting of a single turn of wire of area A situated in a uniform magnetic field B and each of which is rotating with the same constant angular velocity ω about a common axis as illustrated schematically in Fig.3.2(a); each coil is connected in series with a circuit of resistance R via a slip-ring commutator. At a time t the magnetic fluxes Φ_1 and Φ_2 linked with Coils 1 and 2 respectively are given by $\Phi_1 = BA \sin \omega t$ and $\Phi_2 = BA \sin(\omega t + \phi)$ and the magnitudes of the induced e.m.f.s are (see eqn (1.27))

$$E_1 = \omega BA \cos \omega t \quad \text{and} \quad E_2 = \omega BA \cos(\omega t + \phi)$$

The resultant 'sinusoidal' currents are

$$I_1 = \hat{I}_1 \cos \omega t \quad \text{and} \quad I_2 = \hat{I}_2 \cos(\omega t + \phi) \quad (3.8)$$

where $\hat{I} = \omega BA/R$ is the amplitude, or 'peak' value, of the currents

46 Alternating currents

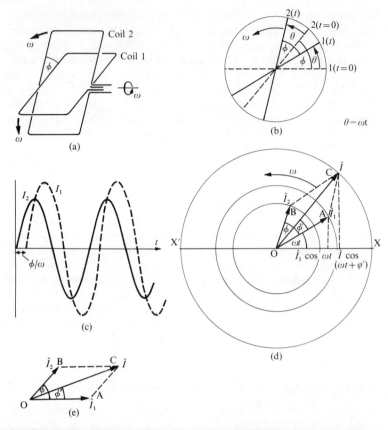

FIG. 3.2. The addition of phasors of the same frequency: (a) two coaxial rotating coils; (b) 'end-view' of the coils; (c) the time-dependent induced currents; (d) the phasor diagram; (e) the 'frozen' phasor diagram.

(the current in a coil reverses direction twice in each revolution, or cycle, of the coil). For the sense of rotation chosen for the illustration Coil 2 'leads' Coil 1 and I_2 is said to lead I_1 by a phase angle ϕ (or simply by phase ϕ).

I_1 and I_2 can be represented on a 'circle', or polar, diagram as shown in Fig.3.2(d) where, for the sake of generality, they are now taken to have different peak values (due to the coils having different areas say). The lines OA and OB, which are both taken to rotate in an anticlockwise sense with angular velocity ω, have lengths proportional to $\hat{I}_1$ and $\hat{I}_2$. The instantaneous values of I_1 and I_2 of eqns (3.8) are given by the projections of OA and OB, respectively, on to the 'horizontal' diameter, or base line X′OX.

Alternating currents

The period of revolution T of a coil, and hence of the induced sinusoidal current, equals $2\pi/\omega$ and the number of cycles per second, or frequency, $f \equiv 1/T = \omega/2\pi$: The unit of frequency is the hertz (Hz); 1 Hz = 1 cycle per second. The angular frequency ω is often called the *pulsatance* and is measured in radians per second. In the ensuing text ω will often occur in contexts where 'frequency' is being discussed; ω will always be 'angular frequency' where $\omega = 2\pi f$.

If the coils are connected in series to an external circuit then the current I will be the sum of I_1 and I_2, namely

$$I = \hat{I}_1 \cos \omega t + \hat{I}_2 \cos (\omega t + \phi). \tag{3.9}$$

This resultant current can be written as $I = \hat{I} \cos (\omega t + \phi')$ if

$$\hat{I}^2 = \hat{I}_1^2 + \hat{I}_2^2 + 2\hat{I}_1\hat{I}_2 \cos \phi \tag{3.10}$$

and

$$\tan \phi' = (\hat{I}_2 \sin \phi)/(\hat{I}_1 + \hat{I}_2 \cos \phi)^\dagger. \tag{3.11}$$

The form of this expression for $\hat{I}^2$ suggests that the resultant of two sinusoidal currents I_1, I_2 of the same frequency can be obtained by treating the currents as vectors, of magnitudes proportional to the peak values of the currents and inclined at an angle equal to their phase difference, and then completing the parallelogram of vectors; the resultant yields $\hat{I}$ and $\cos \phi'$ (see Fig.3.2(d)). The instantaneous value of the resultant is found from its projection on the common base line, as for the components I_1 and I_2.

Obviously three or more sinusoidally varying quantities (or *phasors* as they are called) of the same frequency can be combined by the same procedure; the same kind of phase-amplitude diagram is often used to illustrate the formation of interference and diffraction patterns in optics.

Since all of the current 'vectors' in Fig.3.2(d) are rotating at the same angular velocity, the diagram can be 'frozen' as shown in Fig.3.2(e). Only differences of phase are of interest and so any direction can be chosen as the base line although usually, for convenience, the phase differences are measured relative to a reference current, for example I_1 (or a reference sinusoidal voltage if it is voltages that are being combined).

The fact that the instantaneous values of the phasors are given by the projections of their amplitude vectors on to a base line leads

---

†To show this, expand the expressions for $(I_1 + I_2)$ and $(\hat{I} \cos (\omega t + \phi'))$ using standard trigonometrical rules. Then square the expressions so obtained and compare the coefficients of the terms in $\sin^2 \omega t$, $\cos^2 \omega t$, and $\sin \omega t \cdot \cos \omega t$. From the three resultant equations the desired relationships can be obtained.

naturally, as will now be seen, to a representation of phasors by complex numbers.

3.2.2. The use of complex numbers

The symbol j is used to represent an operation which if carried out twice in succession on a number reverses the sign of that number, i.e. if a real number is denoted by y then $j(jy) = j^2 y = -y$. As well as being an operation j can also be thought of as the number resulting from j operating on the real number one. Hence j^2 can be thought of as the result of j operating twice on the number one yielding $j^2 = -1$ so that j can be thought of also as 'the square root of minus one'; j is said to be an *imaginary* number.

An imaginary number jy (y being real) can be combined with a real number x to form a *complex* number w where

$$w = x + jy. \qquad (3.12)$$

Complex numbers are represented conveniently in graphical form in

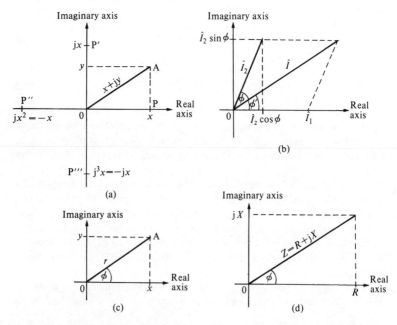

FIG. 3.3. The Argand diagram: (a) the representation of a complex number $w = x + jy$; (b) the addition of phasors I_1 and I_2 of the same frequency; (c) the polar form of a complex number; (d) the representation in the complex plane of a complex impedance $Z = R + jX$.

an Argand diagram (Fig.3.3). Since j^2 acting on the real number x translates P to P″ the operation can be imagined as a rotation of the line OP through an angle of 180°. Hence j itself can be thought of as an operator which rotates OP through 90° to P′ and the operations j^3 and j^4 cause rotations of 270° and 360° respectively (N.B. $1/j = j/j^2 = -j$). So if real numbers are represented to scale along the x-axis and imaginary numbers are represented likewise along the y-axis then the complex number w is represented by a point such as A and OA is referred to as 'a vector in the complex plane' of which the amplitude, or modulus, is equal to $(x^2 + y^2)^{1/2}$ and whose argument is $\phi = \tan^{-1}(y/x)$.

A function $y = a^z$ is called an exponential function and the inverse relation is $z = \log_a y$. An exponential function of particular interest is $y = e^z$ where

$$e^z = 1 + z + \frac{z^2}{2!} + \frac{z^3}{3!} + \ldots . \tag{3.13}$$

and $e^1 = 2.718\ldots$. The number e is the base of natural logarithms but a more useful property of exponential functions in the present context, which will be stated without proof, is

$$e^{jz} = \cos z + j \sin z. \tag{3.14}$$

(Memorize this relationship!)

So for a complex number

$$w = x + jy \tag{3.15a}$$

$$\text{mod } w = |w| = (x^2 + y^2)^{1/2} \tag{3.15b}$$

$$\arg w = \tan^{-1}(y/x). \tag{3.15c}$$

Now the position of a point in the complex plane can also be expressed in terms of polar coordinates. For instance the point A (see Fig.3.3(c)) has polar coordinates (r, ϕ) where

$$\begin{aligned} x &= r \cos \phi \\ y &= r \sin \phi. \end{aligned} \tag{3.16}$$

On substituting for x and y in eqn (3.12) we find

$$w = r(\cos \phi + j \sin \phi).$$

Thus on using eqn (3.14) it follows that

$$w = re^{j\phi}. \tag{3.17}$$

Further, it can be seen from Fig.3.3, or from eqns (3.16) that

$$r = (x^2 + y^2)^{1/2}.$$

Eqn (3.17) represents the complex number w in polar form; r and ϕ are the modulus and argument of w respectively.

It should be noted that for two complex numbers

$$w_1 = r_1 e^{j\phi_1}, \quad w_2 = r_2 e^{j\phi_2},$$

it follows that

$$w_1 w_2 = (r_1 r_2) e^{j(\phi_1 + \phi_2)}$$

$$\frac{w_1}{w_2} = \frac{r_1}{r_2} e^{j(\phi_1 - \phi_2)}.$$
(3.18)

We will now assume for the moment that a phasor of the general form $I = \hat{I} \cos(\omega t + \alpha)$ can be represented by $I = \hat{I} \exp j(\omega t + \alpha)$. Then eqn (3.9), which can be written as $\hat{I} \cos(\omega t + \phi') = \hat{I}_1 \cos \omega t + \hat{I}_2 \cos(\omega t + \phi)$ becomes

$$\hat{I} e^{j(\omega t + \phi')} = \hat{I}_1 e^{j\omega t} + \hat{I}_2 e^{j(\omega t + \phi)}.$$
(3.19)

The common factor $\exp(j\omega t)$ cancels out (this is the equivalent of going to the 'frozen' phasor diagram Fig.3.2(e)) and so

$$\hat{I} e^{j\phi'} = \hat{I}_1 + \hat{I}_2 e^{j\phi}.$$
(3.20)

Now in any equation involving complex numbers, such as $(a + jb) = (c + jd)$ the real parts on both sides are equal $(a = c)$ and so also are the imaginary parts $(b = d)$. An alternative, equivalent, pair of statements is that the moduli are equal $(\sqrt{(a^2 + b^2)}) = (\sqrt{(c^2 + d^2)})$ as also are the arguments $(\tan^{-1}(b/a) = \tan^{-1}(d/c))$. Hence, on equating moduli in eqn (3.20),

$$\hat{I}^2 \cos^2 \phi' + \hat{I}^2 \sin^2 \phi' = \hat{I}_1^2 + \hat{I}_2^2 \cos^2 \phi + 2\hat{I}_1 \hat{I}_2 \cos \phi + \hat{I}_2^2 \sin^2 \phi$$

or

$$\hat{I}^2 = \hat{I}_1^2 + \hat{I}_2^2 + 2\hat{I}_1 \hat{I}_2 \cos \phi.$$
(3.21)

Equating the arguments of eqn (3.20) yields

$$\tan \phi' = (\hat{I}_2 \sin \phi)/(\hat{I}_1 + \hat{I}_2 \cos \phi).$$
(3.22)

It can be seen that eqns (3.21) and (3.22) are identical to eqns (3.10) and (3.11) (thus justifying our earlier assumption of the exponential representation of a phasor) and that they relate the peak values and relative phases (i.e. phase differences) of the phasors. The *instantaneous* value of a phasor is given by the *real* part of an expression of the general form $\hat{I} \exp\{j(\omega t + \alpha)\}$ namely $\hat{I} \cos(\omega t + \alpha)$; for example in Fig.3.3(b) the real part of $\hat{I}_2 \exp\{j(\omega t + \phi)\}$ is the projection on to the real axis

Alternating currents

in the complex plane of a vector of modulus $\hat{I}_2$ and argument ϕ.† Thus the problem of adding two phasors is reduced, by using the complex number representation, to the problem of finding the resultant of two vectors in the complex plane (Fig.3.3(b)).

Bearing in mind the possible conceptual difficulties introduced by the use of complex numbers it may be thought that, for the problem just considered, there has been no significant simplification! However, in more complicated a.c. circuit problems the use of the frozen phasor diagram (as a useful conceptual aid) coupled with the well-developed theory of complex numbers does simplify the analyses greatly; this will become apparent later in this, and other, chapters.

As a slightly more complicated illustrative example consider the problem of finding the resultant of three current phasors $\hat{I}_1 \cos \omega t$, $\hat{I}_2 \cos(\omega t + \phi_2)$, and $\hat{I}_3 \cos(\omega t + \phi_3)$. The resultant $\hat{I} \cos(\omega t + \phi)$ can be expressed in terms of the three component currents using the exponential representation

$$\hat{I} e^{j(\omega t + \phi)} = \hat{I}_1 e^{j\omega t} + \hat{I}_2 e^{j(\omega t + \phi_2)} + \hat{I}_3 e^{j(\omega t + \phi_3)}$$

or

$$\hat{I} e^{j\phi} = \hat{I}_1 + \hat{I}_2 e^{j\phi_2} + \hat{I}_3 e^{j\phi_3}. \quad (3.22)$$

Equating the real and imaginary parts of this equation yields

$$\hat{I} \cos \phi = \hat{I}_1 + \hat{I}_2 \cos \phi_2 + \hat{I}_3 \cos \phi_3$$
$$\hat{I} \sin \phi = \hat{I}_2 \sin \phi_2 + \hat{I}_3 \sin \phi_3. \quad (3.23)$$

Hence, the modulus of the resultant is given by

$$\hat{I}^2 = (\hat{I}_1 + \hat{I}_2 \cos \phi_2 + \hat{I}_3 \cos \phi_3)^2 + (\hat{I}_2 \sin \phi_2 + \hat{I}_3 \sin \phi_3)^2 \quad (3.24)$$

and the argument by

$$\tan \phi = (\hat{I}_2 \sin \phi_2 + \hat{I}_3 \sin \phi_3)/(\hat{I}_1 + \hat{I}_2 \cos \phi_2 + \hat{I}_3 \cos \phi_3). \quad (3.25)$$

Again the instantaneous value of the resultant is given by the real part of $\hat{I} \exp\{j(\omega t + \phi)\}$ namely $\hat{I} \cos(\omega t + \phi)$.

---

†Equally well we could have chosen the instantaneous value of a phasor to be given by the imaginary part of an expression of the form $\hat{I} \exp\{j(\omega t + \phi)\}$, namely $\hat{I} \sin(\omega t + \phi)$, i.e. the projection on the imaginary axis of a vector of modulus $\hat{I}$ and argument ϕ.

However, once a choice has been made at the beginning of a particular calculation with regard to using the real or imaginary part then you must adhere to that choice when, at a later stage, you may wish to evaluate the instantaneous values of phasors appearing in that calculation.

It is common practice to write the real part of a complex number w as Re (w) and the imaginary part as Im (w).

3.3. Impedance

Suppose that a current $I = \hat{I} \cos \omega t$ is flowing through a 'pure' resistor R; then the voltage V across the resistor is given by

$$V = (R\hat{I}) \cos \omega t, \qquad (3.26)$$

i.e. V is in phase with I and has a peak value of $(R\hat{I})$.

Now suppose that a current $I = \hat{I} \cos \omega t$ is flowing through a 'pure' inductor L; by eqn (1.28) the voltage across the inductor is given by

$$V = L\,dI/dt = -\omega L \hat{I} \sin \omega t = \omega L \hat{I} \cos(\omega t + 90°), \qquad (3.27)$$

i.e. the voltage 'leads' the current by a phase angle of $90°$ and has a peak value of $(\omega L \hat{I})$.

If a current $I = \hat{I} \cos \omega t$ flows through a 'pure' capacitor C then

$$V = \frac{q}{C} = \frac{1}{C} \int I\,dt = \frac{\hat{I}}{\omega C} \sin \omega t = \frac{\hat{I}}{\omega C} \cos(\omega t - 90°), \qquad (3.28)$$

i.e. the current leads the voltage by $90°$.

A query might be raised at this point as to how the 'response' (the current) can lead the 'excitation' (the voltage). However it must be noted and remembered that this analysis using phasors is applicable only when the transient currents and voltages associated with the 'starting-up' process have decayed to negligible proportions and the amplitudes and phases of the remaining currents and voltages have *steady* values. Thus in the foregoing arguments based on e.m.f.s generated in rotating coils it was assumed implicitly that the coils had been rotating at constant angular velocity for a long enough time for transient e.m.f.s to be negligible.

Now consider a circuit element which can be represented by a pure inductor L in series with a pure resistor R. If a current $I = \hat{I} \cos \omega t$ is flowing through this circuit element then the voltage V across the element is, from eqns (3.26) and (3.27),

$$V = RI \cos \omega t + \omega L \hat{I} \cos(\omega t + 90°). \qquad (3.29)$$

Using the rules for combining phasors which are embodied in eqns (3.10) and (3.11) it follows that, if V is written as $\hat{V} \cos(\omega t + \phi)$ then, since $\cos 90° = 0$,

$$V^2 = R^2 \hat{I}^2 + \omega^2 L^2 \hat{I}^2 = \hat{I}^2 (R^2 + \omega^2 L^2) \qquad (3.30)$$

and

$$\tan \phi = \omega L / R. \qquad (3.31)$$

Similarly, for a circuit element consisting of a pure capacitor and a pure resistor in series, then for a current $I = \hat{I} \cos \omega t$ the voltage V across the element is

$$V = \hat{V} \cos(\omega t + \phi) = R\hat{I} \cos \omega t + \frac{\hat{I}}{\omega C} \cos(\omega t - 90°)$$

where

$$\hat{V}^2 = R^2\hat{I}^2 + \hat{I}^2/(\omega^2 C^2) = \hat{I}^2(R^2 + 1/(\omega^2 C^2)) \quad (3.32)$$

and

$$\tan \phi = -\frac{1}{\omega CR}. \quad (3.33)$$

It can be seen from eqns (3.27), (3.28), and (3.29) that for a.c. currents the ratio of the voltage across a circuit element to the current through it it is not as simple to interpret as in the d.c. case because, in general, there is a component of voltage which is 90° out of phase with the current (phasors which are 90° out of phase with each other are said to be 'in quadrature'). The ratio of the peak values of the voltage and current can be found (see eqn (3.30) for example) as also can the phase difference between them, but only through a separate equation.

It is at this juncture that the usefulness of the complex number representation becomes apparent; the ratio of the voltage across a circuit element to the current through it is called the impedance Z where

$$Z \equiv \frac{V}{I} = \frac{\hat{V}e^{j(\omega t + \phi)}}{\hat{I}e^{j\omega t}} = \frac{\hat{V}e^{j\phi}}{\hat{I}} \quad (3.34)$$

Thus impedance as so defined is a complex quantity and if $Z = (a + jb)$, say, then $(a^2 + b^2) = \hat{V}^2/\hat{I}^2$ and $b/a = \tan \phi$. On comparing these relationships with eqns (3.30) and (3.31) it can be seen that for an element consisting of an inductor L in series with a resistor R, the real part of Z is to be identified with R and the imaginary part with ωL. For a capacitor C and resistor R in series the real part of Z is identified again with R and the imaginary part with $\{-1/(\omega C)\}$. The imaginary part of an impedance is called the reactance and is usually denoted by the symbol X, i.e. $Z = R + jX$ in general (see Fig.3.3(d)).

From the definition of Z, the voltage V across the circuit element of impedance $Z = R + jX$ through which the current $I = \hat{I} \exp(j\omega t)$ is flowing is given by

$$V = ZI = (R + jX)\hat{I}e^{j\omega t} \quad (3.35)$$

and the instantaneous value of V is given by the real part of this expression which is

$$V = R\hat{I} \cos \omega t - X\hat{I} \sin \omega t. \quad (3.36)$$

54 Alternating currents

If $X = \omega L$, say, then this equation becomes identical with eqn (3.29) since $\cos(\omega t + 90°) = -\sin \omega t$.

It has been seen that the basic laws governing d.c. networks are $V = RI$ for a (resistive) circuit element coupled with $\Sigma(E + V) = 0$ for a closed loop and $\Sigma I = 0$ at a node. Analogous relationships are valid in a.c. networks namely $V = ZI$ coupled with $\Sigma(E + V) = 0$ and $\Sigma I = 0$ where V, E, Z, and I are represented now by complex functions of course. Thus for impedances in series and parallel

$$Z = Z_1 + Z_2 + Z_3 + \ldots, \qquad \frac{1}{Z} = \frac{1}{Z_1} + \frac{1}{Z_2} + \frac{1}{Z_3} + \ldots. \qquad (3.37)$$

For a capacitor C in parallel with a resistor R

$$\frac{1}{Z} = \frac{1}{R} + j\omega C = (1 + j\omega CR)/R$$

or

$$Z = R/(1 + j\omega CR).$$

This expression has a complex denominator and so we *rationalize* by multiplying numerator and denominator by the conjugate complex† of the denominator, viz.

$$Z = \frac{R(1 - j\omega CR)}{(1 + \omega^2 C^2 R^2)}.$$

Hence,

$$|Z| = \frac{R(1 + \omega^2 C^2 R^2)^{1/2}}{(1 + \omega^2 C^2 R^2)}$$

or

$$|Z| = R(1 + \omega^2 C^2 R^2)^{-1/2}$$

and

$$\tan \phi = -\omega CR.$$

The circuit theorems introduced in Chapter 2 for d.c. circuits (Kirchhoff's laws, superposition theorem, Thévenin's and Norton's theorems) apply to instantaneous values of currents, voltages, and e.m.f.s in the a.c. case providing that the impedances are linear.

A circuit element with a current–voltage characteristic of the form $I = aV + bV^2$, for instance, would be said to be 'non-linear' because of the term proportional to V^2. If $V = \hat{V} \exp(j\omega t)$ then

†If $w = (a + jb)$, then the conjugate complex of w, denoted by w^*, is given by $w^* = (a - jb)$. Note that

$$ww^* = (a + jb)(a - jb) = (a^2 + b^2) = |w|^2.$$

Alternating currents 55

$$I = a\hat{V}e^{j\omega t} + b(\hat{V}e^{j\omega t})^2 = a\hat{V}e^{j\omega t} + b\hat{V}^2 e^{j2\omega t}$$

and I contains a harmonic at a frequency 2ω.

Consider again a pure inductor L through which a current $I = \hat{I}\exp(j\omega t)$ is flowing. The important information concerning the current and voltage across the inductor is contained in the following two statements:

(a) $\dfrac{\text{peak value of voltage, } \hat{V}}{\text{peak value of current, } \hat{I}} = \omega L$ (eqn (3.27));

(b) V leads I in phase by $90°$.

If the phase of the current is chosen as the reference phase, that is the current is to be represented as a vector directed along the positive real axis in the complex plane, then the above two statements can be symbolized by the compact mathematical statement $\hat{V} = j\omega L \hat{I}$.

If the phase of the voltage is chosen as reference phase then the mathematical statement is $\hat{I} = j^3 \hat{V}/\omega L$ or $\hat{I} = -j\hat{V}/\omega L$. It follows that $\hat{V} = -\omega L \hat{I}/j = j\omega L \hat{I}$ as before. So it is immaterial whether the phase of the voltage or the current is chosen as a reference for phase (this applies to circuit elements in general and not just to pure inductors).

Analogous arguments for a pure capacitor yield the mathematical statement $\hat{V} = -j\hat{I}/\omega C = \hat{I}/j\omega C$ and for a pure resistor $\hat{V} = R\hat{I}$ of course.

Henceforth the symbols e, v, i will be used to represent sinusoidal e.m.f.s, voltages, and currents where it will be assumed, unless explicitly stated otherwise, that the quantities are vectors in the complex plane; the peak value of e.m.f., voltage, or current is given by the modulus of e, v, or i respectively and the phase angle, relative to whichever of e, v, or i is chosen to denote a reference for phase, by the argument of e, v, or i.

For a general two-terminal circuit element of impedance $Z = (R + jX)$ through which current i is flowing $v = Zi = Ri + jXi = v_R + v_X$, where v_R is in phase with i and v_X is $90°$ out-of-phase with i (a phase lead or lag depending on whether X is positive or negative, i.e. on whether X is an inductive or a capacitive reactance).

As an example consider a source of e.m.f. of 50 V (peak value) at 50 Hz, which has negligible internal impedance, connected in series with a circuit element which can be represented by a resistance $10\,\Omega$ in series with an inductance of 0.1 H. The current is given by

$$i = \frac{e}{Z} = \frac{e}{(R + j\omega L)} = \frac{e(R - j\omega L)}{(R^2 + \omega^2 L^2)}.$$

So

$$i(\text{peak}) = \frac{e(\text{peak})(R^2 + \omega^2 L^2)^{1/2}}{(R^2 + \omega^2 L^2)} = \frac{e(\text{peak})}{(R^2 + \omega^2 L^2)^{1/2}}$$

$$= 1.5 \text{ A (peak value of current)}$$

and $\tan \phi = -\omega L/R$ where in this case $\phi = -72°$ approximately (phase of i relative to e).

3.4. Admittance

In Section 2.5 it was seen that it is convenient to discuss parallel d.c. circuits by using the concept of conductance G so that for resistors $R_1, R_2, R_3, \ldots$ in parallel

$$\frac{1}{R} = \frac{1}{R_1} + \frac{1}{R_2} + \frac{1}{R_3} + \ldots \quad \text{or} \quad G = G_1 + G_2 + G_3 + \ldots \quad (3.38)$$

For a.c. circuits the concept of admittance $Y (\equiv 1/Z)$ is introduced where the real part of Y is a conductance G and the imaginary part is a susceptance B, i.e. if $Z = R + jX$ then

$$Y = \frac{1}{(R+jX)} = \frac{(R-jX)}{(R+jX)(R-jX)}$$

or

$$Y = \frac{R-jX}{(R^2 + X^2)} = G + jB \quad (3.39)$$

where $G = R/|Z|^2$ and $B = -X/|Z|^2$. For R and jX in parallel then $Y = G + jB$ where $G = 1/R$ and $B = 1/X$.

From eqns (3.37) it can be seen that the resulting impedance of a number of two-terminal networks in series is equal to the sum of the component impedances and for a number of such networks in parallel the resulting admittance is the sum of the component admittances.

It will have been noticed by now, probably, that there are formal similarities between pairs of equations which relate voltage and current in circuit elements. For instance,

$$v = (R)i \qquad i = (G)v$$

$$v = (j\omega L)i \qquad i = (j\omega C)v$$

$$v = \left(\frac{-j}{\omega C}\right)i \qquad i = \left(\frac{-j}{\omega L}\right)v.$$

Elements in series: Elements in parallel:
$v = (Z_1 + Z_2)i$ $i = (Y_1 + Y_2)v.$

In each case the elements represented by the terms in parentheses in each pair of equations are said to be the duals of one another. Networks may also be duals of one another as we shall see in later chapters.

3.5. Power in a.c. circuits

If there is a phase difference ϕ between the current $i = \hat{I} \cos \omega t$ through a circuit element of impedance Z {where $Z = (R + jX)$} and the voltage $v = \hat{V} \cos(\omega t + \phi)$ across it then the power dissipated in the element is given by $p = iv$ or

$$p = \hat{I}\hat{V} \cos \omega t \cos(\omega t + \phi)$$
$$= \hat{I}\hat{V} \cos \phi \cos^2 \omega t - \hat{I}\hat{V} \sin \phi \sin \omega t \cos \omega t. \qquad (3.40)$$

The second term in this expression is oscillatory ($\sin \omega t \cos \omega t = \frac{1}{2} \sin 2\omega t$) and represents a flow of energy into and out of the source of e.m.f.. However, there is a net flow of energy from the source represented by the first term and this is the term of practical interest. The average power P delivered by the source is given by the average of p over a complete cycle of period T:

$$P = \frac{\hat{I}\hat{V} \cos \phi}{T} \int_0^T \cos^2 \omega t \, dt - \frac{\hat{I}\hat{V} \sin \phi}{T} \int_0^T \sin \omega t \cos \omega t \, dt$$

$$= \frac{\hat{I}\hat{V} \cos \phi}{\omega t} \int_0^{2\pi} \cos^2 \omega t \, d(\omega t) - \frac{\hat{I}\hat{V} \sin \phi}{\omega t} \int_0^{2\pi} \sin \omega t \cos \omega t \, d(\omega t)$$

or
$$P = \frac{\hat{I}\hat{V} \cos \phi}{2} \qquad (3.41)$$

since
$$\int_0^{2\pi} \cos^2 x \, dx = \int_0^{2\pi} \sin^2 x \, dx = \pi \quad \text{and} \quad \int_0^{2\pi} \sin x \cos x \, dx = 0.$$

Using these relationships the mean squared value of the current, $\overline{I^2}$, is given by

$$\overline{I^2} = \frac{1}{T} \int_0^T i^2 \, dt = \frac{\hat{I}^2}{T} \int_0^T \cos^2 \omega t \, dt = \frac{\hat{I}^2}{2}$$

and the 'root mean square' value $I_{\text{r.m.s.}}$ is given by

$$I_{\text{r.m.s.}} = \hat{I}/\sqrt{2}. \qquad (3.42)$$

An analogous relationship holds for voltages and so

$$P = I_{\text{r.m.s.}} V_{\text{r.m.s.}} \cos \phi. \qquad (3.34)$$

Now since $\cos \phi = R/\sqrt{(R^2 + X^2)}$ and $\hat{V} = \hat{I}\sqrt{(R^2 + X^2)}$ it follows that eqn (3.41) can be rewritten as

58   Alternating currents

$$P = \hat{I}^2 R/2 = I_{\text{r.m.s.}}^2 R$$

or, alternatively,

$$P = (\hat{V}^2 \cos^2\phi)/2R = (V_{\text{r.m.s.}}^2 \cos^2\phi)/R. \qquad (3.44)$$

In the expression for P, $\cos\phi$ is known as the power factor of the circuit element and is equal to 1 for a pure resistor and 0 for a pure inductor or capacitor. Thus only the real part of a complex impedance is used in calculating the power dissipated in the circuit element.

Remember that since P involves the integrals of products of terms such as $\exp\{j(\omega t + \phi)\}$, and so is not a linear function of voltage or current, it cannot be calculated by using the complex number representation (see Section 3.3), i.e.

$$P \neq \frac{1}{T} \int_0^T \operatorname{Re}[\hat{I}\hat{V}e^{j\omega t}e^{j(\omega t + \phi)}]\,dt = \frac{\hat{I}\hat{V}}{T} \int_0^T \operatorname{Re}[e^{j(2\omega t + \phi)}]\,dt = 0.$$

To re-emphasize this point and to illustrate further the exponential representation of phasors let us calculate in a slightly different way the power dissipated in a circuit element of impedance $Z = R + jX$. If the instantaneous value of the current through the element is represented by $i = \operatorname{Re}[\hat{I}e^{j\omega t}]$ then the voltage across the element is represented by $v = i(R + jX)$.

Thus the power P dissipated in the element is given by

$$P = \frac{1}{T}\int_0^T iv\,dt = \frac{1}{T}\int_0^T \operatorname{Re}(\hat{I}e^{j\omega t}) \operatorname{Re}\{\hat{I}e^{j\omega t}(R + jX)\}\,dt$$

$$= \frac{\hat{I}^2}{T}\int_0^T \cos\omega t \operatorname{Re}\{(\cos\omega t + j\sin\omega t)(R + jX)\}\,dt$$

$$= \frac{\hat{I}^2}{T}\int_0^T \cos\omega t\{R\cos\omega t - X\sin\omega t\cos\omega t\}\,dt$$

Using the results for the integrals which we obtained above we have $P = \hat{I}^2 R/2$ which agrees with our previous result.

Consider the circuit shown in Fig. 3.4; the power delivered to the load by the generator is given by

$$P = I_{\text{r.m.s.}}^2 R_L = \frac{E_{\text{r.m.s.}}^2 R_L}{(R_g + R_L)^2 + (X_g + X_L)^2}. \qquad (3.45)$$

In practical situations the impedance of the generator is not variable

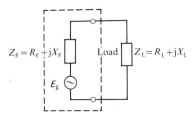

FIG. 3.4. A generator of internal impedance Z_g connected in series with a load of impedance Z_L.

usually and hence the power delivered to the load is a maximum when $X_L = -X_g$ and, as can be shown easily by differentiation of the expression for P with respect to R_g, when $R_L = R_g$. If these conditions are met then the load is said to be 'matched' to the generator.[†]

3.6. Real capacitors and inductors

'Pure' resistors, capacitors, and inductors cannot be realized in practice; even a straight piece of wire exhibits a small self-capacitance and a small self-inductance (see Section 5.4.2). Also, for example, a metal oxide 1 MΩ resistor may well have a self-capacitance ≈ 0.3 pF; this may seem to be very insignificant but at 1 MHz the reactance of the capacitance is only 500 kΩ approximately which, since the self-capacitance is effectively in parallel with the resistance, means that the impedance of the resistor as a whole has a value of about 300 kΩ only. Obviously, the impedance will decrease with increasing frequency; the existence of such self-capacitances, and of 'stray' capacitances in general, means that high frequency circuits are characterized by low levels of impedance.

The stray capacitance and mutual inductance between adjacent wires and components of a circuit may be significant even in the audio frequency range and thus, in general, great thought must be given to the design of the lay-out of the components of a circuit.

3.6.1. Capacitors

Energy losses can occur both by leakage currents between the plates of the capacitor and by loss of energy from the alternating electric field in the dielectric between the plates ('dielectric losses'). The real part of the impedance of a capacitor may be represented by either a resistance

[†] Strictly speaking, the load is said to be 'conjugately' matched to the generator under these conditions since $Z_L = Z_g^*$. Certain considerations may lead to other matching criteria being adopted in practical situations (see Section 8.4).

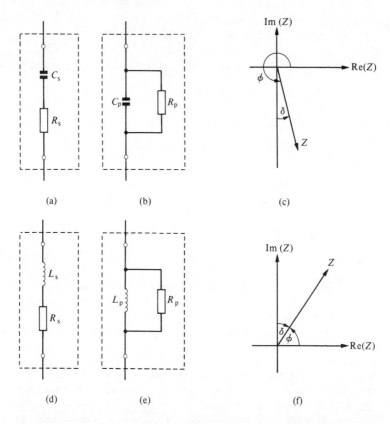

FIG. 3.5. Equivalent circuits and complex impedance vectors for real capacitors and inductors showing the effect of energy losses. In (c) the loss angle δ is given by $\delta \approx \tan \delta = \omega C_s R_s$ and in (f) $|\tan \phi| = \omega L_s/R_s$.

in parallel or in series (see Fig.3.5). The appropriate value of the parallel or series resistance will, in general, be frequency dependent although a constant value of resistance may be used over limited ranges of frequency (see Section 5.2).

For the series equivalent circuit the impedance is represented in the complex plane as shown in Fig.3.5(c) and since for most practical situations capacitors are nearly ideal, and R_s is small, $\tan(\phi - 3\pi/2)$ is very small; the dissipative processes in the capacitor are represented commonly by the *loss angle* δ where $\tan \delta = R_s/(\omega C_s)^{-1}$ and since δ is very small, so that $\tan \delta \approx \delta$, we have $\delta \approx \omega C_s R_s$. For the parallel equivalent circuit, for which R_p will be very large, it can be shown easily that the loss angle δ is given by $\delta \approx \tan \delta = (\omega C_p R_p)^{-1}$.

Alternating currents   61

3.6.2. Inductors

Although the resistivity of the wire of the coil of the inductor is an obvious source of energy losses it is not the sole source nor, in general, is it even the most important source. For instance if the inductor has a ferromagnetic core there will be power dissipation associated with both eddy currents and hysteresis in the core (see Section 5.5.3). Also the current in the wire tends to be restricted increasingly to the outer regions as the frequency is increased (the 'skin' effect; see Section 5.4.1) and so the a.c. resistance is higher than the d.c. resistance. In addition, the coil acts as an aerial, however inefficiently, and so power is lost also by radiation; this effect increases with frequency. As for a capacitor an inductor can be represented by equivalent circuits containing either a series or a parallel resistance (see Fig.3.5); the series resistance corresponds physically to losses in the windings of the coil ('copper' losses) and the parallel resistance to eddy current and hysteresis losses in the core ('iron' losses).

For the series equivalent circuit $Z = (R_s + j\omega L_s)$ and it is usual to specify the losses through a 'quality factor', Q, defined by $Q = |\tan \phi| = \omega L_s / R_s$. N.B. $\tan \delta = 1/Q$. In the parallel equivalent circuit $Q = R_p / \omega L_p$. For the majority of practical cases $Q \geqslant 10$ and for the high quality coils, used in radiofrequency circuits for instance, $Q \sim 100$.

At frequencies greater than 20–30 kHz 'iron' losses become prohibitive, even for laminated core materials (see Section 5.5.3) and so ferrite coils are used. At very high frequencies air-cored coils are used.

3.7. Transformers

As mentioned in Section 1.4 a transformer is an example of a mutual inductor and, for the purposes of this introductory discussion, may be imagined to be a pair of coils wound on a core of ferromagnetic material. For convenience, at this stage, the two coils are shown in Fig.3.6(a) as continuations of one another, with a common terminal. Also we will assume, for simplicity, that there is no 'leakage' of magnetic flux, i.e. all of the flux due to the currents in the two coils is confined to the material of the core.

It is necessary to exercise some care in assigning polarities to the induced e.m.f.s. Imagine that i_p is increasing and generating an increasing magnetic flux ϕ_p. The e.m.f. induced in the secondary coil, through the mutual coupling with the primary, will be in such a sense that the resultant secondary current generates a flux ϕ_s in the *opposite sense* to ϕ_p. In turn, the changing secondary flux will induce an e.m.f. in the primary circuit in such a sense as to *reinforce* ϕ_p. Hence, in drawing the equivalent circuit of Fig.3.6(b), where R_p is the resistance of the primary coil, the mutual inductive effect is represented by a source of

62 Alternating currents

the same polarity as the primary source v_p. The induced e.m.f. due to the self-inductance of the primary coil opposes the primary source, of course.

On applying Kirchhoff's voltage law to the primary circuit and assuming that the voltages and currents have an $e^{j\omega t}$ dependence

$$i_p R_p + j\omega L_p i_p = v_p + j\omega M i_s$$

or
$$i_p R_p + j\omega L_p i_p - j\omega M i_s = v_p. \tag{3.46}$$

Similarly, for the secondary circuit

$$i_s Z_L + i_s R_s + j\omega L_s i_s - j\omega M i_p = 0 \tag{3.47}$$

where R_s is the resistance of the secondary coil. If we write $(R_p + j\omega L_p) = Z_p$ and $(R_s + j\omega L_s + Z_L) = Z_s$, then eqn (3.47) can be written as

$$i_s = \frac{j\omega M i_p}{Z_s} \tag{3.48}$$

and on substituting for i_s in eqn (3.46) we find

$$i_p \left(Z_p + \frac{\omega^2 M^2}{Z_s} \right) = v_p$$

so that
$$Z_{in} \equiv \frac{v_p}{i_p} = Z_p + \frac{\omega^2 M^2}{Z_s}. \tag{3.49}$$

Also, with the configuration we have adopted, $v_L = -i_s Z_L$ and so

$$\frac{v_L}{v_p} = \frac{-j\omega M Z_L}{Z_s \left(Z_p + \dfrac{\omega^2 M^2}{Z_s} \right)}$$

or
$$\frac{v_L}{v_p} = \frac{-j\omega M Z_L}{(Z_s Z_p + \omega^2 M^2)}. \tag{3.50}$$

If we write $Z_s = R_s + jX_s$ then eqn (3.49) becomes

$$Z_{in} = R_p + j\omega L_p + \frac{\omega^2 M^2 (R_s - jX_s)}{|Z_s|^2}$$

or
$$Z_{in} = \left(R_p + \frac{\omega^2 M^2 R_s}{|Z_s|^2} \right) + j\omega \left(L_p - \frac{\omega M^2 X_s}{|Z_s|^2} \right).$$

Thus the effective primary resistance (real part of Z_{in}) is increased and, if X_s is inductive, the effective primary inductance is decreased (imaginary part of Z_{in}). The increase in the effective resistance in the primary circuit represents the dissipation of energy in the secondary circuit.

If we assume, for the present, that there are no energy losses in the transformer then $R_p = R_s = 0$ and from eqns (3.46) and (3.47) it follows that

$$i_s = \frac{j\omega M v_p}{\{\omega^2(M^2 - L_p L_s) + j\omega L_p Z_L\}}$$

and

$$i_p = \frac{(j\omega L_s + Z_L)v_p}{\{\omega^2(M^2 - L_p L_s) + j\omega L_p Z_L\}}.$$

Now if the secondary of the transformer is shorted ($Z_L = 0$) then as M is increased from a small value ($M^2 < L_s L_p$) then the above equations show that both i_p and i_s increase and would become infinite for $M^2 = L_s L_p$; hence this condition sets an upper bound to M of $\sqrt{L_s L_p}$. In general we can write $M = k\sqrt{L_s L_p}$ where the coefficient of coupling k lies in the range $0 \leqslant k \leqslant 1$.

In an ideal transformer there would be no energy losses, no flux leakage, perfect coupling ($k = 1$), and the inductances of the primary and secondary coils would be so large that they would swamp all other impedances in the circuit. Many transformers, especially at power (or 'mains') frequencies, approach quite closely to this ideal, with efficiencies as high as 95 per cent, and so the ideal model serves as a useful basis for further discussions of real transformers.

If ωL_p, ωL_s are large enough then i_p, i_s are vanishingly small; further $L_p/L_s = (n_p/n_s)^2$ where n_p, n_s are the number of turns on the primary and secondary coils respectively (see eqn (1.32)). Under these assumptions we have, from eqn (3.50)

$$\frac{v_L}{v_p} = \frac{-M}{L_p} = -\sqrt{\frac{L_s}{L_p}} = -\frac{n_s}{n_p}. \tag{3.51}$$

Thus for the ideal transformer the secondary voltage v_B (using the labelling of Fig.3.6(a)) is $180°$ out-of-phase with the primary voltage v_A and is 'stepped up' or 'stepped down' depending on the magnitude of the ratio n_s/n_p.

Under the conditions of no loss and perfect coupling, only, then

$$\frac{i_s}{i_p} = \frac{j\omega M}{(j\omega L_s + Z_L)}$$

and, if $\omega L_s \gg Z_L$, then

64   Alternating currents

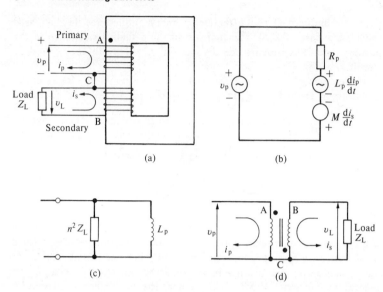

FIG. 3.6. (a) A schematic diagram of a transformer. (b) The equivalent primary circuit. (c) The equivalent primary circuit under the assumption of no losses and perfect coupling. (d) An illustration of the 'dot' convention.

$$\frac{i_s}{i_p} = \frac{M}{L_s} = \sqrt{\frac{L_p}{L_s}} = \frac{n_p}{n_s}. \tag{3.52}$$

Thus the current is stepped up if the voltage is stepped down and vice versa and $v_p i_p = v_L i_s$: this is a manifestation of the assumed perfect power transmission, of course.

Now from eqn (3.49), and remembering that we are assuming that $\omega L_p \gg R_p$, we have

$$Z_{in} = j\omega L_p + \frac{\omega^2 M^2}{(j\omega L_s + Z_L)}$$

or

$$\frac{1}{Z_{in}} = \frac{(j\omega L_s + Z_L)}{\{j\omega L_p (j\omega L_s + Z_L) + \omega^2 M^2\}}.$$

Whence,

$$\frac{1}{Z_{in}} = \frac{L_s}{L_p Z_L} + \frac{1}{j\omega L_p}.$$

Thus,

$$\frac{1}{Z_{in}} = \frac{1}{n^2 Z_L} + \frac{1}{j\omega L_p}. \tag{3.53}$$

where $n \equiv n_p/n_s$. From eqn (3.53) an equivalent circuit for the transformer can be drawn as indicated in Fig.3.6(c); the term $n^2 Z_L$ is called the 'reflected impedance' of the secondary circuit at the primary side of the transformer. If $\omega L_p \gg n^2 Z_L$ then $Z_{in} = n^2 Z_L$.

From the discussion of Section 3.5 it follows that maximum power transfer between the source and load is effected if $n^2 X_L = -X_g$; $n^2 R_L = R_g$; this shows that a transformer can be used as a matching device. Examples of the use of transformers in this role which are commonly met, are in the matching of moving coil record player 'pick-ups', ribbon microphones, and other low impedance transducers, to the input stage of an amplifier.

Other common uses of transformers, apart from mains transformers, are in the coupling between stages in multi-stage amplifiers and as isolating transformers; in this latter use the coils have no common connection and so provide d.c. isolation.

A 'dot' convention is used to establish a relationship between the circuit diagram (including the senses of the voltages and currents) and the real transformer 'on the bench'. This is necessary because in a circuit diagram such as Fig.3.6(d) the senses of i_p and i_s have no physical significance. The 'dot' convention is as follows: if i_p and i_s both flow into (or both flow out of) the dotted terminals of the primary coils, respectively, then the magnetic flux due to i_s reinforces that due to i_p with the consequence that the terms j$\omega M i_p$, j$\omega M i_s$ in eqns (3.46) and (3.47) would have the same sign as the terms j$\omega L_s i_s$, j$\omega L_p i_p$ respectively. By referring to the physical diagram (Fig.3.6(a)) we have seen that the flux due to i_s opposes that due to i_p and we notice that whereas i_p flows into the dotted terminal A, i_s flows out of the dotted terminal C. Hence the terms j$\omega M i_p$, j$\omega M i_s$ are of opposite sign to the terms j$\omega L_s i_s$, j$\omega L_p i_p$ respectively, in accordance with the convention.

The assignments of dots to the terminals of real transformers can be made as follows. If a constant voltage with the polarity of v_p, as shown in Fig.3.6(a), is switched on then there will be a transient voltage between the terminals of the secondary coil. If one dot is assigned to the positive terminal of the primary coil then the other dot should be assigned to that terminal of the secondary coil which becomes positive. Alternatively the assignments of the dots can be made by tracing the senses of the two windings on the core. The dot convention is illustrated by Figs.3.6(a) and 3.6(d) in conjunction with eqns (3.46) and (3.47).

In the case of transformers operating at audio and radio-frequencies their characteristics depart much further from the ideal. Referring to Fig.3.7(a) and applying Kirchhoff's voltage law to the primary and secondary circuits, we have

$$v_p = i_p(R_p + j\omega L_p) - j\omega M i_s$$
$$0 = -j\omega M i_p + i_s(R_s + j\omega L_s + Z_L).$$

Alternating currents

So

$$Z_{in} = \frac{v_p}{i_p} = (R_p + j\omega L_p) + \frac{\omega^2 M^2}{(R_s + j\omega L_s + Z_L)}.$$

Since the coupling is assumed now to be less than perfect, we take $M = k\sqrt{L_p L_s}$ and so

$$Z_{in} = (R_p + j\omega L_p) + \frac{\omega^2 k^2 L_p L_s}{(R_s + j\omega L_s + Z_L)}. \quad (3.54)$$

Now we want to obtain an equivalent circuit for the transformer as we did for the ideal transformer and a mathematical trick which will help us to arrange eqn (3.54) in a more helpful form is to add and subtract a term $(j\omega k L_p)$ on the right-hand side. We then have

$$Z_{in} = R_p + j\omega L_p(1-k) + \frac{\omega^2 k^2 L_p L_s}{(R_s + j\omega L_s + Z_L)} + j\omega k L_p$$

$$= \{R_p + j\omega L_p(1-k)\} + \frac{j\omega k L_p\{R_s + Z_L + j\omega L_s(1-k)\}}{(R_s + j\omega L_s + Z_L)}$$

$$= \{R_p + j\omega L_p(1-k)\}$$

$$+ \cfrac{1}{\cfrac{(R_s + Z_L)}{j\omega k L_p\{R_s + Z_L + j\omega L_s(1-k)\}} + \cfrac{L_s}{k L_p\{R_s + Z_L + j\omega L_s(1-k)\}}}.$$

Now $(1-k)$ will be considerably smaller than one and so if we assume that $Z_L \gg j\omega L_s(1-k)$ and remember that $L_p/L_s = n^2$ where $n \equiv n_p/n_s$ we have

$$Z_{in} \approx \{R_p + j\omega L_p(1-k)\}$$

$$+ \cfrac{1}{\cfrac{1}{j\omega k L_p} + \cfrac{1}{n^2 k\{R_s + Z_L + j\omega L_s(1-k)\}}}. \quad (3.55)$$

The equivalent circuit which represents this expression is shown in Fig. 3.7(b). In physical terms the leakage flux is described, in the approximation we have adopted, by an independent primary inductance $(1-k)L_p$.

Additional amendments can be made to the equivalent circuit to account for the 'magnetizing' current and the 'iron' losses. In an ideal transformer, and if $Z_L = 0$, then the flux produced by i_s exactly cancels that due to the primary current; from eqn (3.48) $i_s = ni_p$. The flux which is actually present in a real situation, and which causes the

Alternating currents

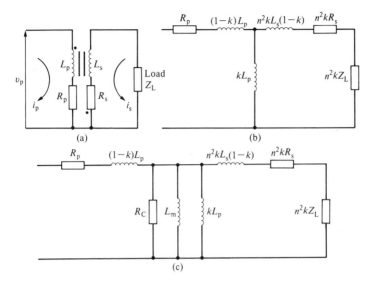

FIG. 3.7. A circuit diagram for a non-ideal transformer is shown in (a) where R_p, R_s represent 'copper' losses in the primary and secondary windings respectively. In the equivalent circuit (b) the imperfect coupling is represented by the factor k and in (c) the magnetizing current in the primary winding and the 'core' losses are represented by L_m and R_c respectively.

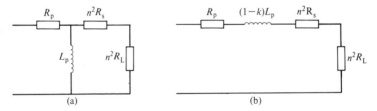

FIG. 3.8. (a) The equivalent circuit used in the definition of the efficiency of a transformer (under the assumption $k \approx 1$). (b) 'High' frequency equivalent circuit.

induced e.m.f. which is the source of transformer action, is generated by the so-called magnetizing current flowing in the primary winding; this situation is represented by a shunt inductance L_m in the primary circuit (see Fig.3.7(c)). 'Iron', or 'core', losses due to hysteresis and eddy currents can be represented by a component current in the primary which is in phase with the primary voltage; so we insert a shunt resistor R_c as shown in Fig.3.7(c). The copper losses are represented by the resistance R_p.

For an audio power transformer the load is usually resistive with a magnitude much less than ωL_s and, assuming $k \approx 1$, we can draw the equivalent circuit of Fig.3.8(a). So the efficiency of the transformer, which is defined as the ratio of the power dissipated in the load to the total power delivered to the transformer, is given by

$$\text{Efficiency} = n^2 R_L/(R_p + n^2 R_s + n^2 R_L).$$

For an efficiency of 90 per cent, and for $R_p \approx n^2 R_s$, this means that for efficient performance R_L should be equal to about $20 R_s$.

At frequencies well below the designed operating range the shunt inductance L_p causes the voltage across the primary winding to fall and at high frequencies the series leakage inductance $(1-k)L_p$ again causes the voltage applied to the primary to fall (see Fig.3.8(b)); for these reasons the efficiency of the transformer falls off outside its designed operating range. By inserting a series capacitor of suitable value in the primary circuit and a shunt capacitor in the secondary circuit the frequency range of operation can be extended. Some other aspects of transformers will be discussed in Sections 5.5.2 and 5.5.3.

3.8. a.c. networks

3.8.1. Examples of mesh and nodal analysis

The analysis of a.c. networks proceeds according to the basic principles which were established for d.c. networks as described in Chapter 2. However it will be useful to consider a few examples involving a.c. currents and voltages.

Suppose we wish to calculate the power delivered by the source in the network shown in Fig.3.9. First, you should note that a capacitor or inductance is labelled by the value of its impedance at the frequency of operation. The mesh equations are

$$i_1(10-j2) - 6i_2 = 50$$
$$-6i_1 + (13+j5)i_2 - 7i_3 = 0$$
$$-7i_2 + (10+j2)i_3 = 0.$$

Since the voltage of the source is given we need to know the modulus of i_1, and its phase angle ϕ, in order to calculate the power delivered i.e.

$$\text{power delivered} = 1/2 \times 50 \times |i_1| \cos \phi$$

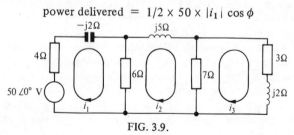

FIG. 3.9.

Alternating currents

Using the rules given in Section 2.4 we have

$$i_1 = \frac{\begin{vmatrix} 50 & -6 & 0 \\ 0 & (13+j5) & -7 \\ 0 & -7 & (10+j2) \end{vmatrix}}{D}$$

where

$$D = \begin{vmatrix} (10-j2) & -6 & 0 \\ -6 & (13+j5) & -7 \\ 0 & -7 & (10+j2) \end{vmatrix}$$

$$= (10-j2)\{(13+j5)(10+j2)-49\}$$
$$+ 6\{-6(10+j2)\}$$

or
$$D = 502 + j546.$$

Thus,
$$i_1 = 50\{(13+j5)(10+j2)-49\}/(502+j546)$$

or
$$i_1 = \frac{50(71+j76)}{(502+j546)}.$$

On rationalizing we have

$$i_1 = \frac{50(71+j76)(502-j546)}{5.50 \times 10^5}.$$

After expanding we find

$$i_1 = \frac{50(7.71 \times 10^4 - j6.14 \times 10^2)}{5.50 \times 10^5}$$

or
$$i_1 = 7.01 \underline{/\tan^{-1}(-6.14/771)},$$

i.e.
$$i_1 = 7.01 \underline{/-0.456°} \text{ ampere}.$$

Thus,
$$\text{Power} = \tfrac{1}{2} \times 50 \times 7.01 \times \cos(0.456)° = 175 \text{ W}.$$

Consider now the network shown in Fig.3.10(a) which can be used as a phase-shifting network, i.e. the phase of v_0 with respect to v_1 varies with the value of R, the ratio $|v_0/v_1|$ remaining constant. The mesh equations are

$$2R_0 i_1 - 2R_0 i_2 = v_1$$

$$-2R_0 i_1 + \left(2R_0 + R - \frac{j}{\omega C}\right) i_2 = 0$$

70   Alternating currents

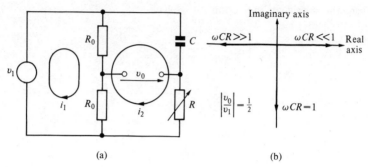

FIG. 3.10. (a) A phase-shifting network. (b) The ratio v_0/v_1 plotted in the complex plane (C kept constant, R varied).

$$i_1 = \frac{\begin{vmatrix} v_1 & -2R_0 \\ 0 & \left(2R_0 + R - \dfrac{j}{\omega C}\right) \end{vmatrix}}{\begin{vmatrix} 2R_0 & -2R_0 \\ -2R_0 & \left(2R_0 + R - \dfrac{j}{\omega C}\right) \end{vmatrix}}$$

$$= \frac{\left(2R_0 + R - \dfrac{j}{\omega C}\right) v_1}{2R_0\left(2R_0 + R - \dfrac{j}{\omega C}\right) - 4R_0^2}$$

$$= \frac{\left(2R_0 + R - \dfrac{j}{\omega C}\right) v_1}{\left(2RR_0 - j\dfrac{2R_0}{\omega C}\right)}$$

Also $i_2 = \dfrac{\begin{vmatrix} 2R_0 & v_1 \\ -2R_0 & 0 \end{vmatrix}}{\begin{vmatrix} 2R_0 & -2R_0 \\ -2R_0 & 2R_0 + \left(R - \dfrac{j}{\omega C}\right) \end{vmatrix}} = \dfrac{2R_0 v_1}{\left(2RR_0 - j\dfrac{2R_0}{\omega C}\right)}.$

Now,
$$v_0 = (i_1 - i_2)R_0 - i_2 R$$
$$= i_1 R_0 - i_2 (R + R_0)$$

and, on substituting for i_1, i_2, we find

$$\frac{v_0}{v_1} = -\frac{1}{2} \frac{\left(R + \dfrac{j}{\omega C}\right)}{\left(R - \dfrac{j}{\omega C}\right)}$$

and, after rationalizing,

$$\frac{v_0}{v_1} = -\frac{\left\{\left(R^2 - \dfrac{1}{\omega^2 C^2}\right) + j\dfrac{2R}{\omega C}\right\}}{2\left(R^2 + \dfrac{1}{\omega^2 C^2}\right)}.$$

Whence,

$$\left|\frac{v_0}{v_1}\right| = \frac{1}{2}$$

and

$$\tan \phi = \left(\frac{-2R}{\omega C}\right) \bigg/ \left(\frac{1}{\omega^2 C^2} - R^2\right)$$

or

$$\tan \phi = \frac{-2}{\left(\dfrac{1}{\omega CR} - \omega CR\right)}.$$

For $\omega CR \ll 1$, $\phi \approx 0°$ (see Fig.3.10(b)); for $\omega CR = 1$, $\phi = -90°$, and for $\omega CR \gg 1$, $\phi \approx -180°$. Hence, if ω is constant, it is possible to vary the phase of v_0 relative to v_1 through a range of $180°$, by varying R, without change of amplitude of v_0/v_1. It is implicit, note, that the impedance of the source is negligible and that the impedance of whatever network or device that is connected to the output terminals is large enough so that it doesn't load the phase-shifting network.

As a final example in this section consider the problem of finding v_2 in the network shown in Fig.3.11(a). We will use nodal analysis to solve this problem; the voltage at node B is v_2 and we will denote the voltage at node A by v_1. First, we will convert the generator voltage into the cartesian form $(a + jb)$; we have that $\sqrt{(a^2 + b^2)} = 10$ and

72 Alternating currents

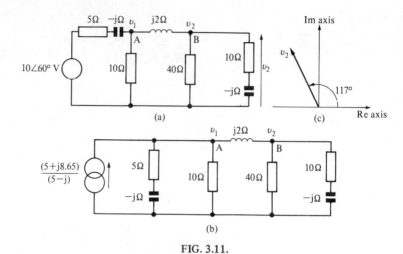

FIG. 3.11.

$b/a = \tan 60°$ which give $a = 5.00$ and $b = 8.65$. Hence the circuit of Fig.3.11(a) can be redrawn as in Fig.3.11(b). The nodal equations are

$$v_1 \left\{ \frac{1}{(5-j)} + \frac{1}{10} + \frac{1}{j2} \right\} - \frac{v_2}{j2} = \frac{(5+j8.65)}{(5-j)}$$

$$-\frac{v_1}{j2} + v_2 \left\{ \frac{1}{j2} + \frac{1}{40} + \frac{1}{(10-j)} \right\} = 0$$

or

$$\frac{v_1(19-j30)}{65} + j\frac{v_2}{2} = \frac{16.35+j48.2}{26}$$

$$j\frac{v_1}{2} + \frac{v_2(2-j7.92)}{16.2} = 0.$$

Thus,

$$v_2 = \frac{\begin{vmatrix} \dfrac{(19-j30)}{65} & \dfrac{(16.35+j48.2)}{26} \\ \dfrac{j}{2} & 0 \end{vmatrix}}{\begin{vmatrix} \dfrac{(19-j30)}{65} & \dfrac{j}{2} \\ \dfrac{j}{2} & \dfrac{(2-j7.92)}{16.2} \end{vmatrix}}$$

Alternating currents 73

$$= -\frac{(9.25 - j3.15)}{(1.90 + j2.00)} = -\frac{(9.25 - j3.15)(1.90 - j2.00)}{7.61}$$

or, finally,

$$v_2 = -1.48 + j2.85.$$

In polar form (see Fig.3.11(c)) we have

$$v_2 = 3.21 \tan^{-1}\left(\frac{2.85}{-1.48}\right)$$

or

$$v_2 = 3.21\, \underline{/117°}.$$

3.8.2. Coupling between networks

This topic fits naturally into Section 7.3 dealing with two-part networks but it is of such great practical importance that it is useful to introduce it at this earlier stage.

Consider Fig.3.4. which depicts a generator coupled to a load. Now the 'generator' could be the Thévenin (or Norton) equivalent of a more complicated network (e.g. the first stage of an amplifier) and the 'load' could be the input impedance of a more complicated network such as the next stage of the amplifier. We have discussed in Section 3.5 the conditions required to maximize the coupling of power from a generator to a load. Let us now consider a multi-stage amplifier which amplifies a small signal from a voltage source (such as a microphone); see Fig.3.12(a). Each stage of the amplifier is represented, at its input, by its effective input impedance Z_{in_1}, Z_{in_2}, etc. At its output a stage of the amplifier is represented by a Thévenin equivalent generator.

In Fig.3.12(a) we can see that

$$i_1 = \frac{v_g}{(Z_g + Z_{in_1})}$$

and

$$v_1 = i_1 Z_{in_1} = \frac{Z_{in_1}}{(Z_g + Z_{in_1})} \cdot v_g, \quad (3.56)$$

i.e. Z_{in_1} and Z_g act as a 'potential divider' (see Fig.2.4). If no current is drawn from the output of the first stage then $v_2 = A_v v_1$; hence A_v is the 'open-circuit voltage gain', or 'no load voltage gain', of the first stage. Since the signal to be amplified is a voltage signal we wish to maximize the voltage coupling between stages of the amplifier, i.e. for the coupling between the signal generator and the first stage of the amplifier we wish to maximize

74 Alternating currents

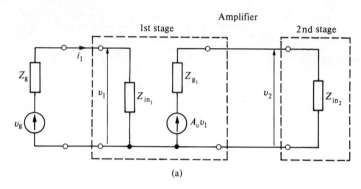

(a)

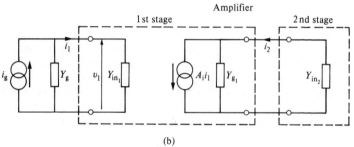

(b)

FIG. 3.12. Equivalent circuits for an amplifier represented in terms of (a) constant voltage generators and (b) constant current generators.

$$\frac{v_1}{v_g} = \frac{Z_{in_1}}{(Z_g + Z_{in_1})}. \qquad (3.57)$$

Obviously, (v_1/v_g) tends towards its maximum value of unity as Z_{in_1} becomes very much larger than Z_g; this describes ideal voltage coupling i.e. the impedance of the 'generator' is negligible in comparison with that of the 'load': for many practical purposes $Z_{in_1}/Z_g \geqslant 20$ is good enough although in some specialized instruments values of Z_{in_1}/Z_g as high as 10^6 are required. You should note that 'Z_g' includes any series element in the interstage coupling such as a d.c. blocking capacitor.

For the two-stage amplifier of Fig.3.12(a) it can be seen that

$$\frac{v_2}{A_v v_1} = \frac{Z_{in_2}}{(Z_{g_1} + Z_{in_2})}$$

and so

$$\frac{v_2}{v_g} = \frac{Z_{in_1}}{(Z_g + Z_{in_1})} \times A_v \times \frac{Z_{in_2}}{(Z_{g_1} + Z_{in_2})}. \qquad (3.58)$$

This argument can be extended, simply, for a multi-stage amplifier.

Alternating currents 75

In some circumstances we may wish to amplify a small current from a signal generator; in this case we want good 'current coupling' between the signal generator and its load. Since we are interested in currents we represent the generator, and the outputs of the stages of the amplifier, by their Norton equivalent networks; see Fig.3.12(b). In this case we have $i_1 = v_1 Y_{in}$ and $i_g = (i_1 + v_1 Y_g)$; thus

$$\frac{i_1}{i_g} = \frac{Y_{in_1}}{(Y_g + Y_{in_1})}. \tag{3.59}$$

This equation is identical, formally, to eqn (3.57), because the network of Fig.3.12(b) is the dual of that of Fig.3.12(a) and Y_g and Y_{in_1} act as a 'current divider' network.

If $Y_{in_2} = 0$ then $i_2 = A_i i_1$ where A_i is the short-circuit current gain of the stage. We have, in general,

$$\frac{i_2}{i_g} = \frac{Y_{in_1}}{(Y_g + Y_{in_1})} \times A_i \times \frac{Y_{in_2}}{(Y_g + Y_{in_2})}.$$

It can be seen that for good current coupling the admittance of the 'load' should be very much greater than the admittance of the 'generator'.

An example of a voltage-coupled situation is where a transistor is used in the 'common-emmitter' configuration to amplify a voltage, and a transistor connected in the 'common-base' configuration is a good current amplifier (see Section 7.4).

As an example of coupling consider the network shown in Fig.3.13(a).: we could imagine that the capacitor couples an amplifier

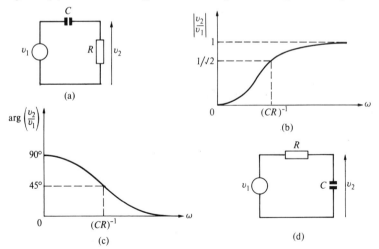

FIG. 3.13. CR networks which act (a) as a high-pass network and (d) as a low-pass network.

stage, which is represented by a generator v_1 of negligible output impedance, to a load R, which could represent the input resistance of the next stage of the amplifier. It can be seen that

$$v_2 = \frac{v_1}{\left(R - \dfrac{j}{\omega C}\right)} \cdot R$$

or

$$\frac{v_2}{v_1} = \frac{R\left(R + \dfrac{j}{\omega C}\right)}{\left(R^2 + \dfrac{1}{\omega^2 C^2}\right)}$$

so that finally

$$\left|\frac{v_2}{v_1}\right| = \frac{R\left(R^2 + \dfrac{1}{\omega^2 C^2}\right)^{1/2}}{\left(R^2 + \dfrac{1}{\omega^2 C^2}\right)}$$

$$= \frac{\omega CR}{(1 + \omega^2 C^2 R^2)^{1/2}}.$$

Now the argument of (v_2/v_1) is $\tan^{-1}(1/\omega CR)$ and so we can write

$$\frac{v_2}{v_1} = \frac{\omega CR}{(1 + \omega^2 C^2 R^2)^{1/2}} \; \underline{/\tan^{-1}(1/\omega CR)}.$$

The variation with frequency of $|v_2/v_1|$ and $\arg(v_2/v_1)$ are shown in Figs.3.13(b) and 3.13(c). Note that for $\omega = (CR)^{-1}$ then $|v_2/v_1| = 1/\sqrt{2} = 0.707$ and $\tan^{-1}(1/\omega CR) = 45°$, so that this network acts as a simple high-pass filter. It is left as an exercise to carry out similar calculations for the network of Fig.3.13(d), which acts as a low-pass filter.

For a practical situation of coupling between stages of an amplifier the 'generator' will have a non-zero output impedance and there may well be significant stray capacitance in parallel with the 'load'.

3.9. a.c. bridges

The measurement of the values of inductors and capacitors is made most conveniently and precisely by means of an a.c. bridge, although resonance methods are useful at high frequencies. For the general form

Alternating currents

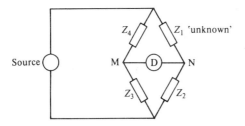

FIG. 3.14. A general form of a.c. bridge where D is the detector.

of bridge, as shown in Fig.3.14, the same arguments as those used for the Wheatstone bridge lead to the balance condition

$$Z_1 Z_3 = Z_2 Z_4 \quad \text{or} \quad R_1 + jX_1 = \frac{(R_2 + jX_2)(R_4 + jX_4)}{(R_3 + jX_3)}. \quad (3.60)$$

For balance the potentials of M and N must be equal at all times which implies equality of amplitude and of phase. These two conditions for balance are exemplified, of course, by the equality of the real and the imaginary parts of eqn (3.60). The combinations of choices of circuit elements for $R_2, R_3, R_4, X_2, X_3, X_4$ are limited, for practical purposes, by the following considerations:

(i) For convenience only two elements should be adjustable in meeting the two balance conditions.

(ii) These two balancing operations should be independent in their effects upon the balance conditions.

(iii) Whilst high precision pure resistors and pure capacitors are commonly available this is not the case where 'pure' inductors are concerned. So the adjustable elements in most a.c. bridges are resistors and capacitors.

Conditions (i) and (ii) imply that it is desirable to be able to write $R_1 + jX_1 = (A + jB)$ where A and B are real and where one of the adjustable elements should appear in A but not in B and the other in B but not in A. If the arms of a bridge are assumed to consist of inductances, capacitances, and resistances in *series* then it can be shown from eqn (3.60), by straightforward (but tedious!) analysis, that the desired conditions can be met if R_2 and X_2 are the adjustable elements and the *ratio* Z_4/Z_3 is either pure real or pure imaginary (obviously Z_2 and Z_4 are equivalent in this analysis and hence are interchangeable).

The arms of a bridge may be made up also of inductances, capacitances, and resistances in *parallel* in which case the balance condition (eqn (3.60)) can be written in terms of admittances as

Alternating currents

$$\frac{1}{Y_1 Y_3} = \frac{1}{Y_2 Y_4} \quad \text{or} \quad R_1 + jX_1 = \frac{Y_3}{Y_2 Y_4}. \tag{3.61}$$

In this case $(R_1 + jX_1)$ can be written in the form $(A + jB)$ if R_3 and X_3 are the adjustable elements and the *product* $Z_2 Z_4$ is either pure real or pure imaginary.

Thus an a.c. bridge may be balanced by adjusting series elements in an arm (Z_2) adjacent to the unknown impedance (Z_1), the 'fixed' arms (Z_3, Z_4) appearing in the balance equation as a ratio (a ratio-arm bridge), or by adjusting parallel elements in the arm (Z_3) opposite to the unknown impedance, the 'fixed' arms appearing in the balance equation as a product (a product-arm bridge).

The usefulness of a particular bridge circuit depends critically on the response of the detector to the voltage between M and N arising from a small deviation from the exact balance condition. By using the same method of analysis as was used for the d.c. Wheatstone bridge (see Section 2.7) the out-of-balance voltage between M and N and the associated current in the detector branch can be calculated. However, it will suffice here to point out that, if the impedance of the detector is much greater than that of the bridge arms, maximum sensitivity is obtained if Z_1, Z_2, Z_3, and Z_4 are of equal magnitude; if the impedance of the detector is not large compared to that of the arms then maximum power is available for operating it if its impedance is equal to that of the network which is connected to its terminals.

Although practical bridge circuits will be discussed later, in Chapter 7, one further practical point can be mentioned here. It is desirable that the two balance conditions should be independent of frequency. Although drifting of the frequency of the generator during the course of a measurement should not be a problem, the fact that the waveform of the generator will contain harmonics of the fundamental frequency would be troublesome if the balance conditions were frequency-dependent since the ability to detect small out-of-balance signals would be impaired.

PROBLEMS

3.1. A capacitor C is charged from a source of e.m.f. E which is then disconnected and the capacitor is allowed to discharge through a resistor R. Show that at time t after the beginning of the discharge the charge q on the capacitor is given by

$$q = q_0 \exp(-t/CR)$$

where q_0 is the charge on the capacitor at time $t = 0$.

3.2. For the situation of Problem 3.1 what is the initial value of the discharge current if $E = 100\,\text{V}, R = 100\,k\Omega, C = 1000\,\mu\text{F}$?

Alternating currents 79

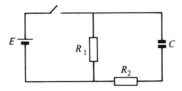

FIG. 3.15.

3.3. What is the time constant for (a) the charging (switch closed) and (b) the discharging (switch open), of the capacitor in the circuit shown in Fig.3.15? *Hint*: Apply Kirchhoff's laws and remember the p.d. across a capacitor C is q/C where q is the charge on the capacitor plates.

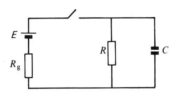

FIG. 3.16.

3.4. What is the time constant for the charging of the capacitor C when the switch is closed in the circuit in Fig.3.16?

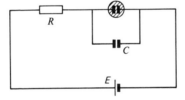

FIG. 3.17.

3.5. A neon lamp is connected across a capacitor C as shown in the circuit in Fig.3.17. The lamp discharges when the voltage across it is V_f, say, ($V_f < E$) and, having once started, the discharge will be maintained at voltages down to V_q ($V_q < V_f$). Show that the period T of the flashing of the neon lamp is given by

$$T = CR \ln \{(E - V_q)/(E - V_f)\}.$$

It may be assumed that the neon lamp has negligible capacitance compared to C and a negligible resistance compared to R when it is discharging.

3.6. Consider the circuit shown in Fig.3.18 and then answer the following questions:
(a) What is the voltage across the inductor at the instant when the switch is closed?
(b) What is the current through the inductor after a long time has elapsed since the switch was closed?

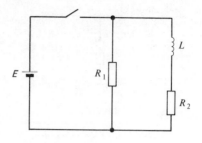

FIG. 3.18.

(c) If the switch is now opened what is the value of the time constant of the decay of the current in the inductor?

3.7. Find the amplitude and phase of the resultant of two phasors $4\cos\omega t$ and $3\cos(\omega t - 3\pi/2)$.

3.8. What is the difference in phase between the two phasors represented by

$$\hat{I}_1 \exp\{j(\omega t + \phi_1)\} \quad \text{and} \quad \hat{I}_2 \exp\{j(\omega t - \phi_2)\}?$$

3.9. A current $i = 10^{-2} \cos \omega t$ A, where $\omega/2\pi = 5774$ Hz, flows in the network shown in Fig.3.19. If $R = 100\,\Omega$ and $C = 1/2\pi\,\mu\text{F}$, express the voltage v as a phasor.

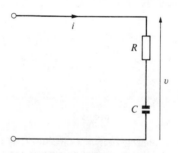

FIG. 3.19.

3.10. A p.d., $v = \hat{V} \sin \omega t$, is applied to the network shown in Fig.3.20. Express the current i as a phasor.

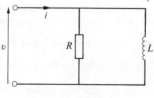

FIG. 3.20.

Alternating currents 81

3.11. A current of amplitude 1 A at 50 Hz is flowing through a circuit element consisting of a 100-Ω resistance in series with a capacitor of 10 μF. Calculate the amplitude of the voltage across the circuit element and the phase of this voltage relative to that of the current.

3.12. A voltage of amplitude 10 V and frequency 1 kHz is applied to a circuit element consisting of a resistance of 100 Ω and an inductance of 0.1 H in series. What is the amplitude of the current flowing through the element and what is the phase angle of the current relative to the applied voltage?

3.13. An e.m.f. of 240 V r.m.s. at 50 Hz is applied to a circuit containing a capacitance of 5 μF in series with a 1-kΩ resistor. Find the rate at which energy is dissipated in the circuit and also find the power factor.

3.14. When an ideal capacitor and a resistor are connected in parallel to a 100-V d.c. supply a steady current of 1 A flows. When, instead of the d.c. supply a 100-V r.m.s., 50 Hz, a.c. supply is connected the current is 2 A r.m.s. Find (a) the value of the resistance and (b) the capacitance of the capacitor.

3.15. A resistor of 500 Ω and a capacitor of 1 μF are connected in series. If the voltage across the resistor is $5 \times 10^{-2} \sin(1000t - \pi/6)$ V express the voltage across the capacitor as a phasor. Now express as a phasor the voltage across the series combination of the resistor and capacitor.

3.16. Obtain expressions for the conductance and susceptance of the network shown in Fig.3.21.

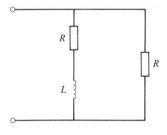

FIG. 3.21.

3.17. Find the amplitude of v and its phase relative to $i = \hat{I} \cos \omega t$ in Fig.3.22.

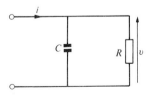

FIG. 3.22.

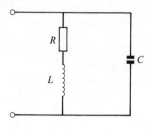

FIG. 3.23.

3.18. Obtain expressions for the conductance and susceptance of the network shown in Fig.3.23.

3.19. An e.m.f. of amplitude 240 V at a frequency of 50 Hz is applied to a circuit element consisting of an inductance of 0.1 H in series with a resistance of 100 Ω. Find the power factor of the circuit element and the power dissipation.

3.20. An e.m.f. of amplitude 240 V at a frequency of 50 Hz is applied to a circuit element consisting of a resistance of 1 MΩ in parallel with a capacitance of 1 μF. Find the power factor of the circuit element and the power dissipation.

3.21. Show that the modulus of the impedance, and the phase angle ϕ, for a network consisting of a capacitance C in parallel with a resistance R are given by

$|Z| = R/(1 + \omega^2 C^2 R^2)^{1/2}$ and $\tan \phi = -\omega RC$ respectively.

3.22. A coil has an inductance of 1 mH, a self-capacitance of 1 pF, and a resistance of 100 Ω. Assuming that the self-capacitance is in parallel with the inductance and resistance, calculate the effective inductance and resistance of the coil at a frequency of $10/2\pi$ MHz.

3.23. Show that the capacitances and resistances of the equivalent networks of Fig.3.5 (a) and (b) are related by

$R_s = R_p/(1 + \omega^2 C_p^2 R_p^2) \quad C_s = (1 + \omega^2 C_p^2 R_p^2)/\omega^2 C_p R_p^2.$

3.24. For the network of Fig.3.13(d) find the modulus and argument of the ratio v_2/v_1.

3.25. For the network shown in Fig.3.24 show that Y is independent of frequency if $R^2 = LC$. What is the value of Y in these circumstances?

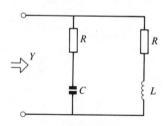

FIG. 3.24.

Alternating currents

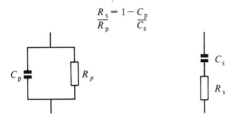

FIG. 3.25.

3.26. Show that for the equivalent series and parallel representations of a lossy capacitor shown in Fig. 3.25 $\dfrac{R_s}{R_p} = 1 - \dfrac{C_p}{C_s}$

3.27. Calculate the primary and secondary currents in the transformer network shown where $L_p = 1\,\text{H}$, $L_s = 0.1\,\text{H}$, $k = 0.95$, $R_p = 10\,\Omega$, $R_L = 1\,\Omega$, and the resistance of the secondary coil can be neglected.

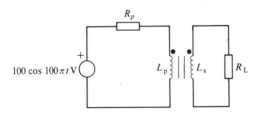

FIG. 3.26.

3.28. What are the duals of the two networks shown in Fig. 3.27?

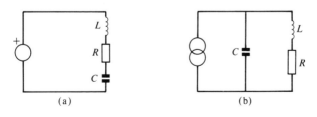

FIG. 3.27.

3.29. Find the voltage transfer function v_o/v_i for the network shown in Fig. 3.28.

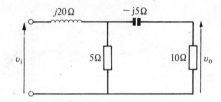

FIG. 3.28.

3.30. Find the Norton equivalent of the network shown in Fig.3.29.

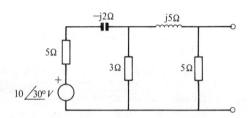

FIG. 3.29.

3.31. Find the Thévenin equivalent of the network shown in Fig. 3.30.

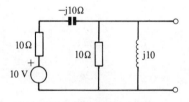

FIG. 3.30.

3.32. Find the current i and the power delivered by source in Fig. 3.31.

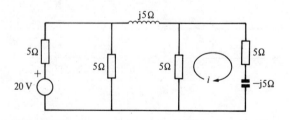

FIG. 3.31.

4. Resonance

4.1. Free, damped oscillations

A 'bob' attached to a horizontal strip of elastic material, as shown in Fig. 4.1(a) is a simple example of a system which, after being displaced from its equilibrium position and then released, will execute oscillations at a characteristic natural frequency. If internal friction in the elastic strip, and friction with air, are negligible then the only force acting on the bob is the elastic restoring force in the strip which, for small enough displacements y, will be assumed to be proportional to y. So from Newton's second law of motion the equation of motion is the second-order differential equation $m\mathrm{d}^2 y/\mathrm{d}t^2 = -ky$ where k is equal to the restoring force per unit displacement for small displacements. This equation can be rewritten as $\mathrm{d}^2 y/\mathrm{d}t^2 = -\omega_0^2 y$ where $\omega_0^2 \equiv k/m$. It can be shown by simple substitution that a function of the form $K \exp(j\omega_0 t)$ satisfies the differential equation but since, for a homogeneous differential equation, the sum of any two solutions is also a solution and also because the general solution of a second-order differential equation contains two arbitrary constants, the general solution can be written in the form

$$y = K_1 \exp(j\omega_0 t) + K_2 \exp(-j\omega_0 t).$$

If the boundary conditions are taken to be $y = \hat{y}$, $\mathrm{d}y/\mathrm{d}t = 0$, at $t = 0$ then it follows that $K_1 = K_2 = \hat{y}/2$ and

$$y = \{\hat{y} \exp(j\omega_0 t) + \hat{y} \exp(-j\omega_0 t)\}/2$$

or

$$y = \hat{y} \cos \omega_0 t.$$

Another idealized physical system which can be described by the same differential equation is an L–C circuit; for this

$$q/C + L\mathrm{d}I/\mathrm{d}t = 0 \text{ (Kirchhoff's voltage law)}$$

or

$$q/C + L\mathrm{d}^2 q/\mathrm{d}t^2 = 0$$

since $I = \mathrm{d}q/\mathrm{d}t$. It can be seen that L and m and c and $1/k$ ('compliance')

86 Resonance

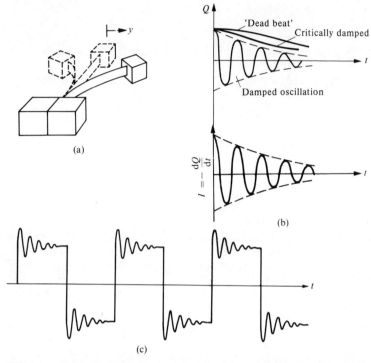

FIG. 4.1. Damped oscillations: (a) an oscillatory mechanical system; (b) various conditions of damping; (c) repetitive shock excitation of a resonant circuit.

are pairs of analogous quantities. Using the boundary conditions $q = \hat{q}, I = 0$, at $t = 0$, then it follows that

$$q = \hat{q}\{\exp(j\omega_0 t) + \exp(-j\omega_0 t)\}$$

or

$$q = \hat{q} \cos \omega_0 t = \text{Re}\{\hat{q} \exp(j\omega_0 t)\}$$

and

$$I = -\omega_0 \hat{q} \sin \omega_0 t = \text{Re}\{j\omega_0 \hat{q} \exp(j\omega_0 t)\}$$

where $\omega_0 = 2\pi f_0 \equiv 1/\sqrt{(LC)}$ is the angular frequency of the free, undamped oscillations.

Since the two systems discussed so far are ideal, lossless systems (i.e. no friction in the mechanical oscillator and no resistance in the $L-C$ circuit) the total energy of the system remains constant and the oscillations continue indefinitely. Note that in the $L-C$ circuit energy is exchanged periodically between the capacitor (potential energy imagined to be stored in the electric field in the capacitor) and the inductor (kinetic energy associated with the current and imagined to be stored in the magnetic field of the Inductor). If dissipative processes

Resonance

are present the energy of motion is converted to other forms of energy, irreversibly, and the oscillation dies away.

Suppose that energy is introduced into a series $L-C-R$ circuit by charging the capacitor to voltage V immediately before it is connected to close the circuit. The capacitor will then discharge and, because of the current I flowing at a subsequent time the energy stored in the magnetic field of the inductor will be $LI^2/2$. Hence the energy balance is described by

Power dissipated = Rate of decrease of stored energy

$$I^2 R = -\frac{d}{dt}\left\{\frac{CV^2}{2} + \frac{LI^2}{2}\right\}, \qquad (4.2)$$

i.e. the power dissipated in the resistor is equal to the rate of decrease of energy associated with the capacitor and inductor. Since $I = dq/dt$ and $C = q/V$, eqn 4.2 can be written as

$$R\left(\frac{dq}{dt}\right)^2 + \frac{1}{2C}\frac{d(q^2)}{dt} + \frac{L}{2}\frac{d}{dt}\left(\frac{dq}{dt}\right)^2 \quad \text{or} \quad R\frac{dq}{dt} + \frac{q}{C} + L\frac{d^2q}{dt^2} = 0. \qquad (4.3)$$

This equation could have been written down by applying Kirchhoff's voltage law to the circuit but it is useful to emphasize the conservation of energy.

If a function of the form $q = K_1 \exp(st) + K_2 \exp(st)$ (note the two arbitrary constants K_1, K_2) is taken as a trial general solution then it will be a solution if s satisfies the auxiliary equation which is obtained after substituting the trial function in the differential equation;

$$Ls^2 + Rs + \frac{1}{C} = 0, \qquad (4.4)$$

ie. if

$$s = \{-R \pm \sqrt{(R^2 - 4L/C)}\}/2L = -\alpha \pm j\omega_r \qquad (4.5)$$

where $\alpha \equiv R/2L$ and $\omega_r^2 \equiv 1/(LC) - \alpha^2 = \omega_0^2 - \alpha^2$.

Taking $q = \hat{q}$, $I = dq/dt = 0$, at $t = 0$, as the boundary conditions and if $R^2 < 4L/C$ (i.e. s is complex) it follows that $K_1 + K_2 = \hat{q}$ and

$$(-\alpha + j\omega_r)K_1 - (\alpha + j\omega_r)K_2 = 0$$

whence $K_1 = K_2 = \hat{q}/2$ and the solution can be written as

$$q = \hat{q}e^{-\alpha t}\cos\omega_r t = \hat{q}e^{-\alpha t} \times \text{Re}\,(e^{j\omega_r t}). \qquad (4.6)$$

This equation represents a damped oscillation of frequency ω_r, the damping being exemplified by the exponential decay factor $\exp(-\alpha t)$ (see Fig. 4.1(b)).

The situation is different if s is real, i.e. if $R^2 > 4L/C$; now $s = -$

$\alpha \pm \omega_r$ and using the same boundary conditions as above it follows that $K_1 = \hat{q}(\alpha + \omega_r)/2\omega_r$ and $K_2 = \hat{q}(\omega_r - \alpha)/2\omega_r$ and

$$q = \frac{\hat{q}e^{-\alpha t}}{2\omega_r}\{\alpha(e^{\omega_r t} - e^{-\omega_r t}) + \omega_r(e^{\omega_r t} + e^{-\omega_r t})\}. \quad (4.7)$$

The important point about this rather formidable looking expression is that it represents a non-oscillatory time dependence for q; the corresponding L–C–R circuit is said to be overdamped or 'dead beat'. For t very small (αt, $\omega_r t \ll 1$) then eqn (4.7) reduces to $q = \hat{q}(1 - \alpha^2 t^2)$ and for t large (αt, $\omega_r t \gg 1$) the term $\exp(-\alpha t)$ dominates and $q \to 0$.

A special case arises if $\omega_r = 0$ ($R^2 = 4L/C$); now s is single-valued but a general solution of the differential must still contain two arbitrary constants. It can be checked, by substitution, that $q = (K_1 + K_2 t)$, $\exp(-\alpha t)$ is a solution and the boundary conditions demand that $K_1 = \hat{q}$ and $K_2 = \hat{q}R/2L$. Again, for t small $q \approx \hat{q}(1 - \alpha^2 t^2)$ and for t large the term $\exp(-\alpha t)$ dominates. Under these conditions the circuit is said to be critically damped.

The indicating mechanisms of many physical instruments (such as the pointer and suspension system of a galvanometer), which would otherwise constitute lightly damped systems, are deliberately damped to an approximately critically damped condition.

Damped oscillator systems, particularly mechanical ones, are often characterized by a parameter called the logarithmic decrement Λ which is defined as the natural logarithm of the ratio of the amplitudes of successive oscillations. So, since the period of the oscillation equals $2\pi/\omega_r$, it follows from eqn (4.6) that

$$\Lambda = \ln\{e^{(-\alpha + j\omega_r)t}/e^{(-\alpha + j\omega_r)(t + 2\pi/\omega_r)}\} = \frac{2\pi\alpha}{\omega_r} = \frac{2\pi}{\{4L/(CR^2) - 1\}^{1/2}}. \quad (4.8)$$

For the great majority of practical cases $R^2 \ll 4L/C$ and $\Lambda = \pi R/(\omega_0 L)$ very closely.

Using eqn (4.6), and remembering that $I = -\,dq/dt$, the total stored energy of the system $W = q^2/2C + \frac{L}{2}(dq/dt)^2$ is given by

$$W = \hat{q}^2 e^{-2\alpha t}\{\cos^2 \omega_r t(1/C + \alpha^2 L) + \omega_r^2 L \sin^2 \omega_r t$$
$$+ 2\alpha\omega L \sin \omega_r t \cos \omega_r t\}/2 \quad (4.9)$$

and the energy loss in one cycle of oscillation is found by subtracting the value of W at $t = (t + 2\pi/\omega_r)$ from the above expression. Oscillatory electrical circuits are usually characterized by a 'quality factor' $Q \equiv 2\pi \times$ (energy stored in the circuit)/(energy loss per cycle). Using the above expression for W

$$Q = 2\pi W(t) \bigg/ \left\{ W(t) - W\left(t + \frac{2\pi}{\omega_r}\right) \right\} = 2\pi/\{1 - \exp(-4\pi\alpha/\omega_r)\}$$

which, if $R^2 \ll 4L/C$, and using the approximation $e^{-x} \approx 1 - x$ for $x \ll 1$, gives

$$Q = \frac{\omega_0 L}{R} \quad \text{and} \quad \Lambda = \pi/Q. \quad (4.10)$$

Notice that the expression defining ω_r, given on p. 87, can be written $\omega_r^2 = \omega_0^2\{1 - 1/(4Q^2)\}$ which in turn can be rewritten as $(\omega_0 + \omega_r)(\omega_0 - \omega_r)/\omega_0^2 = 1/(4Q^2)$. Hence for $(\omega_0 - \omega_r)/\omega_0 = 10^{-2}$, say, which means that $\omega_0 + \omega_r \approx 2\omega_0$ then $Q = 3.5$. Since L–C–R circuits in practical use in amplifiers and radio receivers, for instance, have Q-factors much larger than 3.5 (100 approximately in radio receivers, for example) it follows that the difference between ω_r and ω_0 is generally negligible in practice.

Damped oscillations of an L–C–R circuit can be demonstrated simply by using a 'square wave' generator as the source in the circuit of Fig. 4.3(b), thus periodically 'shocking' the circuit into oscillation, and displaying the voltage across the circuit on a cathode ray oscilloscope (C.R.O.). Note that the 'off time' of the wave form must be long enough to allow sufficient cycles of oscillation to occur between the exciting pulses (see Fig. 4.1(c)).

4.2. Forced, damped oscillations

4.2.1. Series L–C–R circuit

Imagine that an ideal constant voltage generator of negligible internal impedance, and angular frequency ω, is connected in series with an inductor L, a capacitor C, and a resistor R (see Fig. 4.3(a)). If the peak value of the e.m.f. of the generator is E then, using eqn (4.3),

$$\frac{q}{C} + R\frac{dq}{dt} + L\frac{d^2q}{dt^2} = \hat{E}e^{j\omega t} \quad (4.11)$$

where the convention that the actual e.m.f. $= \text{Re}(\hat{E}\exp(j\omega t))$ has been used. If it is assumed that q has the same time dependence as the e.m.f., apart from a phase difference (e.g. $q = \hat{q}\exp j(\omega t + \delta)$) then, after substitution in eqn (4.11), it follows that

$$q = \hat{E}\exp(j\omega t)/\{L(\omega_0^2 - \omega^2) + j\omega R\} \quad \text{where} \quad \omega_0^2 \equiv 1/(LC)$$

and

$$i = \frac{dq}{dt} = j\omega\hat{E}\exp(j\omega t)/\{\omega^2 L(\omega_0^2/\omega^2 - 1) + j\omega R\}.$$

Resonance

The actual current, being given by the real part of this expression, is

$$i = \text{Re}\,[j\omega\hat{E}\exp(j\omega t)\{\omega^2 L(\omega_0^2/\omega^2 - 1) - j\omega R\}/\{\omega^4 L^2(\omega_0^2/\omega^2 - 1)^2 + \omega^2 R^2\}]$$

or

$$i = \frac{-\omega L(\omega_0^2/\omega^2 - 1)\hat{E}\sin\omega t + R\hat{E}\cos\omega t}{\{\omega^2 L^2(\omega_0^2/\omega^2 - 1)^2 + R^2\}}$$

$$= \frac{\hat{E}\cos(\omega t + \phi)}{\{\omega^2 L^2(\omega_0^2/\omega^2 - 1)^2 + R^2\}^{1/2}} \quad (4.12)$$

where the phase angle ϕ is given by

$$\tan\phi = \omega L(\omega_0^2/\omega^2 - 1)/R. \quad (4.13)$$

i can be considered as the resultant of two currents of angular frequency ω which are $\pi/2$ out-of-phase (since $\sin\omega t = \cos(\omega t - \pi/2)$; thus, using eqn (3.10),

$$\hat{I} = \frac{\hat{E}\{\omega^2 L^2(\omega_0^2/\omega^2 - 1)^2 + R^2\}^{1/2}}{\{\omega^2 L^2(\omega_0^2/\omega^2 - 1)^2 + R^2\}} = \frac{\hat{E}}{\{\omega^2 L^2(\omega_0^2/\omega^2 - 1)^2 + R^2\}^{1/2}}. \quad (4.14)$$

Of course the general solution of eqn (4.11) also includes transient terms associated with a free damped oscillation of the circuit due to the shock excitation as the generator is first switched on.[†] However, as has been seen, these transient terms decay exponentially with a time constant $\tau = 1/\alpha = 2L/R$ and $\exp(-\alpha t) = 1/e$ for $t = 1/\alpha$. The transient oscillation decays to 5 per cent, approximately, of its initial amplitude in $t = 3\tau$: For $Q = 10$ and $f = 10$ kHz then $3\tau \approx 1$ ms and for $Q = 100$ and $f = 500$ kHz then $3\tau \approx 200\,\mu s$. Thus in many practical situations the transient oscillations decay quickly and can be ignored and eqn (4.12) is the solution for the current in the steady state situation. For this situation the conventions established for a.c. theory in Chapter 3 can be used i.e. the time dependence in $e^{j\omega t}$ will be assumed and terms like $\hat{I}e^{j\omega t}$, $\hat{E}e^{j\omega t}$, ... etc. will be written simply as $i, e, \ldots$ etc. Hence the current could have been written down immediately as $i = e/Z$ where $Z = R - j/\omega C + j\omega L$ or

---

[†]The solution of an inhomogeneous second-order linear differential equation, such as eqn (4.11), can be written as the sum of the 'complementary function' and a 'particular integral'. The former function is the solution of the homogeneous differential equation formed by setting the right-hand side of eqn (4.11) equal to zero; this situation has just been treated in Section 4.1. The particular integral is any particular function which can be found to be a solution of the inhomogeneous equation; in our case we chose $q = \hat{q}\exp j(\omega t + \delta)$. The complementary function is a transient term, as we have just seen in Section 4.1.

Resonance

$$i = \frac{e\{R + j(1/\omega C - \omega L)\}}{\{(1/\omega C - \omega L)^2 + R^2\}} = \frac{e\{R + j\omega L(\omega_0^2/\omega^2 - 1)\}}{\{\omega^2 L^2(\omega_0^2/\omega^2 - 1)^2 + R^2\}}.$$
(4.15)

This means that the peak value of the current is given by

$$\hat{I} = \frac{\hat{E}\{R^2 + \omega^2 L^2(\omega_0^2/\omega^2 - 1)^2\}^{1/2}}{\{R^2 + \omega^2 L^2(\omega_0^2/\omega^2 - 1)^2\}}$$

$$= \frac{\hat{E}}{\{R^2 + \omega^2 L^2(\omega_0^2/\omega^2 - 1)^2\}^{1/2}} = \frac{\hat{E}}{|Z|} \qquad (4.16)$$

and the phase angle of the current relative to the e.m.f. by $\tan \phi = \omega L(\omega_0^2/\omega^2 - 1)/R$. These results agree with eqns (4.14) and (4.13) respectively.

From eqn (4.16) it can be seen that if one of either ω, L, or C is varied then $\hat{I}$ is a maximum when the condition $\omega^2 = 1/LC \equiv \omega_0^2$ is satisfied; this is an example, in an electrical context, of resonance (sometimes called 'current' resonance). This is also, obviously, the condition for maximum voltage across R since $\hat{V}_R = R\hat{I}$. Now for large Q-factors R is small and this means that $\hat{V}_R$ is small, even at resonance. Hence, in the experimental detection of the resonance, it is usual to monitor the voltage v_C across the capacitor or the voltage v_L across the inductor. It should be noted that the values of the variables ω, L, or C to give maximum $\hat{V}_L$ or $\hat{V}_C$ differ from those satisfying $\omega_0^2 = 1/(LC)$ by factors like $\{1 \pm 1/(nQ^2)\}$ where $n = 1, 2$, or 4. Since, in most cases of practical interest, $Q > 10$ the condition for resonance of $\omega^2 = \omega_0^2 = 1/(LC)$ holds true to a high degree of accuracy.

The variations of $|Z|/R$ and ϕ with respect to ω/ω_0 are shown in Fig. 4.2., which illustrates that $Z = R$ and $\phi = 0$ at resonance i.e. the impedance of the circuit is purely resistive. It should be noticed that for $\omega > \omega_0$, ϕ is negative and the current lags the applied e.m.f. and, for $\omega \gg \omega_0$, $\phi \to -\pi/2$; hence the circuit is said to be 'inductive' for $\omega > \omega_0$. From an analogous argument the impedance is said to be 'capacitive' for $\omega < \omega_0$. From a qualitative standpoint it can be seen that the reactance of the inductance, ωL, is much greater, at high enough frequencies, than that of the capacitor and hence gives the dominant contribution to the impedance of the circuit. The converse holds true at low enough frequencies.

The power P delivered to the circuit by the generator is given by $P = (\hat{E}\hat{I} \cos \phi)/2$ or, using eqns (4.14) and (4.13)

$$P = \hat{E}^2/[2R\{Q^2(\omega_0^2/\omega^2 - 1)^2 \omega^2/\omega_0^2 + 1\}]. \qquad (4.17)$$

Thus maximum power is delivered at resonance and is given by $P_{\max} = \hat{E}^2/2R$. For $\tan \phi = 1$ ($\phi = 45°$) then from eqn (4.13)

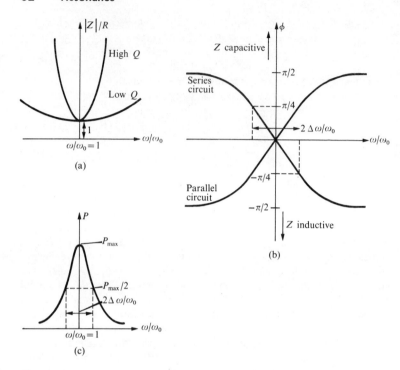

FIG. 4.2. Impedance, phase angle, and power in a resonant circuit.

$Q(\omega_0^2/\omega^2 - 1)\omega/\omega_0 = 1$ whence $(\omega_0^2 - \omega^2) = \omega\omega_0/Q$. Now assuming that $(\omega_0 - \omega)/\omega_0$ is small, which is true even for only moderately large Q-factors (see Fig. 4.2), this relationship can be written as

$$(\omega_0 - \omega)(\omega_0 + \omega) = \omega\omega_0/Q$$

or
$$Q = \omega_0/(2\Delta\omega) \qquad (4.18)$$

where $\Delta\omega \equiv (\omega_0 - \omega)$, and since $(\omega_0 + \omega) \approx 2\omega_0$.

Further, it follows from eqn (4.17) that $P = P_{max}/2$ for $\tan\phi = 1$ (i.e. for $\omega = \omega_0 \pm \Delta\omega$) and thus the 'sharpness' of a resonance is specified, commonly, in terms of the frequency at which the power absorption by the circuit is half the maximum.

To summarize, the quality factor Q of a resonance is equal to $\omega_0 L/R = \omega_0/(2\Delta\omega)$ where $2\Delta\omega$ is the 'half-width' of the power absorption curve, i.e. the width between the half-power points. It is left as an exercise to show that at resonance the voltages across the capacitor and inductor are out-of-phase by 180° and have magnitudes Q times the e.m.f. of the generator. Thus the Q-factor can be thought of also as a 'magnification' factor. If a resonance is detected by measuring the voltage V_C across the capacitor, the frequencies corresponding to $(\omega_0 \pm \Delta\omega)$ are determined by $\hat{V}_C = \hat{V}_{max}/\sqrt{2}$.

4.2.2. Parallel L–C–R circuit

Imagine now that an ideal constant-current generator is connected to a capacitor C and an inductor L as shown in Fig. 4.3(b); the resistor R is placed in series with L since it will usually correspond to the resistance of the inductor. It is convenient to analyse such a parallel circuit in terms of admittances rather than impedance, as outlined in Chapter 2, and so

$$Y = \frac{1}{(R + j\omega L)} + j\omega C = \frac{R}{(R^2 + \omega^2 L^2)} + j\omega \left\{ C - \frac{L}{(R^2 + \omega^2 L^2)} \right\}.$$

(4.19)

The resonance condition for this circuit is defined, for convenience, as the condition for which Y is real i.e. $R^2 + \omega_0 L^2 = L/C$ which leads to an expression for the resonance angular frequency of the form

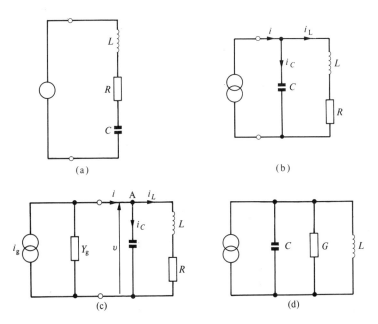

FIG. 4.3. (a) A series L–C–R circuit with an ideal voltage source. (b) A parallel L–C–R circuit with an ideal current source and (c) with a source of admittance Y_g. (d) The dual of the series circuit of (a).

$$\omega_0^2 \approx \left(\frac{1}{LC}\right)\left\{1 - \frac{1}{Q^2}\right\} \qquad (4.20)$$

where, as for the series resonant circuit, $Q = \omega_0 L/R = (1/R)\sqrt{(L/C)}$. Since (usually) $Q > 10$, $\omega_0^2 = 1/(LC)$ again, for practical purposes.

$|Y|$ is a minimum at resonance but, as with the series resonant circuit, slightly different conditions for resonance are obtained depending on whether the variable is ω, L, or C. However, these differences are only $\sim 1/Q^2$ and so can usually be ignored.

From eqns (4.19) and (4.20) it follows that at resonance

$$Y = 1/(RQ^2). \qquad (4.21)$$

Note that for $|Y|$ a minimum then the current drawn from the generator is a minimum; there is a periodic exchange of energy between the inductor and the capacitor due to the circulating currents i_L and i_C. At resonance the amplitude of this exchange is a maximum and i_L and i_C are 180° out-of-phase (see Fig. 4.3(c)).

For an admittance $Y = G + jB$ the phase angle ϕ' is given by $\tan\phi' = B/G$. Now as $Y = 1/Z$, where $Z = R + jX$, then $G = R/(R^2 + X^2)^{1/2}$ and $B = -X/(R^2 + X^2)^{1/2}$ and so, if $\tan\phi = X/R$, then $\tan\phi' = -\tan\phi$.

From eqn (4.19) $\tan\phi' = \omega L^2 C(\omega^2 - \omega_0^2)/R$ assuming $Q^2 \gg 1$ or

$$\tan\phi = \omega L^2 C(\omega_0^2 - \omega^2)/R \approx \{(1 - \omega^2/\omega_0^2)Q\omega\}/\omega_0 \qquad (4.22)$$

where the final approximation is that $\omega_0^2 = 1/(LC)$ and $Q = \omega_0 L/R$. Thus, for $\omega < \omega_0$, i lags v and the circuit is inductive and, for $\omega > \omega_0$, i leads v and the circuit is capacitive (i.e. $\omega L \ll 1/\omega C$ at low enough frequencies and vice versa at high enough frequencies).

4.3. Some practical aspects of resonant circuits

First the effect on the properties of resonant circuits of using real as opposed to ideal generators will be considered. For a series circuit if the voltage generator has internal impedance $Z_g = R_g + jX_g$ then the resonance condition becomes $\{\omega L - 1/(\omega C) + X_g\} = 0$ and so X_g should be small in order that the measured resonance frequency is close to $\{1/(LC)\}^{1/2}$. Also R_g should be small in order not to degrade the Q-factor which will be equal to $\omega_0 L/(R + R_g)$.

For a parallel circuit if the current generator has a shunt admittance Y_g (see Fig. 4.3(c)) then the admittance of the circuit as 'seen' by the ideal constant current generator is

$$Y = Y_g + j\omega C + \frac{1}{(R + j\omega L)} = G_g + jB_g + j\omega C + \frac{R - j\omega L}{(R^2 + \omega^2 L^2)} \qquad (4.23)$$

Consequently Y is real if $B_g + \omega C - \omega L/(R^2 + \omega L^2) = 0$ and so B_g should be small in order that the measured resonance frequency shall be close to $\{1/(LC)\}^{1/2}$.

In practice the current source will be obtained, very often, by using a commercial signal generator (of output impedance between 60 and 600 Ω, say) with a large enough resistor R_g connected in series with its output. If the impedance of the detector (a C.R.O. say) is assumed to be resistive and denoted by R_d then, in order that the resonance shall not be damped significantly due to currents circulating through R_g and R_d, which are in parallel with the resonant circuit, we must have $R_g R_d/(R_g + R_d) \gg Z_{\text{resonance}}$ ($= Q^2 R$). For instance if $Q = 32$ and $R = 50\,\Omega$ then $Q^2 R = 50\,\text{k}\Omega$ and we should have both R_g and R_d greater than 100 kΩ.

For the circuit of Fig. 4.3(c) by equating the voltages across the capacitor C and applying Kirchhoff's current law at node A the following two equations are obtained

$$-\frac{ji_C}{\omega_0 C} = i_L(R + j\omega_0 L) \tag{4.24}$$

$$i_L + i_C = vR/(R^2 + \omega_0^2 L^2) = i. \tag{4.25}$$

From these equations it follows that $i_C = vR(1 + jQ)/(R^2 + \omega_0^2 L^2)$, or $i_C \approx jvRQ/(R^2 + \omega_0^2 L^2)$, and $i_L = -jvRQ/(R^2 + \omega_0^2 L^2)$, i.e. that i_L and i_C are 180° out-of-phase as we deduced earlier. Furthermore $\hat{I}_C = \hat{I}_L = Q\hat{I}$; thus, in the parallel circuit current magnification occurs.

Because of the magnification property of resonant circuits care must be taken to ensure that the capacitors and inductors which are used are rated to withstand the anticipated voltages and currents.

4.4. Coupled resonant circuits

A general description of the properties of a pair of coupled resonant circuits (see Fig. 4.4) is quite complicated because of the number of parameters involved. For illustrative purposes we shall consider the situation where the primary and secondary circuits are identical; this will serve to indicate the most important features of coupled circuits in general.† On applying Kirchoff's voltage law to the network of Fig. 4.4 we have

$$i_p(R + jX) - j\omega M i_s = v_p$$

$$-j\omega M i_p + i_s(R + jX) = 0$$

---

†For a more detailed treatment see, for instance, Harnwell, G. P. *Principles of electricity and electromagnetism.* McGraw-Hill, New York (1949).

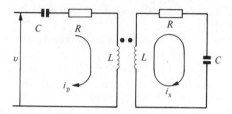

FIG 4.4.   An example of coupled resonant circuits.

where $X = (\omega L - 1/\omega C)$. Thus,

$$i_s = \frac{j\omega M v_p}{(R^2 - X^2 + j2XR + \omega^2 M^2)}.$$

If we consider the situation near resonance such that $\omega = \omega_0 \pm \Delta\omega$ where $\Delta\omega \ll \omega_0$ then

$$X = \left(\omega L - \frac{1}{\omega C}\right) = \frac{L(\omega^2 - \omega_0^2)}{\omega}$$

or

$$X \approx 2L\Delta\omega$$

where we have assumed that $(\omega + \omega_0) \approx 2\omega_0$.

Using this result we find

$$\hat{I}_s = \omega_0 M v_p \{[R^2 - 4L^2(\Delta\omega)^2 + \omega_0^2 M^2] + 16R^2 L^2(\Delta\omega)^2\}^{-1/2}.$$

In order to find the condition for $\hat{I}_s$ to be a maximum we differentiate this expression with respect to $\Delta\omega$ and find that the condition is given by

$$\Delta\omega [4L^2(\Delta\omega)^2 + (R^2 - \omega_0^2 M^2)] = 0.$$

Remembering that $M^2 = k^2 L^2$ and that $Q \equiv \omega_0 L/R$ we find that for $\hat{I}_s$ to be a maximum

$$\Delta\omega = 0 \quad \text{or} \quad \Delta\omega = \pm\frac{\omega_0}{2Q}\sqrt{\{(kQ)^2 - 1\}}.$$

We are only interested in real roots: If $(kQ)^2 > 1$ there are three real roots, there being a minimum of $\hat{I}_s$ for $\Delta\omega = 0$ and maxima for the other two roots. If $(kQ)^2 < 1$ there is a single real root only, at $\Delta\omega = 0$, this being the condition for a maximum. For $(kQ)^2 = 1$ the coupling is said to be critical and the three roots coincide.

The most important feature which emerges from this cursory discussion is that as the coupling is made tighter the resonance is split into two peaks: this is a feature of coupled physical oscillators in general, of course, including such diverse examples as coupled pendulums and interacting atoms in solids.

There is a wide range of variations upon this theme depending on whether both, or only one, of the primary and secondary circuits are tuned circuits, and depending on whether or not they are tuned to the same resonant frequency. In addition the primary and/or secondary circuit may be a parallel circuit and not a series circuit such as we have used for illustrative purposes.

The critically coupled condition is often used in radio-frequency amplifiers to give a 'band-pass' response (see Chapter 7); the 'roll-off' in the stop-bands is steepest for the critically coupled condition.

PROBLEMS

4.1. Show that for forced resonance in a series $L-C-R$ network the voltage across the inductance and capacitance are $180°$ out-of-phase and have amplitudes Q times that of the source.

4.2. Find the admittance of the parallel $L-C-R$ network

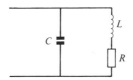

FIG. 4.5.

shown in Fig. 4.5. and hence show that the resonant frequency ω_r is given by
$$\omega_r^2 = \frac{1}{LC} - \frac{R^2}{L^2}, \text{ i.e. } \omega_r^2 = \omega_0^2\left(1 - \frac{1}{Q^2}\right).$$

4.3. Calculate the values of the capacitances which would resonate with a 1-mH inductance at frequencies of 100 kHz and 10 MHz. What values of resistance would give a Q-value of 100 at these two frequencies?

4.4. Find the reactance and the resonant frequency of the network shown in Fig. 4.6.

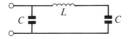

FIG. 4.6.

Resonance

4.5. For the network shown in Fig. 4.7. calculate the values of C, R to give a resonant frequency of 200 kHz and a Q-factor equal to 20.

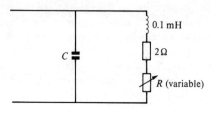

FIG. 4.7.

What is the impedance of this network at resonance?

5. Electromagnetics

THIS chapter is concerned with some of the electrical and magnetic properties of physical media insofar as they are of immediate significance to the material components of electrical circuits. The range of physical phenomena which underly the technical features of modern circuit components is very wide, too wide to be covered exhaustively in a text with the aims of this one. Thus the choice of topics is highly selective and the treatments are largely phenomenological in nature.

5.1. Static electric fields in solids

5.1.1. Conductors, semiconductors, and dielectrics

The d.c. conductivities of solids at 300 K range from 10^7 S m^{-1} in some metals through values in the range $10-10^4$ S m^{-1}, roughly, for semiconductors and down to 10^{-18} S m^{-1} in good insulators. The term dielectric is reserved usually for materials having conductivities less than 10^{-12} S m^{-1} although the divisions between dielectrics, so-called semi-insulators, and semiconductors, and also between semiconductors, semi-metals, and metals are very blurred indeed.

Since the predominant interest in this text is in linear, passive circuit components the effects of electric fields in conductors and semiconductors are covered adequately by the definition of resistance and the statement of Ohm's law given in Section 1.2.2. Thus the prime consideration of Sections 5.1 and 5.2 is with static and alternating fields in dielectrics.

As we shall see in Section 5.2 energy losses occur when alternating electric fields are applied to dielectrics due to a number of physical mechanisms other than 'Ohmic' losses arising in the straightforward transport of charge through the material. For the present we will assume that for a dielectric-filled capacitor the energy loss in the dielectric can be represented by a series resistor R_s.

Now the relative permittivity ϵ of a dielectric is defined as C/C_0 where C_0 is the capacitance of a capacitor before being filled with the dielectric. For a parallel plate capacitor having plates of area A and plate separation t,

$$C = \frac{\epsilon\epsilon_0 A}{t} \quad \text{and} \quad R_s = \frac{t}{\sigma A}$$

and so

$$CR_s = \frac{\epsilon\epsilon_0}{\sigma}. \tag{5.1}$$

If $\epsilon \sim 5$ and $\sigma < 10^{-12} \, \text{S m}^{-1}$, then the time constant CR_s of the dielectric is greater than 50 s roughly and may be as high as 10^7 s in some materials (see Table 5.1).

In their principal areas of use, namely for insulation and as the filling for capacitors, dielectrics also have to possess the property of being able to withstand electric fields of $10^6 \, \text{V m}^{-1}$ at least.

Table 5.1
Typical electrical properties of a selection of dielectrics

Material	ϵ_s	tan δ at 1 kHz	Time constant (s)
Paper	3.7	$< 10^{-2}$	10^3
Mica	6	10^{-3}	5×10^2
Polyethylene	2.2	5×10^{-5}	10
Polypropylene	2.3	10^{-4}	10^4
P.T.F.E.	2.1	5×10^{-5}	10^7
Polystyrene	2.5	$< 5 \times 10^{-5}$	10^7
Polycarbonate	1.5	3×10^{-3}	10^4
Corundum (Al$_2$O$_3$)	7	2×10^{-3}	10^2
Silica (SiO$_2$)	4	10^{-4}	10^5
Ceramics	5.5–7.5	10^{-3}	10^3

N.B. 10^6 s ≈ 12 days.

5.1.2. Polarization and the relative permittivity

Imagine that a slab of an isotropic and homogeneous solid insulating material is subjected to a uniform applied static electric field E_0 as depicted in Fig. 5.1. The end faces of the slab which are perpendicular to the x-axis will acquire positive and negative surface charges, i.e. the slab will become polarized. In general terms there are two sources of this polarization.

(a) *Induced electric dipole moments.* Consider an atom which is electrically neutral (otherwise it would be an ion!); it consists of a positively charged nucleus which is surrounded by a cloud of negative electrons. Normally the centres of the positive and negative charge distributions are coincident. The effect of the applied electric field is to cause the centres of positive and negative charges to be displaced relative to one another, equilibrium being reached when the attractive

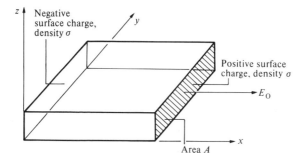

FIG. 5.1. Induced surface charges on opposite faces of a slab of an insulating material in a polarizing field.

force between the displaced charges balances the displacing force due to the field. Now the separated positive and negative charges constitute an electric dipole; if the charges are of magnitude q and their separation at equilibrium is x then the dipole moment p, which is a vector pointing in the direction of the displacement x, is defined by

$$p = qx. \qquad (5.2)$$

The displacements x are very small ($\sim 10^{-16}$ m typically) but are of very great physical significance. It so happens that, in the frequency range which is of relevance for applied alternating fields in electrical circuits (below 10^{11} Hz, broadly speaking), the polarization of atoms as just described is usually swamped by the polarization arising from other microscopic systems. However an atom is the simplest system to describe and the polarization due to other types of microscopic polarizable system arises in a broadly similar way, i.e. the centres of positive and negative charge of an initially neutral charge distribution are displaced by the effect of an applied electric field and so give rise to a dipole moment. For instance, the most significant source of polarization in ionic solids (or in molecules containing ions) is the displacement of the positive and negative ions in opposite senses when a field is applied e.g. the alkali halides such as lithium fluoride, sodium chloride, and potassium bromide.

(b) *Permanent dipole moments.* Many solid materials of practical importance contain microscopic systems which have permanent dipole moments, for instance:

(i) Polar molecules, of which water is probably the most familiar example;

(ii) Permanent dipoles associated with polymer chains;

(iii) Associated impurity ion- vacancy structures of varying degrees of complexity;

(iv) Permanent dipoles associated with interfaces either between grains of the material or between the material and electrodes which are in contact with it.

As far as this section is concerned the important common feature of assemblies of such permanent dipoles is their behaviour under the influence of an applied field. In the absence of an applied field assembly of dipoles will be disordered and the average polarization of a macroscopic sample of the material will be zero (except for the small class of materials called ferroelectrics where co-operative interactions cause the self-polarization of macroscopic regions of the material, a phenomenon akin to ferromagnetism). Due to an applied field E there is a couple of magnitude pE acting on a dipole P in a sense which tends to orient it parallel to the field and the over-all effect of this is to cause a partial ordering of the constituent dipoles of the assembly and hence induce a net macroscopic polarization; the degree of ordering reflects the competition between the disordering effect of thermal vibrations and the ordering effect of the applied field.

On a macroscopic scale the polarization P of a sample is defined as the dipole moment per unit volume, i.e., if the concentration of microscopic dipoles of moment p is N, then

$$P = Np. \tag{5.3}$$

Thus the surface of a sample of a polarized material becomes charged and, referring to Fig. 5.1, if these surface charges have an area density σ then

$$P = \frac{\text{dipole moment of whole slab}}{\text{volume of the slab}}$$

$$= \frac{(\sigma A)l}{Al}$$

or

$$P = \sigma. \tag{5.4}$$

It is instructive to consider a parallel plate capacitor in which the space between the plates is filled with an insulating material (see Fig. 5.2): it will be assumed that the lateral dimensions of the plates are much greater than the separation of the plates so that end effects can be neglected and the electric field can be assumed to be uniform between the plates. If C_0 is the capacitance before the insulating material is present then $\sigma = q/A$ where q is the total net charge on a capacitor plate and A is the area of the plate, $C_0 = \epsilon_0 A/t$ and $E_0 = \sigma/\epsilon$ (see Section 1.1.2.). The slab of material acquires surface charge densities $\pm \sigma'$, say, after it has been inserted between the plates and

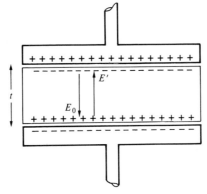

FIG. 5.2. The polarization of an insulating material between the plates of a capacitor.

these are the source of an electric field E' as shown. So

$$E_0 - E' = \frac{-\sigma' + \sigma}{\epsilon_0}$$

whence

$$E' = \frac{P}{\epsilon_0}. \tag{5.5}$$

Hence the potential difference V between the plates is given by

$$V = E_0\left(-\frac{P}{\epsilon_0}\right)t$$

and

$$C = \frac{Q}{V}$$

$$= \frac{C_0 E_0 t}{(E_0 - P/\epsilon_0)t}.$$

Now the average macroscopic field inside the material is $E = E_0 - E'$ (i.e. the electric field between the plates is less than what it was before the material was introduced)

$$C = \frac{C_0(E + E')}{E} \quad \text{or} \quad C = C_0\left(1 + \frac{E'}{E}\right)$$

ie.

$$C = C_0\left(1 + \frac{P}{\epsilon_0 E}\right).$$

The electric susceptibility χ_e of the material is defined through

$$\chi_e \equiv \frac{P}{\epsilon_0 E} \qquad (5.6)$$

(where we are assuming that the material is isotropic so that χ_e does not depend on the direction of E) and so

$$C = C_0(1 + \chi_e) \qquad (5.7)$$

Eqn (5.6) is the link between the atomic/molecular, or microscopic, properties of the material and its macroscopic properties since, in principle at least, χ_e can be calculated from a knowledge of the electronic structure of the constituent atoms and the calculated value can be compared with the experimentally measured macroscopic property, the electric susceptibility. Note that the total electric susceptibility is the sum of the contributions from the various species of polarizable system which exist in a particular material.

From eqn (5.7) it can be seen that $C/C_0 = (1 + \chi_e)$ and this ratio is called the relative permittivity of the material being denoted usually by the symbol ϵ, i.e.

$$\epsilon \equiv 1 + \chi_e. \qquad (5.8)$$

The dielectric constant of a material is $\epsilon_0 \epsilon$ and, as mentioned earlier insulating materials are generally referred to as 'dielectrics'. A discussion of the range of values of the relative permittivity which are encountered in practice, and the classification of dielectrics, is given later in Section 5.3 after an analysis of the polarization due to time-dependent electrical fields. However the static relative permittivity ϵ_s of several materials is given in Table 5.1.

5.1.3. The electric displacement

Imagine a metal sphere carrying a net charge q which is totally immersed in a body of fluid which may be considered to be infinite in extent. The fluid will be polarized under the action of the radial electric field due to the sphere but the electric field (see Fig. 5.3) at any point is reduced by the factor $1/\epsilon$ compared to the field E_0 which would have existed in the absence of the medium; this can be seen as follows

$$\mathbf{E} = \mathbf{E}_0 - \mathbf{E}'; \qquad \mathbf{E}' = \frac{P}{\epsilon_0} \quad \text{and} \quad \chi_e = \frac{P}{\epsilon_0 E}$$

Therefore

$$E(1 + \chi_e) = E_0 \quad \text{i.e.} \quad E = \frac{E_0}{\epsilon}$$

Hence the surface integral of the electric field **E** can be expressed in terms of an integral over a hypothetical spherical surface S centred on the sphere as

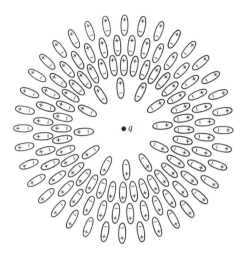

FIG. 5.3. A schematic representation of the polarization of the molecules of a fluid by a charge q.

$$\int_S \mathbf{E} \cdot \mathbf{ds} = \int \frac{q}{4\pi\epsilon_0 \epsilon r^2} \cdot 4\pi r^2 \, dr$$

or

$$\int_S \epsilon \mathbf{E} \cdot \mathbf{ds} = \frac{q}{\epsilon_0}. \qquad (5.9)$$

The reason for this apparent breakdown in Gauss's law lies in the fact that the charge q on the sphere is not the only charge enclosed within the hypothetical surface; there is also a net charge arising from the polarization of the molecules of the fluid. This effect may be difficult to picture but in crude qualitative terms the reason is as follows: If q is positive then naturally the negative ends of the microscopic dipole moments are closest to q. For any hypothetical surface those molecules which lie completely within it do not contribute to the net enclosed charge whilst those molecules which bestride the surface naturally give a net negative charge inside the surface. In analytical terms this is shown by starting from the differential form of Gauss's law:

$$\nabla \cdot \mathbf{E} = \frac{\rho}{\epsilon_0}$$

or

$$\nabla \cdot \mathbf{E} = \frac{\rho_f + \rho_b}{\epsilon_0} \qquad (5.10)$$

where the total charge density ρ is made up of a contribution ρ_f from 'free' (or conduction) charges and a contribution ρ_b from 'bound' charges; the latter are the net charges arising from the polarized molecules of the medium. Now using the divergence theorem, eqn (5.9) can be written as

$$\int_v \nabla \cdot (\epsilon E) dv = \frac{q}{\epsilon_0}$$

where v is the volume enclosed by the hypothetical spherical surface S. Now assuming that the fluid is a perfect insulator then the only conduction charges contained within the surface are those on the metal sphere, i.e.

$$\int_v \rho_f dv = q.$$

Thus, we have

$$\int_v \nabla \cdot (\epsilon E) dv = \frac{1}{\epsilon_0} \int_v \rho_f dv \qquad (5.11)$$

or

$$\nabla \cdot \epsilon E = \frac{\rho_f}{\epsilon_0}.$$

On substituting for ρ_f in eqn (5.10) we obtain

$$\nabla \cdot (\epsilon - 1) E = -\frac{\rho_b}{\epsilon_0}$$

and using $\epsilon - 1 = P/\epsilon_0 E$ from eqns (5.6) and (5.8) it follows that

$$\nabla \cdot P = -\rho_b. \qquad (5.12)$$

By using the relationship in eqn (5.10) we find

$$\nabla \cdot E = \frac{\rho_f}{\epsilon_0} - \nabla \cdot \frac{P}{\epsilon_0}$$

or

$$\nabla \cdot (\epsilon_0 E + P) = \rho_f.$$

The electric displacement **D** is defined through

$$D \equiv \epsilon_0 E + P \qquad (5.13)$$

and so

$$\nabla \cdot D = \rho_f. \qquad (5.14)$$

Also

$$D = \epsilon_0 \epsilon E. \qquad (5.15)$$

Hence, eqn (5.11) becomes $\int \nabla \cdot D dv = \int \rho_f dv$ or, using the divergence theorem again

$$\int_S D \cdot ds = \int_v \rho_f dv \qquad (5.16)$$

which is a statement of Gauss's law for material media.

In the foregoing it has been assumed that χ_e is independent of the magnitude of **E** i.e. that the medium is linear. Also it has been assumed in addition that the medium is isotropic and homogeneous: eqn (5.15) is true only if these assumptions hold.

The behaviour of dielectrics could be described without invoking this auxiliary vector, the displacement, but it turns out to be a useful quantity largely because of its property, exemplified by eqn (5.14), of being related to the distribution of free charges only.

As a simple example of an application of eqn (5.16) consider a coaxial cable or transmission line in which the inner and outer conductors have radii a and b respectively and the space between them is filled with a dielectric of relative permittivity ϵ (see Fig. 5.4). Imagine a hypothetical cylindrical surface of radius r and apply eqn (5.16):

$$\int_s \mathbf{D} \cdot d\mathbf{s} = D \cdot 2\pi r l$$

for a length l of line since D is constant over the surface. If there is charge q per unit length of the inner conductor (and by implication charge $-q$ per unit length of the outer conductor) then using eqn (5.16)

$$D \cdot 2\pi r l = ql.$$

Since $D = \epsilon \epsilon_0 E$ this means that

$$E = \frac{2q}{4\pi \epsilon_0 \epsilon r}$$

and so the potential difference V between the inner and outer conductors is given by

$$V = \int_a^b E \, dr \quad \text{or} \quad V = \frac{2q}{4\pi \epsilon_0 \epsilon} \int_a^b \frac{dr}{r}.$$

Thus

$$V = \frac{2q}{4\pi \epsilon_0 \epsilon} \ln\left(\frac{b}{a}\right)$$

and, since $C = q/V$, we have

$$C = \frac{4\pi \epsilon_0 \epsilon}{2 \ln(b/a)} \quad \text{per metre.}$$

Typical values of ϵ and (b/a) are 3 and 5 respectively which gives $C \approx 100 \, \text{pF m}^{-1}$.

Consider the electric field at a plane interface between two dielectrics (see Fig. 5.4) which has a density of free charges. Imagine a hypothetical Gaussian surface of cylindrical shape which spans the interface as shown in Fig. 5.4(b). In applying eqn (5.16) to this closed surface the contributions to $\int \mathbf{D} \cdot d\mathbf{s}$ from the curved surface become

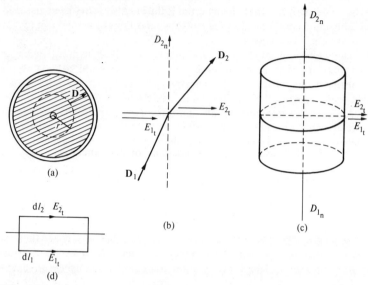

(a)

(b)

(c)

(d)

FIG. 5.4. (a) A cross-section of a dielectric-filled coaxial line. (b) The electric displacement and the normal electric displacement and tangential electric field at an interface between two dielectrics. (c) A hypothetical Gaussian 'pill-box' spanning the interface between two dielectrics. (d) A closed elementary path of integration spanning the interface.

negligible if the height of the cylinder is imagined to shrink to an infinitely small magnitude. Thus we have

$$(D_{2_n} ds - D_{1_n} ds) = \rho_f ds$$

or

$$D_{2_n} - D_{1_n} = \rho_f. \tag{5.17}$$

Now consider a closed path of integration which spans the interface, as shown in Fig. 5.4(c). The contribution to the line integral $\oint \mathbf{E} \cdot d\mathbf{l}$ from the 'end' sections of the path become negligible in the imagined circumstances and so

$$\oint \mathbf{E} \cdot d\mathbf{l} = E_{1_t} dl - E_{2_t} dl = 0.$$

Hence,

$$E_{1_t} = E_{2_t}. \tag{5.18}$$

Thus the boundary conditions at the interface are that:
 (i) The normal component of the electric displacement is discontinuous by ρ_f across the interface.
 (ii) The tangential component of electric field is continuous across the interface.

5.2. Alternating electric fields and dielectric losses

In section 5.1.2 it emerged that the static relative permittivity of a material, which we will now call ϵ_s, is related to the polarization p through

$$\epsilon_s = 1 + \chi_e \qquad \text{or} \qquad \epsilon_s - 1 = \frac{P}{\epsilon_0 E} \qquad (5.19)$$

where P is the sum of the contributions from induced and permanent microscopic dipoles. For alternating electric fields the effective relative permittivity is related to the extent to which the displacement of the charges, or the ordering of the assembly of permanent dipoles, as the case may be, can follow the time-dependent field; the responses of displaced charges and permanent dipoles are markedly different.

A simple mechanical analogy to an assembly of permanent dipoles would be a paddle in water; it is fairly easy to move it backwards and forwards slowly but very difficult to move it quickly. In an analogous fashion the response to an applied alternating field of the net macroscopic polarization of an assembly of permanent dipoles decreases with the frequency of the field until at high enough frequencies the response is very small and the contribution to $(\epsilon - 1)$ is negligible.

The induced dipoles exhibit a resonance type of response to alternating fields since the attractive force between the displaced positive and negative charges is in the nature of a restoring force; the resonant frequencies are higher than 10^{14} Hz which is well above the range $(0-10^{11}$ Hz) which is of interest for electrical circuits. In the frequency range well below resonance the response of the induced dipoles to the alternating field is virtually instantaneous: a mechanical analogy would be a mass on a spring which was subjected to a periodic displacing force. If the response to an alternating electric field can be assumed to be instantaneous over the range of frequencies of interest then the value of the relative permittivity will be independent of frequency and will have a magnitude equal to that of the static value. The essential features of this qualitative description are shown in Fig. 5.5.

The total polarization of a sample can be written as

$$P = P_a + P_i + P_d \qquad (5.20)$$

where P_a is the contribution due to induced atomic dipoles, P_i is the contribution from induced ionic dipoles, and P_d represents the contribution from permanent dipoles. Now from eqns (5.5) and (5.7) it follows that

$$\epsilon - 1 = \frac{P}{\epsilon_0 E}$$

and so

$$\epsilon - 1 = \frac{(P_a + P_i) + P_d}{\epsilon_0 E}.$$

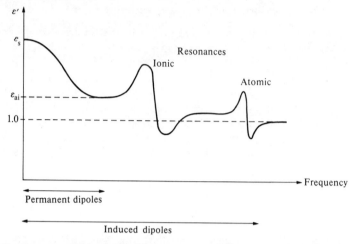

FIG. 5.5. A qualitative representation of the frequency dependence of the real part ϵ' of the dielectric constant of a dielectric material.

For applied alternating fields having frequencies below 10^{11} Hz it will be assumed that the atomic and ionic polarizations are independent of frequency and a contribution ϵ_{ai} to the total relative permittivity is defined through

$$\epsilon_{ai} - 1 \equiv \frac{(P_a + P_i)}{\epsilon_0 E}. \quad (5.21)$$

The dipolar contribution to the total polarization will be assumed to have a 'step function' response (see Fig. 5.6) of the form

$$P_d(t) = P_d^\infty (1 - e^{-t/\tau}) \quad (5.22)$$

where P_d^∞ is the limiting value which P_d would attain after an infinitely long time interval (see eqn 3.3 for the growth of charge on a capacitor in a $C-R$ circuit).

For an assembly of dipoles having such a response the polarization will decay exponentially if the polarizing field is switched off. The exponential decay towards the disordered situation at thermal equilibrium, with $P_d = 0$, is characterized by the time constant τ, i.e. $P_d(t) \propto e^{-t/\tau}$ and τ is often called the relaxation time.

Since the atomic and ionic induced dipoles have been assumed to respond instantaneously to an applied field it follows that after an infinite lapse of time
$$P = P_a + P_i + P_d^\infty$$
where P is the polarization due to a static field. Thus

$$P = (\epsilon_s - 1)\epsilon_0 E. \quad (5.23)$$

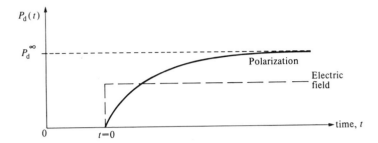

FIG. 5.6. The response to the application of an electric field of the total polarization of an assembly of permanent dipoles.

Now from eqn (5.22)

$$\frac{dP_d(t)}{dt} = \frac{P_d^\infty e^{-t/\tau}}{\tau}$$

$$= \frac{P_d^\infty - P_d(t)}{\tau}. \qquad (5.24)$$

The crucial assumption at this stage of the development of this theory is that, in the situation where there is an alternating applied field $\hat{E}e^{j\omega t}$, P_d^∞ can be replaced by $P_d^\infty(t)$ where this quantity is the limiting value of polarization corresponding to the instantaneous value of the applied field at time t. Hence eqn (5.24) becomes

$$\frac{dP_d(t)}{dt} = \frac{P_d^\infty(t) - P_d(t)}{\tau}$$

$$= \frac{P - (P_a + P_i) - P_d(t)}{\tau}$$

and using eqns (5.20) and (5.21) it follows that

$$\frac{dP_d(t)}{dt} + \frac{P_d(t)}{\tau} = \frac{(\epsilon_s - \epsilon_{ai})}{\tau} \epsilon_0 \hat{E} e^{j\omega t}.$$

This is an example of a well known form of differential equation and the solution is

$$P_d(t) = \frac{(\epsilon_s - \epsilon_{ai})}{(1 + j\omega\tau)} \epsilon_0 \hat{E} e^{j\omega t}$$

Since $(\epsilon - 1) = P/\epsilon_0 E$ and by making use of eqn (5.21), and rationalizing, it follows that

$$\epsilon = \epsilon_{ai} + \frac{(\epsilon_s - \epsilon_{ai})}{(1 + \omega^2\tau^2)} - \frac{j\omega\tau(\epsilon_s - \epsilon_{ai})}{(1 + \omega^2\tau^2)}. \qquad (5.25)$$

So we see that the effective relative permittivity is complex and can be written in the form $\epsilon = \epsilon' - j\epsilon''$ where

$$\epsilon' = \epsilon_{ai} + \frac{(\epsilon_s - \epsilon_{ai})}{(1 + \omega^2 \tau^2)} \tag{5.26}$$

$$\epsilon'' = \frac{\omega \tau (\epsilon_s - \epsilon_{ai})}{(1 + \omega^2 \tau^2)}. \tag{5.27}$$

Now we saw in Section 5.12, eqn 5.7, that the relative permittivity of a material is defined in terms of the capacitance of a capacitor which is filled with that material, i.e.

$$C = \epsilon C_0.$$

Thus the admittance Y of the capacitor is given by

$$Y = j\omega \epsilon C_0$$

or

$$Y = j\omega \epsilon' C_0 + \omega \epsilon'' C_0. \tag{5.28}$$

The second term in the expression for Y is real and can be written as $1/R_p$ where R_p is a resistor in parallel with the capacitance as was anticipated in Section 3.6. The energy losses in the material ('dielectric' losses) are represented by the loss angle δ of the capacitor, where $\tan \delta \approx (\omega C R_p)^{-1}$, and so

$$\tan \delta \approx \frac{\omega \epsilon'' C_0}{\omega \epsilon' C_0}$$

or

$$\tan \delta \approx \frac{\epsilon''}{\epsilon'}. \tag{5.29}$$

Eqns (5.26) and (5.27) are known as the Debye equations and describe well the electrical properties of most dielectrics at frequencies below 10^{11} Hz. The frequency dependence of ϵ' and ϵ'' is shown in Fig. 5.7.

A complication which is encountered with many dielectrics is that there is more than one of these relaxation mechanisms each with its own characteristic relaxation time τ. Some idea of the range of relaxation time encounted in practice can be gained from Table 5.1.

5.3. Types of dielectric

Dielectrics are used in a wide variety of roles in the general fields of electronic and electrical engineering but there are two areas of usage which are of particular importance in the context of electrical circuits, namely in capacitors and in transmission lines.

In the case of capacitors the dielectric material has a dual function:

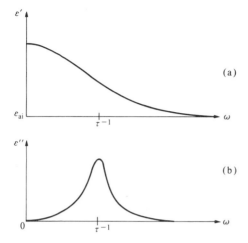

FIG. 5.7. The frequency dependence of (a) the real part and (b) the imaginary part, of the complex dielectric constant.

on the one hand by its presence it increases the capacitance per unit volume and on the other hand it serves to separate and insulate the capacitor plates which, of course, should be as close together as possible bearing in mind the spatial tolerances and the electrical breakdown strength of the material.

Dielectrics fulfil the same general purposes in transmission lines and waveguides; the cross-sectional area of the line or waveguide can be reduced, compared to the air-filled situation, and a solid dielectric also provides mechanical strength, particularly in the case of a coaxial line. The most widely used materials are polyethylene, polystyrene, and polytetrafluoroethylene (P.T.F.E.) for which ϵ' is constant to better than 1 per cent over the frequency range from 50 Hz to 10 GHz. Dielectric windows are used to separate evacuated and air-filled sections of waveguide and dielectric vanes are employed in phase shifters.

A physical property of dielectrics which is important in all of these contexts is the electrical breakdown strength. This property is specified by the electric field strength at which a runaway current starts to flow through the material; in much the same way a spark is a manifestation of breakdown in a gas or vapour. The physical mechanism of breakdown is very complex but the features which are of most direct importance from an engineering point of view are that the value of the field at breakdown is temperature dependent and frequency dependent being higher, very often, for alternating than for steady fields.

The important technical properties of a selection of commonly used dielectrics are given in Table 5.1 but it must not be overlooked that

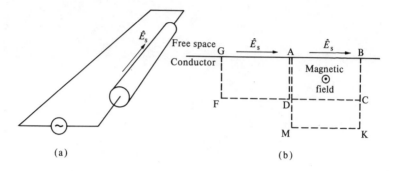

FIG. 5.8. (a) The alternating electric field at the surface of a conductor. (b) The magnetic field in a conductor associated with the currents due to the alternating electric field directed parallel to the surface of the conductor.

material costs and fabrication costs are factors of crucial importance in the choice of material for mass-produced components.

In the most common situations in electrical circuits using discrete components capacitors have to withstand d.c. voltages up to perhaps 10 kV, but more usually much lower values, and alternating voltages up to 250 kV. Breakdown strengths of dielectrics are typically of the order of 10^6–10^8 V m^{-1} and the thickness of dielectric layers are usually such that the operating voltage is about one-tenth of that required to give the breakdown field.

Some typical specifications of commerically available capacitors are:

Polystyrene	10 pF–47 000 pF	630 V working
Ceramic	10 pF–0.5 μF	up to 400 V working
Silvered mica	10 pF–0.01 μF	up to 400 V working
Metallized paper	0.05 μF–2 μF	up to 400 V working
Metallized polycarbonate	0.05 μF–2 μF	up to 400 V working
Metallized polyester	0.01 μF–10 μF	up to 400 V working
Metallized polypropylene	0.01 μF–1 μF	up to 400 V working

For situations where a large value of capacitance is required and relatively large losses can be tolerated then electrolytic capacitors can be used. The basic features of the construction of such capacitors are as follows. Under the action of the polarizing voltage electrochemical action causes the formation of an insulating film on an anode; a film of Al_2O_3 for an aluminium anode and Ta_2O_5 for tantalum. This process, which is known as anodizing, produces extremely thin films, 0.02 μm in thickness typically, which give rise to a large capacitance. Such capacitors are usually constructed by rolling up an aluminium or tantalum foil anode together with an aluminium cathode and a paper separator, the whole being soaked in a liquid electrolyte or packed with a solid one such as MnO_2. Electrolytic capacitors having capacitances as large as $47\,000\,\mu F$ at working voltages of 16 V and with an equivalent series resistance of about $0.03\,\Omega$ are available currently.

5.4. Alternating electromagnetic fields in conductors

5.4.1. The 'skin' effect

Imagine that a signal generator produces an alternating electric field of amplitude $\hat{E}_s$ parallel to the surface of a conductor (see Fig. 5.8(a)). Before we attempt to consider analytically the problem of determining the distribution of current over the cross-section of the wire it will be useful to make some qualitative observations which will serve to give a useful impression of the physical situation. The alternating electric field will cause a current flow in the conductor in a direction parallel to $\hat{E}_s$ and this current will generate an alternating magnetic field which will be directed out of the plane of the paper in the situation depicted in Fig. 5.8(b). The laws of electromagnetic induction tell us therefore that in a closed path, such as ABCD, there will be an induced e.m.f. in such a sense as to drive a current (a so-called eddy current) around the closed path, the sense of the current being from C to D i.e. opposing the current produced by the applied electric field. A path such as ABKM has more magnetic flux linked with it than path ABCD, the induced e.m.f. will be greater and so the eddy current flowing from K to M will be greater than that along CD.

From these qualitative arguments it can be deduced that when an alternating electric field is applied to a conductor the resultant current flow is not disturbed uniformly over the cross-section of the conductor but is most dense at the surface and decreases with depth of penetration into the body of the conductor. A succinct statement is that the current density is greatest in those regions of a conductor which are encircled by the smallest number of magnetic flux lines. This phenomenon is known as the skin effect. There are two consequences which are of particular interest here. First, for a long straight wire, of circular cross-section say, the resistance R at an angular frequency ω will be greater than the value R_0 for direct current since the effective area of cross-section of the wire is smaller: We can write

$$R = R_0(1 + \alpha). \qquad (5.30)$$

Secondly the eddy currents must depend on the angular frequency ω of the alternating field (see Section 1.4), and upon the permeability $\mu\mu_0$ and conductivity σ of the material of the wire, as well as the radius a of the wire. Since the function α describes the effects of the eddy currents and since, furthermore, it is dimensionless, it follows that it must be a function of the dimensionless variable $u \equiv a\sqrt{(\mu\mu_0 \omega \sigma)}$; the proof of this is left as an exercise in the use of dimensional arguments.

It can be shown that the amplitude of an electromagnetic field decays exponentially with depth of penetration z into a conductor, e.g.

$$\hat{E}(z) = \hat{E}_s e^{-z/\delta} \qquad (5.31)$$

where

$$\delta \equiv \sqrt{\left(\frac{2}{\mu\mu_0 \sigma \omega}\right)} = \frac{a\sqrt{2}}{u} \qquad (5.32)$$

is known as the skin depth of the material at angular frequency ω.

The function $\alpha(u)$ cannot be expressed in simple form for general values of u but, to within 10 per cent discrepancy limits roughly:

$$\text{Low frequency approximation } (u < 2); \alpha(u) \approx \frac{u^4}{192} \qquad (5.33)$$

$$\text{High frequency approximation } (u > 8); 1 + \alpha(u) \approx \frac{u}{2\sqrt{2}}. \qquad (5.34)$$

The results of measurements of the resistance of a small loosely wound air-cored coil made up of copper wire of 0.5-mm diameter are displayed in Fig. 5.9. From eqn (5.34) it is predicted that $R \propto \sqrt{\text{frequency}}$ for high enough frequencies and this behaviour is found in the experimental results.

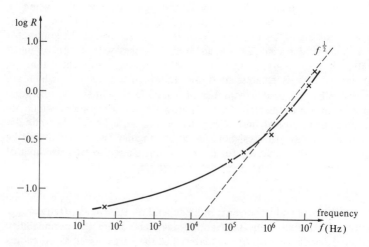

FIG. 5.9. The frequency dependence of the resistance of a loosely wound air-cored coil.

5.4.2. The inductance of lines and coils

Generally speaking if specific values of self-inductance or mutual inductance are required of coils then use would be made of tabulated data[†]. However, it is of interest to review particular physical situations which are of importance to electrical circuits in general.

A long straight wire has self-inductance in association with the energy stored in the magnetic field inside the wire itself. For a wire of circular cross-section of radius a the magnetic field at radius r, where $r < a$, can be found by applying Ampère's circuital law. (see Section 1.4) to the circular path of radius r along which the magnetic field is constant. If the total current I is assumed to be distributed uniformly over the cross-section of the wire then the application of Ampère's circuital law gives

$$B(r) 2\pi r = \mu \mu_0 \frac{r^2}{a^2} I$$

or

$$B(r) = \frac{\mu_0}{4\pi} \frac{2rI}{a^2} \qquad (5.35)$$

where it has been assumed that $\mu = 1$ which is true, to a good approximation, for non-ferromagnetic materials.

So using the expression for the stored energy per unit volume in a magnetic field, namely $B^2/2\mu_0$ J m^{-3} (see Problem 1.5), the total energy W stored per unit length of wire is

$$W = \frac{\mu_0 I^2}{8\pi^2 a^4} \int_0^a r^2 2\pi r \, dr$$

or

$$W = \frac{\mu_0 I^2}{16\pi}.$$

Now since $W = LI^2/2$ also (see eqn (1.34)) it follows that

$$L = \frac{\mu_0}{8\pi} \text{H m}^{-1}. \qquad (5.36)$$

This is usually called the internal self-inductance of the wire; because of the assumption of uniform current distribution this result applies accurately only at low frequencies where the skin depth is large compared to the diameter of the wire; for high frequencies it can be shown that $L = (4\pi^2 a \delta \sigma f)^{-1}$ H m^{-1}.

For a coaxial line there is self-inductance associated with the energy

---

[†] See, for instance, Terman, F. E. *'Radio Engineers Handbook'*. McGraw-Hill, (1943).

stored in the magnetic field which exists between the two conductors. It is left as an exercise to show that the self-inductance L per unit length is given by

$$L = \left(\frac{\mu_0}{4\pi}\right) 2 \ln(b/a) \qquad (5.37)$$

where a, b are the radii of the inner and outer conductors respectively. However it should be noted that the current in the outer conductor produces no magnetic field in the space between the conductors since this current does not link a closed path which lies entirely in this space.

It is also of interest to note that the expression for L (eqn (5.37)) is the inverse of that for the capacitance per unit length of a coaxial line (eqn (1.14)), apart from the fundamental electric and magnetic constants of course. The inductance L per unit length of two parallel cylindrical conductors can be calculated in a entirely analogous method to that given in Section 1.1.2 for the calculation of the capacitance per unit length through the definition and use of a magnetic vector potential. However it will suffice for our purposes here to extend the notion which was introduced in the previous sentence and say, by comparison with eqn 1.20, that

$$L = \left(\frac{\mu_0}{4\pi}\right) 4 \cosh^{-1}(b/2a). \qquad (5.38)$$

Since $\mu_0 = 4\pi \times 10^{-7}\,\mathrm{H\,m^{-1}}$ it follows from eqn (5.38) that $L = 0.4 \cosh^{-1}(b/2a)\,\mu\mathrm{H\,m^{-1}}$ and for $b/2a = 3$ say then $L \approx 0.8\,\mu\mathrm{H\,m^{-1}}$.

The expressions of eqns 5.37 and 5.38 give the external self-inductances of the coaxial line and twin line respectively and to these must be added the internal self-inductances of the conductors involved in order to obtain the total self-inductance.

If the space between the conductors of a coil or line is filled with a material of relative permeability μ then, to a good approximation, the self-inductance is increased by a factor equal to the value of μ. A discussion of the substances which are useful as core materials in coils will be given in Section 5.5: for the present we note that because of increasing energy losses laminated cores made from ferromagnetic materials are confined to the frequency range below 20 kHz roughly and that ferrite cores can be used up to about 50 MHz.

As hinted above the design of coils to obtain maximum inductance, minimum resistance, and minimum distributed capacitance of the windings is largely an empirical procedure. Apart from tabulated data there are formulae which apply with reasonable accuracy to common shapes of coils. For instance, for a single-layer solenoid with n turns as shown in Fig. 5.10(a), the inductance L is given by

$$L = 3.95 a^2 n^2 K/l \quad \mu\mathrm{H} \qquad (5.39)$$

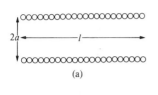
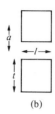
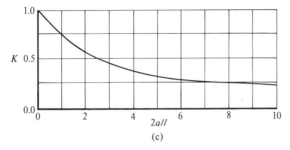

FIG. 5.10. In (a) and (b) are shown the shapes of coils for which the self-inductances are given by eqns (5.39) and (5.40) respectively. The dependence on l and a of the geometrical factor K for a coil of the shape in (a) is shown in (c).

where l and a are measured in metres. The dependence on l and a of the numerical factor K is shown in Fig. 5.10(c). In the case of a multilayer coil as shown in Fig. 5.10(b)

$$L \approx \frac{32 a^2 n^2}{(6a + 9l + 10t)} \mu\text{H}. \qquad (5.40)$$

In the audio-frequency range coils having laminated iron cores have Q-factors which can be as high as 100. At higher frequencies where ferrite-cored or 'air'-cored coils are used the inter-turn capacitance becomes a significant factor as well as the losses arising from the skin effect. The latter effect is compounded with the so-called proximity effect: the current distribution in a turn of a coil is affected not only by the magnetic flux due to the current flowing in that turn but also by the current flowing in neighbouring turns (see the proximity effect discussed in Sections 1.1.2). In single-layer coils a gauge of wire should be chosen so that the coil is loosely wound (diameter of wire $\frac{1}{2}$ to $\frac{3}{4}$ of the spacing of the turns, say): the decrease in losses due to a reduction of the proximity effect will be more than offset the increase in losses via the skin effect due to the smaller area of cross-section of the wire used. The losses due to the skin effect can be reduced by using multi-stranded insulated wire (Litz wire); the conductor is compounded of a number of strands of fine wire which are connected in parallel and interwoven. Thus the current divides equally amongst the strands and, providing the diameter of a strand is less than or of the order of the skin depth, then the current divides fairly uniformly over the cross-section of the compound conductor; the a.c. resistance can be made to approach the d.c. value at moderate radio-frequencies. So in well designed coils a Q-factor of 100–300 can be obtained in the frequency range up to about 1 MHz; above this frequency the capacitance between the strands is a major factor in reducing the effectiveness of Litz wire.

The self-capacitance (or distributed capacitance) of a coil should be as small as possible for two particular reasons. Firstly, the value of the self-capacitance determines the highest frequency at which the coil can be made to resonate. Secondly, the dielectric losses in the materials of the coil (coil former and insulating materials), which are affected by the electric fields existing in the capacitances between windings, can give rise to a significant additional resistance and hence a reduction in the Q-factor of the coil. The general rule to be followed in order to minimize self-capacitance is to wind the coil so that turns which are close to each other are never at widely differing potentials, e.g. use a 'bank' winding.

5.5. Magnetic fields in solids

5.5.1. Basic macroscopic theory

As was mentioned in Section 1.4 the relative permeability μ of a material can be defined on an experimental basis through $\mu \equiv L/L_0$ where L is the self-inductance of a 'long' solenoid when it is filled with the material and L_0 is the inductance when the material is removed. The usefulness of materials having large values of $\mu (\sim 1000)$ is that much larger values of inductance can be obtained than for air-cored coils or, alternatively, for a given value of inductance a coil can be made much smaller. Apart from the advantage from the mechanical point of view of having compact components, another advantage is that the distributed capacitance of a coil is smaller since it is related fairly directly to the dimensions of the coil.

Many parallels can be drawn between the descriptions of the magnetization and the electrical polarization of materials although it is dangerous to push the analogies too far. As with electrical polarization the basic description of magnetization can be given in terms of induced and permanent dipoles: in this case magnetic dipoles, of course. These dipoles arise from the electric currents associated with the orbital motion of electrons[†] about the nuclei of atoms and also, very importantly,

[†] The magnetic dipole moment m of a current I flowing in a path in the form of a 'small' loop of area A is defined to be $m \equiv IA$ A m^2. Of course the magnetic dipole moment is a vector quantity and the convention is to assume that the vector is directed perpendicular to the plane of the loop with its sense derived from the circulation of the current through the right-hand screw rule (see Fig. 5.11(a)).

from the circulation of electric charge associated with the so-called 'spin' of electrons.

We are not concerned here with a detailed description at the atomic level of the origins of the magnetization of solids and so we shall deal with the macroscopic features only. In fact the only classes of solid in which we are interested from a practical point of view are the so-called ferromagnetic and ferrimagnetic materials. In these materials there are interactions which, briefly, cause the magnetic dipole moments of neighbouring atoms to line up parallel to one another despite the disordering effect of thermal vibrations of the atoms. The parallel alignment of atomic magnetic moments occurs usually over limited regions of a sample, called domains, which are of macroscopic dimensions ($\sim 10^{-1}$ mm in lateral extent and often larger); the number, shape, and directions of magnetization of the domains in a particular sample is such as to minimize the free energy. In the absence of an applied magnetic field, or of a 'memory' of a previously applied magnetic field, the magnetization vectors of the individual domains point in all directions so that overall the sample has zero net magnetization.

Consider a small volume element of a cylinder of magnetized material as shown in Fig. 5.11(b); the magnetization M, which is defined as magnetic moment per unit volume, is assumed to be uniform throughout the sample of material. If we assume that M arises from a current I as shown, then the magnetic moment of the current loop is equal to (magnetization) × (volume of the element), i.e.

$$I\,dxdy = M\,dxdydz \quad \text{or} \quad I = M\,dz.$$

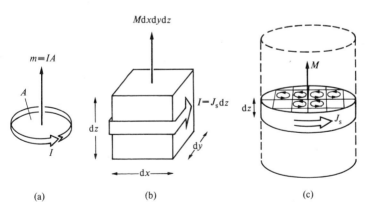

FIG. 5.11. (a) The definition of the magnetic dipole moment m of a 'small' current loop area A carrying current I. (b) The relationship between the magnetization M of a medium and the density J_s of the associated 'bound' current.

It must be emphasized that I is not a conduction current, which would have energy losses associated with it, but a current which is lossless in the same sense that the current due to an electron circulating in an orbit in a atom is lossless according to the precepts of quantum mechanics. In other words I is the large-scale resultant of the microscope electron currents circulating in the interacting atoms of ferromagnetic and ferrimagnetic materials. A lossless current such as I here is usually called an 'Ampèrian' or 'bound' current to distinguish it from the real, conduction, currents which are driven through conductors by electromotive forces. An analogy can be drawn here with the concepts of 'bound' and 'free' electrical charges which were introduced in Section 5.1.3 in the context of electrical polarization.

The bound current I can be expressed in terms of a surface density J_s of bound current; $I = J_s \mathrm{d}z$. So,

$$M = J_s. \tag{5.41}$$

A thin slice of the sample can be imagined to be made up of a large number of volume elements, or cells, such as the one that we have just been considering. The bound currents flowing along the boundaries of adjacent cells cancel each other out so that there is a resultant bound current at the surface of the sample only (see Fig. 11(c); note that this is true only for a uniformly magnetized sample). The magnetic field at any point external to the sample (providing that we don't approach too close to the sample) is identical to that which would be produced by a sheath of current having a surface density $J_s \mathrm{A\,m^{-1}}$. For samples having more complicated shapes and non-uniform magnetization the mathematical analysis is more complex but the physical arguments and conclusions are the same.

We will now consider Ampère's circuital law (see Section 1.4) as applied in material media:

$$\frac{1}{\mu_0} \oint \mathbf{B} \cdot \mathrm{d}\mathbf{l} = \text{total current linked with the closed path}$$

$$= \text{(total conduction current)} + \text{(total bound current)}$$

$$= I_c + I_b.$$

Consider a path element dl of the closed path depicted in Fig. 5.12. For the path element, using the definition of the magnetic moment of a current loop,

$$(M \, \mathrm{d}l \, \mathrm{d}s) \cos \theta = \mathrm{d}I_b \mathrm{d}s.$$

Thus,

$$M \, \mathrm{d}l \cos \theta = \mathrm{d}I_b \quad \text{or} \quad \mathbf{M} \cdot \mathrm{d}\mathbf{l} = \mathrm{d}I_b.$$

So the total bound current I_b linked with the path is given by

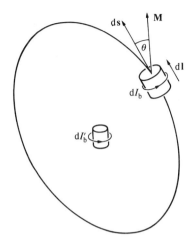

FIG. 5.12. A path element used in the determination of the relationship between the magnetization M of a medium and the bound current density I_b.

$$I_b = \int dI_b = \oint \mathbf{M} \cdot d\mathbf{l}.$$

(Note that there is no net linkage of current with the closed path from current loops such as I_b'.) Hence,

$$\frac{1}{\mu_0} \oint \mathbf{B} \cdot d\mathbf{l} = I_c + \oint \mathbf{M} \cdot d\mathbf{l}$$

or

$$\oint \left(\frac{\mathbf{B}}{\mu_0} - \mathbf{M} \right) \cdot d\mathbf{l} = I_c. \tag{5.42}$$

A very useful auxiliary vector $\mathbf{H}$ is defined through

$$\mathbf{H} \equiv \frac{\mathbf{B}}{\mu_0} - \mathbf{M} \tag{5.43}$$

so that

$$\oint \mathbf{H} \cdot d\mathbf{l} = I_c. \tag{5.44}$$

The great utility of the quantity $\mathbf{H}$ lies in the fact that its line integral round a closed path is simply equal to the total *conduction* current linked with the path; in general the determination of $\mathbf{M}$ and $\mathbf{B}$ in the interiors of magnetized samples is very difficult.

For media other than ferromagnets and ferrimagnets $\mathbf{B}$ and $\mathbf{H}$ are related linearly to one another through the relative permeability μ, i.e.

124 Electromagnetics

Hence,
$$\mathbf{B} = \mu\mu_0 \mathbf{H}. \tag{5.45}$$
$$\mu = 1 + \chi_m$$

where the magnetic susceptibility χ_m is defined through

$$\chi_m \equiv \frac{M}{H}. \tag{5.46}$$

(Note that for such materials $\chi_m \sim 10^{-4}$ or less so that $\mu = 1$ to a very close approximation.)

The relationship between **B** and **H** is much more complex in the case of ferromagnetic and ferrimagnetic materials and the characteristics of some important classes of such materials will be outlined in Section 5.5.3. For the present it will suffice to say that eqn (5.43) still applies but eqn (5.45) should be taken to define a variable quantity μ rather than to imply a linear relationship between **B** and **H**.

5.5.2. Magnetic circuits

Consider a ring made of four ferromagnetic materials and where each material has a different cross-section and length, as shown in Fig. 5.13: it is assumed that the diameter of even the fattest section is small compared to the over-all diameter of the ring so that B, H can be considered to be constant over any radial section of the material. From the circuital law (eqn 5.44) we know that for the indicated closed path

$$H_1 l_1 + H_2 l_2 + H_3 l_3 + H_4 l_4 = NI$$

Where I is the (conduction) current flowing in the winding and N is the total number of turns. Now

$$H_1 l_1 = \frac{B_1 l_1}{\mu_1 \mu_0} \quad \text{or} \quad H_1 l_1 = \frac{\phi_1 l_1}{\mu_1 \mu_0 A}$$

where $\phi_1 \equiv B_1 A_1$ is the magnetic flux in Section 1. Here we are assuming that the lines of **B** are confined to the material of the ring which will be a reasonable approximation for a ferromagnetic material having a relative permeability greater than 10^3. Also since lines of **B** are continuous the magnetic flux will have the same value in each section of the ring.

So we have

$$\left(\phi \; \frac{l_1}{\mu_1 \mu_0 A_1} + \frac{l_2}{\mu_2 \mu_0 A_2} + \frac{l_3}{\mu_3 \mu_0 A_3} + \frac{l_4}{\mu_4 \mu_0 A_4} \right) = NI.$$

This equation can be written

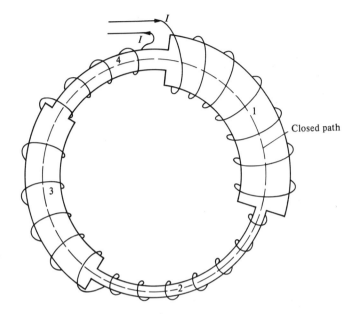

FIG. 5.13. A closed ring composed of four different ferromagnetic materials.

$$\phi \mathscr{R} = \mathscr{H} \tag{5.47}$$

and, by analogy with $IR = E$ in an electrical circuit, $\mathscr{H} (=NI)$ is called the magnetomotive force (m.m.f.), and $\mathscr{R}$ the reluctance, of the magnetic circuit.

It is of interest to examine more closely the analogy between magnetic and electrical circuits: For a conductor of uniform cross-section A, length l, and conducitivity σ the electrical resistance R is

$$R = \frac{l}{\sigma A}$$

and for an element of a magnetic circuit of the same length and cross-sectional area but having relative permeability μ the reluctance is

$$\mathscr{R} = \frac{l}{\mu \mu_0 A}.$$

In the usual electrical circuit situation the conducitvity of the material of the circuit elements is probably at least 10^{20} times greater than that of the surrounding medium. However for most magnetic circuits the relative permeability will be between 10^3 and 10^4 times that of the surrounding medium. Also the length to transverse dimension ratio is

126   Electromagnetics

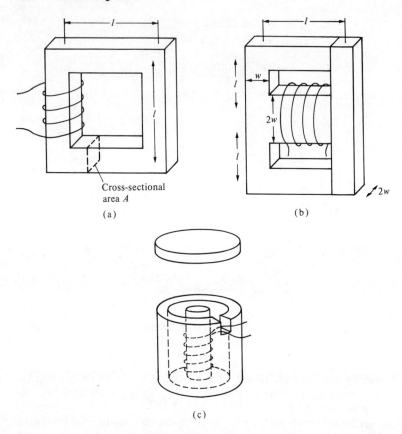

FIG. 5.14. Some forms of magnetic circuit which can be utilized in transformers, for example.

usually much smaller for magnetic than for electrical circuits. The practical significance of these features is that whereas for an electric circuit, for practical purposes, the current flows through the circuit elements only, and not at all through the surrounding medium, in magnetic circuits there is significant magnetic flux outside the elements of the circuit. Such flux leakage is difficult to calculate in practical situations but use can often be made of empirical tables in many circumstances.

A simple example can be used to give you some idea of the order of magnitude of the physical quantities involved: Consider a coil wound on a core of the form shown in Fig. 5.14(a). Using eqn (5.47),

$$\phi = \frac{\mu\mu_0 ANI}{4l} \tag{5.48}$$

where it has been assumed that magnetic flux leakage is negligible and an obvious approximation has been made in taking the effective length of path for the flux to be $4l$.

Assuming $l = 5$ cm, $A = 2$ cm^2, $N = 10^3$, $I = 1$ mA, and $\mu = 10^3$, then $\phi = 4\pi \times 10^{-7}$ Wb and the magnetic field B in the core ($= \phi/A$) is $2\pi \times 10^{-3}$ T or 60 gauss approximately. The self-inductance L of the coil is equal to $d\phi/dI$ (see Section 1.4) and so $L = 4\pi \times 10^{-4}$ H or 1.2 mH approximately.

Now suppose that there is a small gap of width l_g in the core. In this case the flux ϕ' is given by

$$\phi' = \frac{\mu_0 ANI}{\left[\dfrac{4l - l_g}{\mu} + l_g\right]}$$

or

$$\phi' = \frac{\mu\mu_0 ANI}{4l\left(1 + \dfrac{\mu l_g}{4l}\right)} \tag{5.49}$$

where it has been assumed that $l_g \ll 4l$. If we suppose that $l_g = 4l/1000$ then $\phi' = \phi/2$ where ϕ is given by eqn (5.48). This rough calculation illustrates the fact that the magnetic flux in the circuit is reduced dramatically by the existence of even a small air gap, and so also is the inductance of the coil. Another way of looking at this situation is to define an effective relative permeability μ_{eff} for the core such that

$$\phi' = \frac{\mu_{\text{eff}}\mu_0 ANI}{4l}$$

where

$$\mu_{\text{eff}} = \frac{\mu}{\left(1 + \dfrac{\mu l_g}{4l}\right)}. \tag{5.50}$$

Since the self-inductance L of the coil is proportional to μ_{eff}, adjustment to the value of L can be made by altering the magnitude of l_g; this technique and the idea of an effective permeability and some related matters will be discussed in the next section.

Magnetic circuits of a parallel nature can be analysed by direct comparison with the analogous electrical circuits. A commonly used form of transformer core is shown in Fig. 5.14(b) where both primary and secondary windings are on the central limb of the E-section. The advantage of this arrangement is that there is smaller flux leakage than for the simple rectangular type of core depicted in Fig. 5.14(a). It is

left as a problem to show that for a core as shown in Fig. 5.14(b) having a winding of NI ampère-turns the flux ϕ in the central limb is given by the expression
$$\phi = 2\mu\mu_0 w^2 NI/7l.$$

A form of magnetic circuit which is commonly used with ferrites is the 'pot' core which is sketched in Fig. 5.14(c); the coil is wound on the central post of course. The compactness of this arrangement, with the coil enclosed within the ferrite cylinder, gives very small flux leakage.

5.5.3. Properties of magnetic materials

The magnetic properties of materials which are of relevance to the characteristics of circuit elements are described through the relative permeability μ which was defined by eqns (5.43) and (5.45). We will be concerned here with metallic ferromagnetic materials such as iron, steel, and certain alloys thereof, and ferrimagnetic mixed oxides, or ferrites, having the general chemical formula $M^{2+}Fe_2^{3+}O_4^{2-}$ where M^{2+} is a divalent metal ion such as Ni^{2+}, Zn^{2+}, Co^{2+}, Mg^{2+}, Mn^{2+}, or Fe^{2+}. Measurements of B as a function of the magnetizing field H for such materials yield a relationship of the general form shown in Fig. 5.15(a). If the sample is completely unmagnetized, initially, and if a cyclical programme is followed of increasing H to H_{max}, decreasing through zero to $-H_{max}$, followed by increasing through zero to H_{max} again, and so on for 20 or 30 cycles then, eventually, a reproducible symmetrical loop such as curve 1 is obtained. Similar cycles can be followed with larger values of H_{max} and similar 'hysteresis' loops will be obtained. Ultimately for a value of H_{max} denoted as H'_{max} the material becomes 'saturated' i.e. the direction of magnetization of all the domains are parallel to one another and M has its maximum value M_{sat}.

A relative permeability may be defined in several different ways from a hysteresis loop. The 'normal' permeability is usually defined as $B/(\mu_0 H)$ for the tip of a loop and has a dependence on B of the general form shown in Fig. 5.15(b); it should be noted that for some alloys (e.g. Supermalloy) the maximum normal relative permeability can be as high as 10^6 at 0.3 T. The 'differential' relative permeability is defined as $(dB/dH)/\mu_0$ and its magnitude for small values of B, less than about 10^{-3} T, is between 5×10^2 and 10^3 in ferrites and about 10^5 in Supermalloy.

The effective alternating relative permeability for a given amplitude of H is the slope of a line such as PQ in Fig. 5.15(a) which joins the tips of the hysteresis loop which is being followed on the particular cycle; it can be determined by measuring the reactance presented to an alternating current by a coil wound on the material.

In saturable reactors the current in a d.c. winding magnetizes a

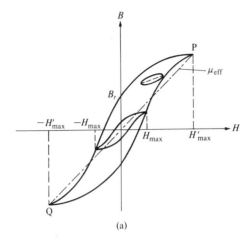

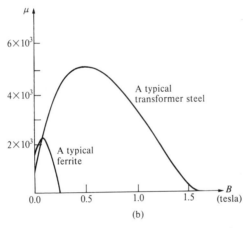

FIG. 5.15. (a) Hysteresis loops showing how the 'normal' and 'incremental' permeabilities are defined. (b) 'Normal' permeability as a function of magnetic field for a typical transformer steel and a typical ferrite.

magnetic core and then the effective permeability for a small hysteresis loop is called the incremental relative permeability as indicated in Fig. 5.15(a). Saturable reactors lie at the heart of so-called magnetic amplifiers; the reactance of the a.c. winding depends on the value of the incremental permeability which is in turn controlled by the bias field produced by the steady current in the d.c. winding.

Energy has to be supplied by the external circuit in driving a sample of material round its hysteresis loop. If ϵ is the back e.m.f. in the magnetic coil then power ϵI has to be expended in driving a current I

through the coil. Since $\epsilon = \mathrm{d}\phi/\mathrm{d}t = nA\,\mathrm{d}B/\mathrm{d}t$ where A is the area, and n the number of turns of the coil, and $I = Hl/n$ from eqn 5.44, where l is the length of the coil, it follows that the energy W expended in one cycle is

$$W = \int \epsilon I\,\mathrm{d}t = V\int H\,\mathrm{d}B \qquad (5.51)$$

where $V \equiv Al$ is the volume of the sample filling the magnetizing coil. In deriving this result that the energy expended per unit volume of sample is equal to the area of the hysteresis loop we have assumed that the resistance of the wire of the coil is negligible and also that eddy current losses in the material are insignificant; so the 'narrower' a loop, the smaller is the hysteresis loss.

'Magnetically soft' materials are characterized by having small values of residual field B_r and coercive force H_c (see Fig. 5.15(a)); such materials are required for alternating current machinery and for power and communications transformers. In the latter devices the faithful reproduction of an imposed waveform requires a $B-H$ characteristic which is as near to linear as possible; this implies that the core material should have a hysteresis loop with as small a curvature and as small an area as possible.

For use in transformers and inductors the electrical resistivity of the core material should be as high as possible since this reduces the amplitude of eddy currents and hence reduces the associated energy losses also. The electrical resistivities of metallic ferromagnetic materials are quite low (10^{-6} to $10^{-7}\,\Omega\,\mathrm{m}$) and so cores are made up from thin laminations, having insulating coatings, which are oriented so that the direction of eddy currents is generally at right angles to the plane of the laminations; hence a comparatively high resistance is presented to the eddy currents.

Since the induced e.m.f. due to a sinusoidally varying magnetic field, $\hat{B} \sin \omega t$ say, is proportional to $\mathrm{d}B/\mathrm{d}t$, and hence has the form $\omega \hat{B} \cos \omega t$, the energy losses arising from eddy currents are proportional to ω^2. For a given frequency eddy current losses can be reduced only by reducing the lamination thickness or increasing the resistivity of the material. There are obvious practical problems in connection with reducing the thickness of laminations, and with stacking them to form cores, quite apart from the increasing time and costs incurred. Also the ratio of the volume of insulating material to ferromagnetic material increases which gives a power stacking factor.

Ferrites are good insulators with resistivities in the range $1\,\Omega\,\mathrm{m}$ to greater than $10^3\,\Omega\,\mathrm{m}$ and while their relative permeabilities are not so high, generally speaking, as metallic ferromagnetic materials, they are high enough nevertheless to make the materials extremely useful. Ferrites are produced by sintering at a temperature between 1000 and

1400°C and the resultant material is homogeneous with no internal air gaps. The saturation flux density in typical ferrites is about 0.3 T which is about five times smaller than in 'soft' metallic ferromagnetic materials; hence ferrites are not very suitable for low-frequency power applications.

The energy losses in coils and transformers are of great significance since they restrict the quality factor Q on the one hand and the efficiency on the other. Energy losses arise from a number of sources of which the principal ones in circuit elements having ferromagnetic cores are eddy current losses, hysteresis losses, and losses due to the resistance of the windings ('copper' losses). The first two named sources of losses are reduced, as we have seen, by using laminated cores and by using materials having narrow hysteresis loops but the upper frequency limit for most practical purposes lies between 20 and 30 kHz. In ferrites, which have been produced to a specification which is designed to optimize their performance in circuit elements, hysteresis losses are small at the low values of B which are commonly used and eddy current losses are negligible because of the high electrical resistivity. However there are also 'residual' losses for which the physical mechanism is not clearly understood although the movement of domain boundaries under the influence of time-dependent magnetic fields seems to be the most favoured candidate..

Now we saw in Section 3.6.2 that as far as its circuit properties are concerned a coil wound on a magnetic core can be represented by a series inductance L_s and a series resistance R_s which represents the total of power losses associated with the coil and magnetic circuit. A relationship for the residual losses which is found to be true approximately for ferrites and which is useful in practice, is

(5.52) $$R_s = K\mu\omega L_s \tag{5.52}$$

where K is a constant. The loss angle δ of the coil is given by

$$\tan\delta = \frac{1}{Q} = \frac{R_s}{\omega L_s}.$$

Now consider a magnetic circuit having a gap of width l_g; if the loss angle is δ' and the corresponding series resistance and inductance are R_s' and L_s' respectively, then

$$\tan\delta' = \frac{R_s'}{\omega L_s'} \tag{5.53}$$

and, using eqns (5.50) and (5.52)

$$\frac{R_s'}{R_s} = \frac{\mu_{\text{eff}} L_s'}{\mu L_s}.$$

132 Electromagnetics

On substituting in eqn (5.53) we have

$$\tan \delta' = \frac{\mu_{\text{eff}}}{\mu} \tan \delta. \quad (5.54)$$

So we see that not only does the inductance of the coil depend on the value of l_g but also that the loss angle depends on l_g through μ_{eff} and decreases as l_g increases; the factor μ_{eff}/μ is usually called the *dilution factor*.

Ferrite 'pot' cores are made available with a range of values of μ_{eff} (typically from 25 to 400) so that by means of the additional provision of an adjustable tuning slug a continuous range of values of inductance can be obtained.

There are numerous specialized texts dealing with the properties of ferrites and also the technical handbooks which are published by the manufacturers.[†]

PROBLEMS

5.1. Show that for the equivalent series network for a lossy capacitor

$$R_s = \frac{\epsilon''}{\omega C_0 (\epsilon'^2 + \epsilon''^2)} \text{ and } \tan \delta = \epsilon''/\epsilon'$$

(See Section 3.6 and Problem 3.26.)

5.2. In a dielectric material in which a single relaxation mechanism is dominant the maximum power loss occurs at a frequency of 1592 Hz. Calculate the value of the relaxation time. If the relative permittivity of the material has the values 5.00 and 1.10 at frequencies 1.6 Hz and 1.6 MHz, respectively, calculate the value of the relative permittivity at frequencies of 160 Hz and 160 kHz.

5.3. For the dielectric material specified in Problem 5.2, calculate the values of $\tan \delta$ at frequencies of 1.6 Hz, 16 Hz, 1.6 kHz, 160 kHz, 1.6 MHz.

5.4. Show that the variable $u \equiv a\sqrt{\mu\mu_0 \omega \sigma}$ is dimensionless. (a is the radius of a wire.)

5.5. Calculate the skin depth for alternating electromagnetic fields in copper (electrical conductivity $\sigma = 5.8 \times 10^7$ Sm^{-1}; magnetic permeability $\mu = 1$) at frequencies of 50 Hz, 5 kHz, 500 kHz, 50 MHz, 5 GHz.

5.6. Show that the distributed self-inductance L of a coaxial line is given by

$$L = \left(\frac{\mu_0}{4\pi}\right) 2 \ln (b/a) \quad \text{H m}^{-1}$$

where a, b are the radii of the inner and outer conductors respectively.

5.7. For the transformer core shown in Fig. 5.14(b) show by making the usual approximation for the path length for magnetic flux, that the magnetic flux ϕ in the central limb is given by

$$\phi = 2\mu\mu_0 w^2 NI/7l$$

where NI is the value of the ampère-turns of the exciting coil.

[†] For example: Schlicke, H. M. *'Dielectromagnetic Engineering'*. J. Wiley (1961).

6. Pulses and transients

6.1. Introduction

SO FAR the discussions of time-dependent e.m.f.s and currents have been concerned almost entirely with the circuit theorems applicable to sinusoidal e.m.f.s and currents of a single frequency and having constant, time-independent amplitudes.

Now according to Fourier's theorem a periodic mathematical function (i.e. a function $v(t)$ such that $v(t + T) = v(t)$ where T is the period) can be represented by a series of sinusoidal 'Fourier components' of the general form

$$v(t) = a_0 + \sum_{n=1}^{\infty}\left(a_n \cos\frac{2\pi n t}{T} + b_n \sin\frac{2\pi n t}{T}\right) \quad (6.1)$$

provided that the function satisfies Dirichlet's criteria.†

For instance the Fourier series for 'half-wave rectified', 'sawtooth', and 'square' waveforms (see Fig. 6.1) can be written as follows (where $\omega \equiv 2\pi/T$):

Half-wave rectified sinusoidal; $v(t) = \hat{V} \sin \omega t$:

$$v(t) = \frac{\hat{V}}{\pi} + \frac{\hat{V}}{2}\sin \omega t - \frac{2\hat{V}}{3\pi}\cos 2\omega t$$

$$-\frac{2\hat{V}}{15\pi}\cos 4\omega t - \frac{2\hat{V}}{35\pi}\cos 6\omega t - \ldots \quad (6.2)$$

Sawtooth:
$$v(t) = \frac{\hat{V}}{\pi}\sin \omega t - \frac{\hat{V}}{2\pi}\sin 2\omega t$$

$$+ \frac{\hat{V}}{3\pi}\sin 3\omega t - \frac{\hat{V}}{4\pi}\sin 4\omega t + \ldots \quad (6.3)$$

†The integral $\int_{-T/2}^{+T/2} |v(t)|dt$ must be finite and $v(t)$ must not have more than a finite number of discontinuities in any finite interval of t.

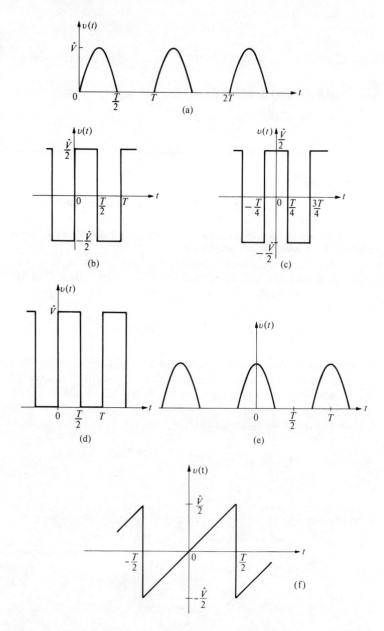

FIG. 6.1. Some examples of periodic, non-sinusoidal waveforms. (a), (e) Half-wave rectified sine waves. (b), (c), (d) 'Square' waves. (f) 'Ramp' waveform.

Square:
$$v(t) = \frac{2\hat{V}}{\pi}\cos\omega t - \frac{2\hat{V}}{3\pi}\cos 3\omega t + \frac{2\hat{V}}{5\pi}\cos 5\omega t - \ldots \quad (6.4)$$

Leaving aside for the moment the reasons for the appearance of sine or cosine terms and plus and minus signs the important feature is the existence of terms whose frequencies are certain integral multiples (i.e. harmonics) of the fundamental frequency $\omega(=2\pi/T)$ and whose amplitudes are a decreasing function of the harmonic number. So, in a determination of the response of an electrical network to a non-sinusoidal, or 'pulsed', e.m.f. it is necessary, in principle, to determine the network responses at the frequencies of each of the Fourier components and then to superpose these to give the resultant response. It is obvious that in order for pulsed signals to be transmitted faithfully by a network (i.e. with their shape preserved) the response of the network must be identical for all of the Fourier components of the incoming signal (see Fig. 6.2). Also notice that the 'sharper' are the 'corners' of a pulse the greater is the number of Fourier components which have significant amplitudes; a comparison of eqns (6.2) and (6.4) gives a rough indication of this.

Although Fourier series, and the related Fourier transformations, have many uses in the description of the responses of networks to non-sinusoidal signals, a more powerful general method uses Laplace transformations: this approach will be described in Section 6.5.

As an example of a Fourier series expansion consider the case of a half-wave rectified sinusoidal wave (see Fig. 6.1(a)): the coefficients a_0, a_n, b_n are given in general by

$$a_0 = \frac{1}{T}\int v(t)\,dt \quad (6.5)$$

$$a_n = \frac{2}{T}\int v(t)\cos n\omega t\,dt \quad (6.6)$$

$$b_n = \frac{2}{T}\int v(t)\sin n\omega t\,dt \quad (6.7)$$

where the integrals are taken over a complete period of $v(t)$. So

$$a_0 = \frac{1}{T}\int_0^T \hat{V}\sin\omega t\,dt$$

or

$$a_0 = \frac{\hat{V}}{T}\int_0^{T/2}\sin\omega t\,dt$$

since $v(t)$ is zero for $T/2 \leqslant t \leqslant T$. Thus,

136 Pulses and transients

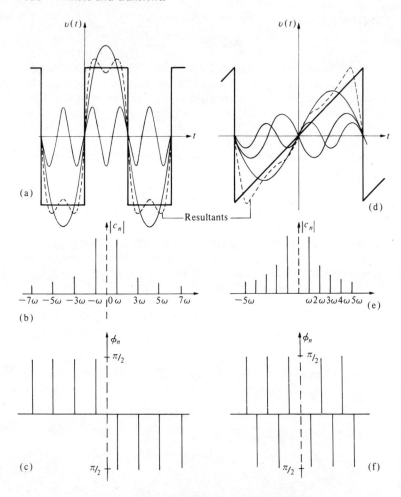

FIG. 6.2. (a) A sketch of the waveform resulting from the addition of the first two Fourier components of a square wave. (b), (c) The amplitude and phase spectra, respectively, of a square wave. (d) The waveform resulting from the addition of the first three Fourier components of a sawtooth wave. (e), (f) The amplitude and phase spectra, respectively.

$$a_0 = \frac{\hat{V}}{\omega T} \int_0^\pi \sin \omega t \, d(\omega t)$$

$$= \frac{\hat{V}}{2\pi} [-\cos \omega t]_0^\pi.$$

Hence
$$a_0 = \frac{\hat{V}}{\pi}.$$

For $n = 1$,
$$a_1 = \frac{2}{T}\int_0^{T/2} \hat{V}\sin\omega t \cos\omega t\, dt$$

$$= \frac{\hat{V}}{\pi}\int_0^{\pi} \sin\omega t \cos\omega t\, d(\omega t)$$

$$= \frac{\hat{V}}{4\pi}\int_0^{\pi} \sin 2\omega t\, d(2\omega t)$$

$$= \frac{\hat{V}}{4\pi}[-\cos 2\omega t]_0^{\pi}.$$

Hence,
$$a_1 = 0.$$

For $n > 1$, we have
$$a_n = \frac{2\hat{V}}{T}\int_0^{T/2} \sin\omega t \cos n\omega t\, dt$$

$$= \frac{\hat{V}}{\pi}\int_0^{\pi} \sin\omega t \cos n\omega t\, d(\omega t)$$

$$= \frac{\hat{V}}{\pi}\int_0^{\pi} \{\sin(n+1)\omega t - \sin(n-1)\omega t\}d(\omega t)$$

$$= \frac{\hat{V}}{\pi}\left[-\frac{\cos(n+1)\omega t}{(n+1)} + \frac{\cos(n-1)\omega t}{(n-1)}\right]_0^{\pi}.$$

Thus,
$$a_n = \frac{-2\hat{V}}{(n^2-1)} \quad (n \text{ even})$$

$$a_n = 0 \quad (n \text{ odd}).$$

For $n = 1$,
$$b_1 = \frac{\hat{V}}{\pi}\int_0^{\pi} \sin^2\omega t\, d(\omega t)$$

$$= \frac{\hat{V}}{2\pi}[\omega t - \sin\omega t \cos\omega t]_0^{\pi}.$$

138 Pulses and transients

So $b_1 = \tfrac{1}{2}\hat{V}$.

For $n > 1$ we have

$$b_n = \frac{\hat{V}}{\pi}\int_0^\pi \sin\omega t\,\sin n\omega t\,d(\omega t)$$

$$= \frac{\hat{V}}{\pi}\int_0^\pi \{\cos(n-1)\omega t - \cos(n+1)\omega t\}d(\omega t)$$

$$= \frac{\hat{V}}{\pi}\left[\frac{\sin(n-1)\omega t}{(n-1)} - \frac{\sin(n+1)\omega t}{(n+1)}\right]_0^\pi$$

Thus $b_n = 0$ for n odd or n even. So, finally, the Fourier series for $v(t)$ is

$$v(t) = \frac{\hat{V}}{\pi}\left[1 + \frac{\pi}{2}\sin\omega t - \frac{2}{3}\cos 2\omega t - \frac{2}{15}\cos 4\omega t - \frac{2}{35}\cos 6\omega t - \ldots\right]$$

As another example consider a square wave as shown in Fig. 6.1(b): since $v(t)$ is an odd function of t $\{v(-t) = -v(t)\}$ it follows that each of the terms of the Fourier series expansion must be an odd function also. Thus the series consists of sine terms only, i.e. $a_n = 0$ $(n > 1)$.

In this example,

$$a_0 = \frac{\omega}{2\pi}\int_0^{T/2}\frac{\hat{V}}{2}dt - \frac{\omega}{2\pi}\int_{T/2}^T\frac{\hat{V}}{2}dt$$

and so, $a_0 = 0$. Also,

$$b_n = \frac{2}{T}\int_0^{T/2}\frac{\hat{V}}{2}\sin n\omega t\,dt - \frac{2}{T}\int_{T/2}^T\frac{\hat{V}}{2}\sin n\omega t\,dt$$

$$= \frac{\hat{V}}{2\pi}\int_0^\pi \sin n\omega t\,d(\omega t) - \frac{\hat{V}}{2\pi}\int_0^\pi \sin n\omega t\,d(\omega t)$$

$$= \frac{\hat{V}}{2\pi}\left[\frac{-\cos n\omega t}{n}\right]_0^\pi - \frac{\hat{V}}{2\pi}\left[-\frac{\cos n\omega t}{n}\right]_\pi^{2\pi}.$$

Therefore $b_n = 0$ (n even) and $b_n = 2\hat{V}/n\pi$ (n odd) and the Fourier series is

$$v(t) = \frac{2\hat{V}}{\pi}\left\{\sin\omega t + \frac{\sin 3\omega t}{3} + \frac{\sin 5\omega t}{5} + \frac{\sin 7\omega t}{7} + \ldots\right\}. \quad (6.8)$$

For the square waves shown in Fig. 6.1(c) and (d) and for the

Pulses and transients

half-rectified sine wave shown in Fig. 6.1(e) it is left for you to show that the Fourier series are, respectively

$$v(t) = \frac{2\hat{V}}{\pi}\left\{\cos \omega t - \frac{\cos 3\omega t}{3} + \frac{\cos 5\omega t}{5} - \frac{\cos 7\omega t}{7} + \ldots\right\} \quad (6.9)$$

$$v(t) = \frac{\hat{V}}{2} + \frac{2\hat{V}}{\pi}\left\{\cos \omega t + \frac{\cos 3\omega t}{3} + \frac{\cos 5\omega t}{5} + \frac{\cos 7\omega t}{7} + \ldots\right\} \quad (6.10)$$

$$v(t) = \frac{\hat{V}}{\pi}\left\{1 + \frac{\pi}{2}\cos \omega t + \frac{2}{3}\cos 2\omega t - \frac{2}{15}\cos 4\omega t + \frac{2}{35}\cos 6\omega t \ldots\right\}. \quad (6.11)$$

Notice in eqns (6.10) and (6.11) that the 'd.c.' term a_0 is equal to the average value of the periodic waveform in question: this is a general feature of the Fourier series expansions of periodic functions.

It can be seen from the examples which we have discussed that the amplitudes and frequencies of the terms in a series corresponding to a particular waveform are independent of the choice of origin; the differences in sign merely represent differences in phase.

6.2. Amplitude and phase spectra

If we modify eqn (6.1) by defining $c_n \equiv (a_n - jb_n)/2$; $c_{-n} \equiv (a_n + jb_n)/2$; $c_0 \equiv a_0$ and utilize the relationships

$$\cos \omega t = (e^{j\omega t} + e^{-j\omega t})/2$$

$$\sin \omega t = (e^{j\omega t} - e^{-j\omega t})/2, \quad (6.12)$$

then we obtain

$$v(t) = \sum_{n=-\infty}^{\infty} c_n e^{jn\omega t} \quad (6.13)$$

where

$$c_n = \frac{1}{T}\int_{\text{period}} v(t) e^{-jn\omega t} dt. \quad (6.14)$$

Eqn (6.13) is the complex form of the Fourier series. Now c_n is complex in general with

(Real part) $\text{Re}(c_n) = \frac{1}{T}\int v(t) \cos n\omega t \, dt$

(Imaginary part) $\text{Im}(c_n) = -\frac{1}{T}\int v(t) \sin n\omega t \, dt.$

The amplitude spectrum of $v(t)$ is defined by

$$|c_n| = \{[\text{Re}\,(c_n)]^2 + [\text{Im}\,(c_n)]^2\}^{1/2} \qquad (6.15)$$

and the phase spectrum by

$$\phi_n = \tan^{-1}\{\text{Im}\,(c_n)/\text{Re}\,(c_n)\}. \qquad (6.16)$$

Consider again the square wave illustrated in Fig. 6.1(b) which we discussed earlier for which $a_n = 0$, $b_n = 0$ for n even and $b_n = 2\hat{V}/n\pi$ for n odd. In this case,

$$\text{Re}\,(c_n) = a_n/2 \qquad \text{Im}\,(c_n) = -b_n/2 = -\hat{V}/n\pi \;\; (n \text{ odd})$$

giving $|c_n| = b_n/2 = \hat{V}/n\pi$ and

$$\phi_n = \begin{cases} \tan^{-1}(-\infty) & (n \text{ positive}) = -\pi/2 \\ \tan^{-1}(+\infty) & (n \text{ negative}) = +\pi/2. \end{cases}$$

For an electrical network, whether it be an active or a passive one, to pass without distortion a non-sinusoidal periodic signal then, ideally, the network should have a response which is independent of frequency. Of course such an ideal is unrealizable in practice and the upper and lower frequency cut-off points of the network's response are of particular significance. The result of combining the first few harmonics only of a square wave, and of a sawtooth wave, are shown in Fig. 6.2; the shape of the original signal is beginning to emerge in each case but there is quite gross distortion.

As a further example of the kind of analysis involved in these situations consider the sawtooth, or 'ramp', waveform shown in Fig. 6.2(d),

and

$$v(t) = \hat{V}t/T$$

$$c_n = \frac{\hat{V}}{T^2} \int_{-T/2}^{T/2} t e^{-jn\omega t} dt$$

or

$$c_n = \frac{\hat{V}}{4\pi^2} \int_{-\pi}^{\pi} \omega t\, e^{-jn\omega t} d(\omega t).$$

It can be shown quite easily, by integrating by parts, that

$$\int x e^{-jnx} dx = \left(\frac{1}{n} + jx\right)\frac{e^{-jnx}}{n}$$

Hence,

$$c_n = \frac{\hat{V}}{4\pi^2}\left[\left(\frac{1}{n} + j\omega t\right)\frac{e^{-jn\omega t}}{n}\right]_{-\pi}^{\pi}$$

and it follows that

$$c_n = \frac{j\hat{V}}{2\pi n} \quad (n \text{ even}); \qquad b_n = -\frac{\hat{V}}{\pi n}$$

$$c_n = -\frac{j\hat{V}}{2\pi n} \quad (n \text{ odd}); \qquad b_n = \frac{\hat{V}}{\pi n}$$

Also

$$c_0 = \frac{\hat{V}}{T^2} \int_{-T/2}^{T/2} t \, dt, \text{ i.e.} \qquad c_0 = 0.$$

The phase spectrum is given by

$$\phi_n = \begin{matrix} +\pi/2 & n \text{ even} \\ -\pi/2 & n \text{ odd} \end{matrix} \qquad \phi_{-n} = \begin{matrix} -\pi/2 & n \text{ even} \\ +\pi/2 & n \text{ odd}. \end{matrix}$$

Displays of amplitude spectra, such as are shown in Fig. 6.2, are useful in that they give useful guidance as to where an infinite Fourier series can be truncated without causing serious distortion to the waveform e.g. a good idea can be obtained of the frequency bandwidth an amplifier must have if it is not to distort seriously the form of the signal.

Incidentally a useful feature of the sawtooth waveform is manifest in Fig. 6.2(e): the amplitude of the harmonics falls off relatively slowly with frequency. This feature is utilized in electronic music synthesizers, for instance, where the harmonics of sawtooth waveforms are filtered appropriately to stimulate the harmonic spectra of musical instruments.

The idea of a negative frequency which has been introduced may be causing you some disquiet but it can be explained quite simply through the representation of $e^{j\omega t}$ on an Argand diagram. We know that

$$e^{j\omega t} = \cos \omega t + j \sin \omega t$$

and this is shown in Fig. 6.3 as a vector in the complex plane having unit magnitude and with its phase angle increasing in the positive sense of rotation at rate ω. Similarly $e^{-j\omega t}$ (note the 'negative' frequency) is represented as a vector rotating in the negative sense.

To recover a 'real', physical cosine or sine variation we combine $e^{j\omega t}$ and $e^{-j\omega t}$:

$$\cos \omega t = (e^{j\omega t} + e^{-j\omega t})/2$$

$$\sin \omega t = -j(e^{j\omega t} - e^{-j\omega t})/2.$$

6.3 Pulse trains

A situation which is often approximated fairly closely in practice is that of an infinite train of narrow pulses. If the pulses are 'rectangular'

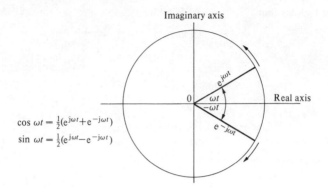

FIG. 6.3. The representations of $e^{j\omega t}$ and $e^{-j\omega t}$ ($=e^{j(-\omega t)}$) as rotating vectors in the complex plane.

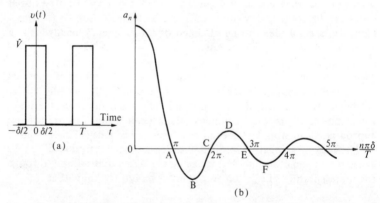

FIG. 6.4. (a) Rectangular pulses; the mark-space ratio is δ/T. (b) The amplitudes a_n of the Fourier components.

having duration δ, period T, and amplitude V, as shown in Fig. 6.4(a) then it can be shown quite easily, by using the standard procedures which have been given in section 6.1, that

$$a_0 = \hat{V}\delta/T$$
$$a_n = (2\hat{V}/n\pi) \sin(n\pi\delta/T) \tag{6.17}$$
$$b_n = 0.$$

The function a_n of eqn (6.17) is sketched in Fig. 6.4(b); if T/δ is an integer then the harmonics given by $n = pT/\delta$ (where p is also an integer) are missing from the line spectrum. For example, if $T/\delta = 2$

Pulses and transients

(a 'square' wave), then the even harmonics are missing as we have seen in our earlier analysis.

If the 'mark to space' ratio is very big (i.e. the 'duty cycle' δ/T is very small), as is often the case in practice, then although the amplitudes of the Fourier components are much less than $\hat{V}$ they do not decrease so rapidly with increasing n. For instance for $\delta/T = 10^{-3}$ the amplitudes of the 3500th harmonic ("F" in Fig. 6.4(b)) is 3/7 of the 1500th harmonic ('B' in Fig. 6.4(b)) and the amplitude of the 6500th harmonic is 3/13 of the 1500th. The amplitude a_1 of the first harmonic is given by $a_1 \approx 2\hat{V}\delta/T$ (using the approximation $\sin \theta \approx \theta$ for θ small) and so $a_1 \approx 2\hat{V} \times 10^{-3}$. Hence for $a_n/a_1 \leq 10^{-1}$ then it follows that $n \geq 3 \times 10^3$ approximately. Thus a rough, but useful indication of the frequency up to which the response of a network ought to be frequency-independent (or 'flat') in order to give reasonably distortionless transmission of a train of 'narrow' pulses ($\delta/T < 10^{-2}$, say) is 3000 times the pulse repetition frequency (p.r.f.) $1/T$, e.g. for $\delta/T = 10^{-3}$ and p.r.f. = 10^3 s^{-1} then the required upper frequency limit is 3 MHz.

6.4. Fourier integrals and transformations

So far we have considered periodic signals only and we have seen that they have discrete, or line, spectra. A non-periodic signal, or transient signal, can be thought of as a periodic signal having an infinitely long period; hence the separation of the Fourier components, which is equal to $2\pi/T$, will tend to zero and the amplitude spectrum will be *continuous*.

In the complex form of the Fourier series (eqn (6.13)), and in eqn (6.14), we shall replace $v(t)$ by $f(t)$ and the fundamental frequency ω by ω_0 to give

$$f(t) = \frac{1}{T} \sum_{n=-\infty}^{\infty} \left\{ \int_{-T/2}^{T/2} f(t)e^{-jn\omega t} dt \right\} e^{jn\omega_0 t}.$$

Since $T = 2\pi/\omega_0$ and writing $n\omega_0 = \omega$, we have

$$f(t) = \frac{1}{2\pi} \sum_{n=-\infty}^{\infty} F(\omega)e^{j\omega t} \frac{\omega}{n}$$

where

$$F(\omega) \equiv \int_{-T/2}^{T/2} f(t)e^{-j\omega t} dt. \qquad (6.18)$$

In the limit of $T \to \infty$ we write the spacing $\Delta\omega (= \omega/n)$ between the Fourier components as $d\omega$ and the summation sign becomes an integral. Hence we have

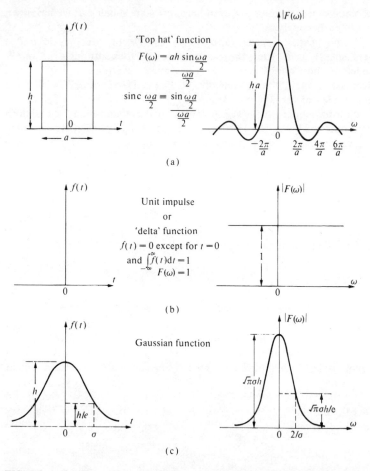

FIG. 6.5. Examples of some functions and their Fourier transforms: (a) the 'top-hat' function,' (b) the Unit impulse or 'delta' function,' (c) the Gaussian function.

$$\lim_{\Delta\omega \to 0} f(t) = \frac{1}{2\pi} \int_{-\infty}^{\infty} F(\omega) e^{j\omega t} d\omega. \qquad (6.19)$$

The functions $f(t)$, $F(\omega)$ are called a Fourier transform pair and eqn (6.19) defines a Fourier transform.

A restrictive condition on the type of function $f(t)$ for which a Fourier transform exists is that $\int_{-\infty}^{\infty} |f(t)|^2 dt$ must be finite. This integral

Pulses and transients   145

is essentially the sum of the squares of the amplitudes of the Fourier components which in turn is related to the total energy in the signal which is being represented by $f(t)$ and which must be finite, of course. For example, the 'step' function (see Fig. 6.7) is very important in the study of the behaviour of electrical networks and systems; but the step function does not satisfy the aforementioned criterion and so we cannot find its Fourier transform (in reality a step function signal, generated by closing a switch, say, cannot persist for ever but it is generally a very satisfactory approximation to assume that it does so). The step function can be handled by using the Laplace transform technique which was alluded to earlier in this chapter.

Some transient functions for which Fourier transforms do exist are shown in Fig. 6.5; note that the unit impulse is the limiting case, as $a \to 0$, of a 'top hat' function for which $ha = 1$.

The particular related pairs of functions shown in Fig. 6.5 illustrate a very important principle namely that of 'reciprocal spreading' between the time duration and related frequency bandwidth of a transient signal. Consider, for example, the top hat function of Fig. 6.5(a): most of the energy in the pulse will be contained in frequencies less than $2\pi/a$, the frequency at which the first zero of $F(\omega)$ occurs, and so the bandwidth Δf of the pulse can be considered to be of the order of $1/a$ Hz. Since the time duration Δt of the pulse is equal to a we have $\Delta t \Delta f = 1$. Bearing in mind the degree of arbitrariness in our definition of the bandwidth we write

$$\Delta t \Delta f \sim 1. \qquad (6.20)$$

This approximate relationship applies to transient signals of any shape of which the Gaussian function and the delta function in Fig. 6.5 are examples; the smaller the extent in time occupied by a signal the wider is the frequency range which it occupies. The delta function is an extreme example of this principle; it is infinitely narrow in time and hence its frequency spectrum is 'flat' from zero to infinity.

6.5. The Laplace transform

The nature of the general problem with which we are concerned is shown in Fig. 6.6(a): we wish to know the response $r(t)$ of a given network to a given signal, or excitation, $e(t)$. If $e(t)$ has a pure sinusoidal time dependence then $r(t)$ can be found by using steady-state a.c. circuit theory, as introduced in Chapter 3, in conjunction with other techniques such as mesh or nodal analysis. No matter how familiar these techniques have become it is worth emphasizing the underlying principles. For example if the network is a series $L-C-R$ network and the required response is the current $i(t)$, then $i(t)$ is found

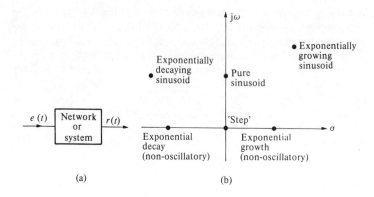

FIG. 6.6. (a) A general schematic representation of a network or system showing the time domain excitation and response $e(t)$ and $r(t)$ respectively. (b) The representation in the complex frequency domain (s-domain) of some non-sinusoidal signal waveforms.

by solving the integro-differential equation

$$Ri(t) + L\frac{di(t)}{dt} + \frac{1}{C}\int i(t)dt = e(t)$$

which is obtained by applying Kirchhoff's laws. Since we are assuming that $e(t)$ has the form $e(t) = \hat{E}e^{j\omega t}$ and that, in the steady state, $i(t)$ has a sinusoidal time dependence also, $i(t) = \hat{I}e^{j(\omega t + \phi)}$, then $di(t)/dt = j\omega i(t)$ and $\int i(t) dt = i(t)/j\omega$ (if the capacitor is uncharged at $t = 0$). So, under these circumstances, the *integro-differential* equation has been transformed to the *algebraic* equation

$$R\hat{I} + j\omega L\hat{I} - \frac{j\hat{I}}{\omega C} = \hat{E}e^{-j\phi}$$

which can be solved for $\hat{I}$ easily.

It is worth emphasizing that the concept of 'reactance', for instance, was defined in the early days when the study of electronics was concerned largely with radio communications and the signals of interest were usually sinusoidal waves. Nowadays we are concerned with signals having a wide variety of forms; 'square', 'ramp', 'sawtooth', etc. It is important to realize that the concept of reactance is meaningless except for steady-state sinusoidal waveforms.

For non-sinusoidal excitations the solution of the integro-differential equation is either extremely difficult or impossible by the techniques used so far whereas the Laplace transform method provides solutions fairly readily. The essence of the method is again to transform the basic integro-differential equation into an algebraic one. However, before considering the Laplace transformations it is necessary to introduce the notion of the complex frequency s.

A sinusoidal wave of angular frequency ω and constant amplitude $\hat{V}$ can be expressed as $v = \hat{V}e^{j\omega t}$ in which it is implicit that the real or imaginary part, as the case may be, represents the physical quantity. Now a 'sinusoidal' wave whose amplitude increases exponentially with time can be expressed as $v = (\hat{V}e^{\sigma t})e^{j\omega t}$ or $v = \hat{V}e^{(\sigma + j\omega)t}$ where σ is real. Similarly a wave having an exponentially damped amplitude is written $v = \hat{V}e^{(-\sigma + j\omega)t}$. The *complex frequency* s is defined through $s \equiv \sigma + j\omega$: waveforms of these, and other, types can be represented by points in the complex s-plane as shown in Fig. 6.6(b).

The Laplace transform $F(s)$ of a function $f(t)$ is defined through

$$F(s) = \mathscr{L}[f(t)] = \int_0^\infty f(t)e^{-st}dt, \qquad (6.21)$$

i.e. the operation transforms a function $f(t)$ in the time domain into a function $F(s)$ in the complex frequency domain or s-domain; $F(s)$ and $f(t)$ constitute a Laplace transform pair.[†]

It should be noted that the commonly used convention of denoting functions in the time domain by lower case characters and functions in the s-domain by upper case characters has been followed here. After an algebraic equation has been solved to give a solution for the voltage or current in the s-domain then the time dependence is obtained by using an *inverse* Laplace transformation, i.e.

$$f(t) = \mathscr{L}^{-1}[F(s)] = \frac{1}{2\pi j}\oint_0^\infty F(s)e^{st}ds. \qquad (6.22)$$

The operation of taking the inverse Laplace transformation of a function $F(s)$ involves an integration around a closed path in the (complex) s-plane. A detailed consideration of this problem would necessitate a long diversion into the mathematical realm of functions of a complex variable so we will take advantage of the extensive tables of Laplace transform pairs which have been drawn up.[‡]

The Laplace transforms of some commonly encountered functions and operations will be derived now:

(i) Unit step function: $f(t) = u(t)$ (see Fig. 6.7(a).)

[†] The lower limit of the integral in the definition of the Laplace transform should be thought of as lying just on the negative side of $t = 0$. This allows the initial conditions in a circuit (e.g. the initial charge on a capacitor) to be introduced explicitly. In mathematical phraseology the lower limit should be written as $t \to 0^-$ which means 'as t approaches ever closer to zero from the negative side.'

[‡] See, for example, Holbrook, J. G. *Laplace transforms for electronic engineers.* Pergamon Press, Oxford (1959).

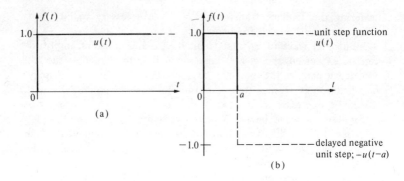

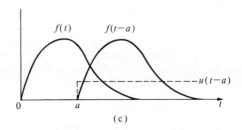

FIG. 6.7. (a) The unit step function $u(t)$. (b) A square pulse generated by a step function and a delayed step function $u(t-a)$. (c) A delayed version $f(t-a)$ of a function $f(t)$ switched on by a delayed step function $u(t-a)$.

$$\mathscr{L}[u(t)] = \int_0^\infty u(t) e^{-st} dt$$

$$= \int_0^\infty e^{-st} dt$$

$$= -\frac{1}{s} \int e^{-st} d(-st)$$

$$= -\frac{1}{s}[e^{-st}]_0^\infty$$

or

$$F(s) = \mathscr{L}[u(t)] = \frac{1}{s}$$

(ii) Exponential function: $f(t) = e^{-\alpha t}$.

$$\mathscr{L}[f(t)] = \int_0^\infty e^{-\alpha t} e^{-st} dt$$

$$= \int_0^\infty e^{-(\alpha+s)t} dt$$

Therefore,
$$F(s) = \frac{1}{(s+\alpha)}$$

Obviously it follows that
$$\mathscr{L}[e^{\alpha t}] = \frac{1}{(s-\alpha)}.$$

(iii) Sine wave: $f(t) = \sin(\omega t + \phi)$. We know that

$$\sin(\omega t + \phi) = \frac{1}{2j}[e^{j(\omega t+\phi)} - e^{-j(\omega t+\phi)}]$$

and so
$$\mathscr{L}[f(t)] = \frac{e^{j\phi}}{2j}\mathscr{L}[e^{j\omega t}] - \frac{e^{-j\phi}}{2j}\mathscr{L}[e^{-j\omega t}].$$

Using the previous result we have, therefore, that

$$\mathscr{L}[f(t)] = \frac{e^{j\phi}}{2j(s-j\omega)} - \frac{e^{-j\phi}}{2j(s+j\omega)}$$

$$= \frac{s(e^{j\phi} - e^{-j\phi}) + j\omega(e^{j\phi} + e^{-j\phi})}{2j(s^2 + \omega^2)}$$

or, finally,
$$\mathscr{L}[\sin(\omega t + \phi)] = \frac{s\sin\phi + \omega\cos\phi}{(s^2 + \omega^2)}.$$

(iv) The Laplace transform of a derivative. The Laplace transforms of derivatives and integrals are of crucial importance since, in circuit analysis, the voltages and currents in capacitors and inductors involve derivatives and integrals with respect to time.

We assume that all the functions we will be dealing with are zero for $t < 0$.

$$\mathscr{L}\left[\frac{df(t)}{dt}\right] = \int_0^\infty \left[\frac{df(t)}{dt}\right] e^{-st} dt.$$

and on integrating by parts we have

$$\mathscr{L}\left[\frac{df(t)}{dt}\right] = [e^{-st}f(t)]_0^\infty + s\int_0^\infty f(t)e^{-st} dt.$$

Now the second term on the right-hand side is the Laplace transform $F(s)$ of $f(t)$ and in the first term $e^{-st} \to 0$ as $t \to \infty$ and $e^{-st} \to 1$ as $t \to 0$. So we have

$$\mathscr{L}\left[\frac{\mathrm{d}f(t)}{\mathrm{d}t}\right] = sF(s) - f(0). \tag{6.23}$$

We will refer later to the significance of $f(0)$ but for the moment it is important to notice that, apart from the 'initial condition' specified by $f(0)$, the process of differentiation in the time domain transforms into the simpler operation of multiplication by s in the s-domain.

(v) The Laplace transform of an integral.

$$\mathscr{L}\left[\int_0^t f(t)\,\mathrm{d}t\right] = \int_0^\infty e^{-st}\left[\int_0^t f(t)\,\mathrm{d}t\right]\mathrm{d}t$$

and on integrating by parts we have

$$\mathscr{L}\left[\int_0^t f(t)\,\mathrm{d}t\right] = \left[\frac{-e^{-st}}{s}\int_0^\infty f(t)\,\mathrm{d}t\right]_0^\infty + \frac{1}{s}\int_0^\infty e^{-st}f(t)\,\mathrm{d}t$$

$$= \frac{F(s)}{s} + \frac{f^{-1}(0)}{s}. \tag{6.24}$$

Here $f^{-1}(0)$ is the conventional way of denoting the value of $\int f(t)\,\mathrm{d}t$ at $t = 0$.

(vi) The shifting theorem; We will find that delayed functions of time are of common occurence in the analysis of the response of networks to transient excitations: For instance, a simple example is that of a square pulse which can be represented as the resultant of a step function beginning at $t = 0$ and a negative-going delayed step function (see Fig. 6.7(b)). A delayed version of a function $f(t)$ is imagined to be 'switched on' by a delayed unit step function $u(t-a)$ where a is the time delay as sketched in Fig. 6.7(c).

$$\mathscr{L}[f(t-a)u(t-a)] = \int_{t=0}^\infty f(t-a)u(t-a)e^{-st}\mathrm{d}t$$

$$= \int_{t=a}^\infty f(t-a)e^{-st}\mathrm{d}t$$

$$= e^{-as}\int_{t-a=0}^\infty f(t-a)e^{-s(t-a)}\mathrm{d}(t-a)$$

since $d(t-a) = dt$, or
$$\mathscr{L}[f(t-a)u(t-a)] = e^{-as}F(s)$$
where
$$F(s) = \mathscr{L}[f(t)].$$

Some of the commonly occurring Laplace transform pairs are tabulated in Table 6.1.

Table 6.1
Some Laplace transform pairs

$f(t)$	$F(s)$
A	$\dfrac{A}{s}$
At	$\dfrac{A}{s^2}$
$e^{-\alpha t}$	$\dfrac{1}{(s+\alpha)}$
$\dfrac{e^{-\alpha t} - e^{-\beta t}}{(\beta - \alpha)}$	$\dfrac{1}{(s+\alpha)(s+\beta)}$
$\sin(\omega t + \phi)$	$\dfrac{s\sin\phi + \omega\cos\phi}{(s^2 + \omega^2)}$
$\dfrac{df(t)}{dt}$	$sF(s) - f(0)$
$\int f(t)\,dt$	$\dfrac{F(s)}{s} + \dfrac{f^{-1}(0)}{s}$

The best way to illustrate the practical significance of these definitions, transformations, and theorems is to examine some simple examples in the context of electrical circuits.

Consider the $R-C$ circuit shown in Fig. 6.8; the switch is closed at time $t = 0$ so that the e.m.f. in the circuit can be represented by a step function of amplitude E_0. In the time domain, Kirchoff's voltage equation for this circuit is

$$Ri(t) + \frac{1}{C}\int i(t)\,dt = E_0 u(t) \tag{6.25}$$

and on transforming these functions to the s-domain we have

$$RI(s) + \frac{1}{C}\frac{I(s)}{s} + \frac{i^{-1}(0)}{Cs} = \frac{E_0}{s}. \tag{6.26}$$

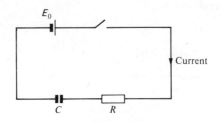

FIG. 6.8.

Now $i^{-1}(0)$ is equal to $\int i(t)dt$ evaluated at $t = 0$ which is the initial charge on the capacitor, say q_0. So on rearranging eqn (6.26) we have

$$I(s) = \frac{(E_0 - q_0/C)}{R\left(s + \dfrac{1}{CR}\right)}. \tag{6.27}$$

In order to obtain an expression for $i(t)$, the current in the time domain, we require $\mathcal{L}^{-1}\left[\left(s + \dfrac{1}{CR}\right)^{-1}\right]$, which is of the form $\mathcal{L}^{-1}[(s + \alpha)^{-1}]$, and is listed in Table 6.1. Thus, on transforming eqn (6.27) back to the time domain, we have

$$i(t) = \frac{(E_0 - q_0/C)e^{-t/CR}}{R}. \tag{6.28}$$

This simple example illustrates the point that the differential equation (eqn (6.25)) for the network has an algebraic equivalent in the s-domain; with experience you will be able to visualize the s-domain algebraic equation without the necessity of first writing down the time domain equation. We summarize the voltage–current relationships for circuit elements in Table 6.2.

Table 6.2.

	Time domain	Steady state a.c.	s-domain
Resistor R	$e(t) = Ri(t)$	$E = RI$	$E(s) = RI(s); Z(s) = R$
Capacitor C	$e(t) = \dfrac{1}{C}\int i(t)dt$	$E = \dfrac{I}{j\omega C}$	$E(s) = \dfrac{I(s)}{sC} + \dfrac{i^{-1}(0)}{sC}$;
			$Z(s) = \dfrac{1}{sC} + \dfrac{i^{-1}(0)}{sC}$
Inductor L	$e(t) = L\dfrac{di(t)}{dt}$	$E = j\omega L$	$E(s) = sLI(s) - sLi(0);$
			$Z(s) = sL - sLi(0)$

Pulses and transients 153

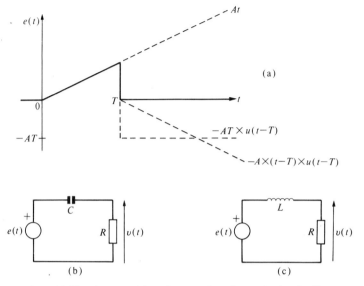

FIG. 6.9. (a) The decomposition of a ramp function excitation for the purpose of determining the response of a network. (b) and (c) see text.

A rather more complicated example is that of a ramp voltage applied to a $C-R$ circuit, as shown in Fig. 6.9(b), where we wish to determine the time dependence of the voltage $v(t)$ across the resistor R. It is important to notice that the ramp voltage has been decomposed into three constituent functions whose Laplace transforms we know; by consulting Table 6.1 and by using the shifting theorem we can write down an expression for $E(s)$, the Laplace transform of $e(t)$:

$$E(s) = \frac{A}{s^2} - \frac{ATe^{-sT}}{s} - \frac{Ae^{-sT}}{s^2}. \tag{6.29}$$

Now

$$V(s) = \frac{E(s)R}{\left(R + \dfrac{1}{sC}\right)}$$

where it should be noted that the s-domain voltages and currents have been treated by the same rules as in steady state a.c. circuit analysis. Thus,

$$V(s) = \frac{sE(s)}{\left(s + \dfrac{1}{CR}\right)} \tag{6.30}$$

154 Pulses and transients

and on substituting for $E(s)$ from eqn (6.29) we have

$$V(s) = A\left\{\frac{1}{s\left(s+\frac{1}{CR}\right)} - \frac{Te^{-sT}}{\left(s+\frac{1}{CR}\right)} - \frac{e^{-sT}}{s\left(s+\frac{1}{CR}\right)}\right\}.$$

From Table 6.1 we have

$$\mathscr{L}^{-1}\left[\left(s+\frac{1}{CR}\right)^{-1}\right] = e^{-t/CR}$$

and using the shifting theorem

$$\mathscr{L}^{-1}\left[\left(s+\frac{1}{CR}\right)^{-1}e^{-sT}\right] = e^{-(t-T)/CR}u(t-T).$$

Also

$$\mathscr{L}^{-1}[\{(s+\alpha)(s+\beta)\}^{-1}] = (e^{-\alpha t} - e^{-\beta t})/(\beta - \alpha)$$

and so

$$\mathscr{L}^{-1}\left[\frac{1}{s\left(s+\frac{1}{CR}\right)}\right] = CR(1-e^{-t/CR})$$

and

$$\mathscr{L}^{-1}\left[\frac{e^{-sT}}{s\left(s+\frac{1}{CR}\right)}\right] = CR(1-e^{-(t-T)/CR})u(t-T).$$

So, finally

$$v(t) = ACR\left\{1 - e^{-t/CR} - u(t-T) - \frac{T}{CR}\cdot e^{-(t-T)/CR}\cdot u(t-T)\right.$$

$$\left. + e^{-(t-T)/CR}\cdot u(t-T)\right\}. \tag{6.31}$$

This expression for $v(t)$ is drawn in Fig. 6.10(a). It is left as an exercise for you to show that, if a ramp voltage is applied to an $L-R$ circuit, the expression for the voltage across the inductor L has the same form as the expression of eqn (6.31) apart from the product CR being replaced by the quotient L/R.

Consider now a ramp voltage applied to an $L-R$ circuit as shown in Fig. 6.9(c) and suppose that we wish to know the voltage across the resistor R.

Pulses and transients 155

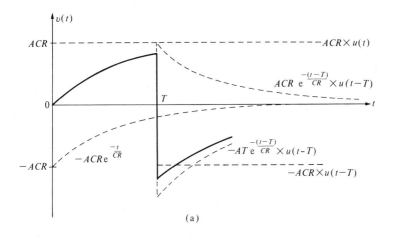

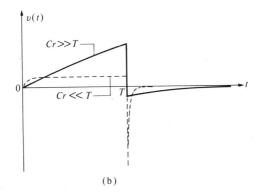

FIG 6.10. The response $v(t)$ of the network of Fig. 6.9(b) (a 'high-pass' C–R network) to a ramp function excitation. (a) The general response (eqn 6.31). (b) The response is reasonably faithful to the excitation if $CR \gg T$ but a good approximation to $(CR)\,\mathrm{d}e(t)/\mathrm{d}t$ is obtained if $CR \ll T$.

$$V(s) = \frac{E(s)R}{(R + sL)}$$

$$= \frac{RE(s)}{L\left(s + \dfrac{R}{L}\right)}.$$

Using eqn (6.29) for $E(s)$ we find that

156 Pulses and transients

$$V(s) = \frac{AR}{L}\left(\frac{1}{s^2\left(s+\frac{R}{L}\right)} - \frac{Te^{-sT}}{s\left(s+\frac{R}{L}\right)} - \frac{e^{-sT}}{s^2\left(s+\frac{R}{L}\right)}\right).$$

The expression $[s^2(s+R/L)]^{-1}$ is not listed in Table 6.1; however using the method of partial fractions (see Appendix 2) we have

$$\frac{1}{s^2\left(s+\frac{R}{L}\right)} = \frac{F}{\left(s+\frac{R}{L}\right)} + \frac{Gs+H}{s^2}$$

$$= \frac{s^2(F+G) + s(GR/L+H) + HR/L}{s^2\left(s+\frac{R}{L}\right)}.$$

Equating the coefficients of the terms in s^2, s^1, s^0 in the numerators of the two sides of the equation we find

$$s^2: \quad 0 = F+G$$

$$s^1: \quad 0 = \frac{GR}{L} + H$$

$$s^0: \quad 1 = \frac{HR}{L}.$$

Thus $F = L^2/R^2$; $G = -L^2/R^2$; $H = L/R$, and

$$\frac{1}{s^2\left(s+\frac{R}{L}\right)} = \frac{L^2/R^2}{\left(s+\frac{R}{L}\right)} - \frac{L^2}{R^2 s} + \frac{L}{Rs^2}. \tag{6.32}$$

It can be seen that all of the terms on the right-hand side of this equation are contained in the list of Laplace transform pairs of Table 6.1. Whilst more comprehensive tables would contain $\mathscr{L}^{-1}\left[\left\{s^2\left(s+\frac{R}{L}\right)\right\}^{-1}\right]$ this simple exercise demonstrates that the partial fraction technique should always be born in mind as a possible means of simplifying the problem of obtaining the inverse Laplace transform of a complicated function in the s-domain.

The techniques of mesh and nodal analysis and the concepts of Thévenin and Norton equivalent circuits can be carried over from a.c. circuit analysis.

6.6. Practical details

The practical problems of the generation, transmission, and shaping of pulses constitute, in themselves, a large sector of the subject of electronics, for instance in telecommunications, radar, computers, and television. Hence we can touch only briefly on one or two interesting and important aspects here. We have already discussed in Section 6.3 the bandwidth problem which is associated with the transmission of a train of pulses; now we will consider the distortion of transient and pulsed signals which occurs when they are transmitted through $C-R$ networks.

It is obvious from eqn (6.31) and Fig. 6.10(a) that the response of a $C-R$ network to a 'ramp' excitation exhibits significant distortion. Let us consider two extreme cases:

(i) $CR \gg T$. For $t \leqslant T$ eqn (6.31) becomes

$$v(t) = ACR(1 - e^{-t/CR}) \tag{6.33}$$

and, using the series expansion for the exponential function, we have

$$v(t) \approx ACR\left[1 - \left(1 - \frac{t}{CR} + \frac{t^2}{2C^2R^2} - \ldots\right)\right]$$

and since if $T \ll CR$ then, in this region, $t \ll CR$ also, we have

$$v(t) \approx At. \tag{6.34}$$

For $t \geqslant T$ eqn (6.31) becomes

$$v(t) = ACR\,e^{-(t-T)/CR}\left[1 - \frac{T}{CR} - e^{-T/CR}\right] \tag{6.35}$$

or

$$v(t) \approx -\frac{AT^2}{2CR}e^{-(t-T)/CR}. \tag{6.36}$$

The response described by eqns (6.34) and (6.36) is sketched in Fig. 6.10(b).

(ii) $CR \ll T$. For $t \gg CR$, but with $t \leqslant T$, eqn (6.33) becomes

$$v(t) \approx ACR \tag{6.37}$$

and for $t \geqslant T$ eqn (6.34) becomes

$$v(t) \approx -AT\,e^{-(t-T)/CR}. \tag{6.38}$$

The response described by eqns (6.37) and (6.38) is also sketched in Fig. 6.10(b).

It will have been noticed, no doubt, from the foregoing analysis and/or from Fig. 6.10(b) that in case (i) the form of $v(t)$ is a reasonable fascimile of the ramp voltage $e(t)$ bearing in mind that $AT \gg AT^2/2CR$. In case (ii), $v(t)$ has a form which approximates closely to a scaled version of the time derivative of the ramp voltage; of course it is impossible to obtain the infinite value of the mathematical derivative at $t = T$ in a real physical network but there is a negative-going spike of comparatively large amplitude. The differentiating property which this high-pass C–R network† possesses under the condition $CR \ll T$ applies to any form of excitation of course.

A 'proof' of the differentiating property of a C–R network may already have been met in which the argument goes as follows: In the network of Fig. 6.9(b), if R is small enough, then all of the excitation voltage $e(t)$ appears across C i.e.

$$e(t) = \frac{1}{C} \int i(t)\,\mathrm{d}t$$

or

$$i(t) = C\frac{\mathrm{d}e(t)}{\mathrm{d}t}.$$

Since $v(t) = Ri(t)$ we have therefore that

$$v(t) \approx CR\frac{\mathrm{d}e(t)}{\mathrm{d}t}.$$

The derivation using Laplace transforms may seem longwinded by comparison but the method is rigorous and it does yield much more information since it gives an analytical form of the response for all values of T and CR.

Such is the utility of the technique that it is worth considering, in detail, one more example. Consider the situation in which the ramp voltage of Fig. 6.9(a) is applied to a C–R low-pass network† as shown in Fig. 6.11(a): in the s-domain we have

$$V(s) = \frac{E(s)}{sC\left(R + \dfrac{1}{sC}\right)} \quad \text{or} \quad V(s) = \frac{E(s)}{CR\left(s + \dfrac{1}{CR}\right)}.$$

Using eqn (6.29) for $E(s)$

---

†Such a C–R configuration is called a 'high-pass' network since if $e(t)$ is sinusoidal then $v(t)/e(t)$ increases from zero towards a value of unity as the frequency of $e(t)$ is increased from zero to infinity. Conversely, for a 'low-pass' network the response decreases from its value at d.c. towards zero as the frequency is increased from zero towards infinity (see Chapter 7).

$$V(s) = \frac{A}{CRs^2\left(s + \frac{1}{CR}\right)} - \frac{ATe^{-sT}}{CRs\left(s + \frac{1}{CR}\right)} - \frac{Ae^{-sT}}{CRs^2\left(s + \frac{1}{CR}\right)}$$

and using eqn (6.32) with CR in place of L/R we have

$$V(s) = ACR\left[\frac{1}{\left(s + \frac{1}{CR}\right)} - \frac{1}{s} + \frac{1}{CRs^2} - \frac{Te^{-sT}}{(CR)^2 s\left(s + \frac{1}{CR}\right)}\right.$$

$$\left. - \frac{e^{-sT}}{\left(s + \frac{1}{CR}\right)} + \frac{e^{-sT}}{s} - \frac{e^{-sT}}{CRs^2}\right].$$

The inverse Laplace transforms are given in Table 6.1 whence

$$v(t) = ACR\left[e^{-t/CR} - u(t) + \frac{t}{CR} - \frac{T}{CR}\{1 - e^{-(t-T)/CR}\}u(t-T)\right.$$

$$\left. - e^{-(t-T)/CR} \cdot u(t-T) + u(t-T) - \frac{(t-T)}{CR} \cdot u(t-T)\right].$$

(6.39)

This response is sketched in Fig. 6.11(b).

Let us consider again two extreme cases:
(i) $CR \gg T$. For $t \leq T$, eqn (6.39) becomes

$$v(t) = ACR\left(e^{-t/CR} - 1 + \frac{t}{CR}\right) \quad (6.40)$$

and using the series expansion of $e^{-t/CR}$ we obtain

$$v(t) \approx \frac{1}{CR} \cdot \frac{At^2}{2}. \quad (6.41)$$

For $t \geq T$, eqn (6.39) becomes

$$v(t) = ACR\, e^{-(t-T)/CR}\left[\frac{T}{CR} - 1 + e^{-T/CR}\right] \quad (6.42)$$

and since $T \ll CR$ we have

$$v(t) \approx \frac{AT^2}{2CR} e^{-(t-T)/CR}. \quad (6.43)$$

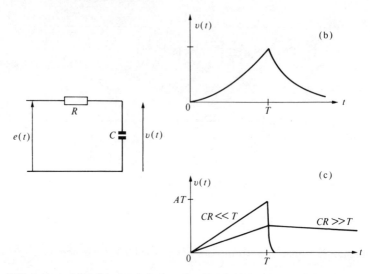

FIG. 6.11. (a) A 'low-pass' C–R network. (b) The general response to a ramp function excitation. (c) The response is reasonably faithful to the excitation if $CR \ll T$ but a good approximation to $(CR)^{-1} \int e(t)\,dt$ is obtained if $CR \gg T$.

The form of response described by eqns (6.40) and (6.42) is sketched in Fig. 6.11(c).

(ii) $CR \ll T$. In the region $t \leqslant T$, but with $t \gg CR$, eqn (6.39) becomes

$$v(t) \approx At. \tag{6.44}$$

For $t \gtrsim T$ we use eqn (6.41) again but since $T \gg CR$ we have

$$v(t) \approx ATe^{-(t-T)/CR}. \tag{6.45}$$

The form of this response is also sketched in Fig. 6.11(c). You will notice that on this occasion a reasonable fascimile of the ramp excitation $e(t)$ is obtained if $CR \ll T$ and that, if $CR \gg T$, then a good approximation to a scaled version of the integral of $e(t)$ is obtained. Again this integrating property of a low-pass C–R network applies to any form of excitation.

For an excitation $e(t)$ in the form of an ideal square, or rectangular pulse of amplitude A and duration δ

$$e(t) = Au(t) - Au(t - \delta)$$

i.e. the sum of a positive step and a delayed negative step and

$$E(s) = \frac{A}{s} - \frac{Ae^{-s\delta}}{s}. \tag{6.46}$$

If this excitation is applied to a high-pass $C-R$ network then using

$$V(s) = \frac{E(s)R}{\left(R + \dfrac{1}{sC}\right)},$$

the response is

$$V(s) = \frac{A}{\left(s + \dfrac{1}{CR}\right)} - \frac{Ae^{-s\delta}}{\left(s + \dfrac{1}{CR}\right)}$$

or

$$v(t) = Ae^{-t/CR} - Ae^{-(t-\delta)/CR} \cdot u(t-\delta). \qquad (6.47)$$

In the case of a train of rectangular or square pulses, if there is sufficient time for the transients to decay to negligible proportions between successive pulses, then the analysis for a single pulse can still be used. The waveform of eqn (6.47) is shown in Fig. 6.12 together with the differentiated and integrated forms. The high-pass $C-R$ network is used very often as an interstage coupling network in electronic systems and this means, in the case of transient or pulsed signals that some distortion is introduced as indicated in Fig. 6.12(a) and (b).

A typical interstage coupling situation is depicted in Fig. 6.13(a) where C_2 represents a shunt capacitance arising at the input to the 'next' stage of the system. It is left as an exercise to show that $v(t)$ has the same general form as eqn (6.47) apart from a multiplying factor $C_1/(C_1 + C_2)$ and with $R_2(C_1 + C_2)$ replacing RC. It is interesting to analyse the network shown in Fig. 6.13(b); you can show easily that

$$\frac{V(s)}{E(s)} = \frac{R_2}{(R_1 + R_2)} \frac{(1 + sC_1R_1)}{\left[1 + s(C_1 + C_2)\dfrac{R_1R_2}{(R_1 + R_2)}\right]} \qquad (6.48)$$

Hence, if

$$sC_1R_1 = s(C_1 + C_2)\frac{R_1R_2}{(R_1 + R_2)},$$

then $V(s)/E(s)$ is independent of s and $v(t)$ will be an undistorted version of $e(t)$. The condition reduces to

$$C_1R_1 = C_2R_2$$

and a network satisfying this condition is usually called a compensated attenuator. Such networks are often used in probes to be used in conjunction with oscilloscopes and other test instruments: In this case R_2 would be the input resistance and C_2 would be the shunt capacitance at the input of an oscilloscope, say. Typical values for the circuit elements would be $R_2 = 1\,\text{M}\Omega$; $C_2 \approx 40\,\text{pF}$; $R_1 = 9\,\text{M}\Omega$. C_2 would be

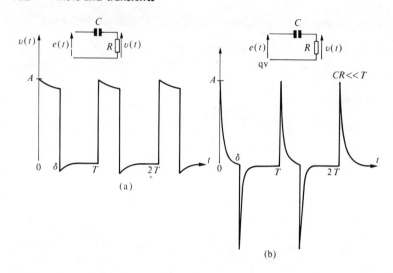

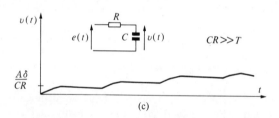

FIG. 6.12. The response of high-pass and low-pass C–R networks to an excitation in the form of a train of rectangular pulses of amplitude A, duration δ, and period T. (a) The general case for a high-pass network. (b) An extreme case (differentiation) for a high-pass network. (c) Another extreme case (integration) for a low-pass network.

an adjustable capacitor which is integral with the probe unit itself. For such values the signal would be attenuated by a factor of ten but an ancillary advantage is that the input impedance at the probe is ten times greater than that at the terminals of the test instrument to which it is connected.

So far we have not considered the characteristics of the source of pulses. The output impedance of a source contains a reactive component, inevitably, which reflects the fact that it is impossible to generate ideal mathematical rectangular pulses in practice; to put it another way we cannot obtain real pulses $e(t)$ which have mathematical discontinuities in $e(t)$ and its derivatives. For reasonably 'clean' rectangular pulses the rise and fall times are usually taken to be the

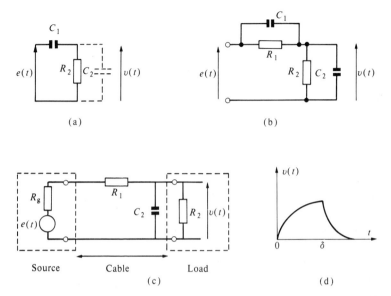

FIG. 6.13. (a) A representation of an inter-stage coupling network where C_2 is the shunt input capacitance of the succeeding stage. (b) This network becomes a so-called compensated attenuator if $C_1 R_1 = C_2 R_2$. (c) The equivalent circuit for a voltage source and a load connected by a cable having an equivalent lumped series resistance R_1; C_2 includes the shunt capacitance of the load.

time for the signal to rise, or fall, as the case may be, between the limits 10 per cent and 90 per cent of the amplitude of the pulse. Pulse generators are commonly available which give 'rectangular' pulses having durations as short as 15 ns with rise times of 5 ns at a pulse repetition rate of 50 MHz.

A situation which may be encountered often, in practice, is that in which the source of pulses is connected to an amplifier via a coaxial cable. If the cable is short enough so that the transmission time is much less than the duration of the pulse then the cable can be represented by an equivalent lumped series resistance and lumped shunt capacitance.† The circuit diagram for such a situation is shown in Fig. 6.13(c) where C_2 includes the shunt capacitance at the input of the amplifier as well as the equivalent lumped shunt capacitance of the cable whose series resistance is represented by R_1; for simplicity the impedance of the source is shown as being purely resistive. It can be shown quite

†Transmission lines, of which coaxial cables are a particular example, are treated in Chapter 8; as a rough guide the transmission time for 2.5 m of coaxial cable is about 1 μs.

Pulses and transients

easily in this case that

$$\frac{V(s)}{E(s)} = 1 \bigg/ \left[C_2(R_g + R_1)\left\{ s + \frac{(R_2 + R_g + R_1)}{C_2 R_2 (R_g + R_1)} \right\} \right]$$

and if $E(s)$ corresponds to a rectangular pulse, as specified by eqn (6.46),

$$e(t) = \frac{AR_2}{(R_2 + R_g + R_1)}[\{1 - e^{-t(R_2 + R_g + R_1)/C_2 R_2 (R_g + R_1)}\}$$

$$- \{1 - e^{-(t-\delta)(R_2 + R_g + R_1)/C_2 R_2 (R_g + R_1)}\} u(t - \delta)]$$

which is sketched in Fig. 6.13(d). So we see that the effect of the shunt capacitance is to distort the input waveform with a tendency towards integration. If the value of $(R_g + R_1)$ can be reduced then

$$e(t) \approx A[\{1 - e^{-t/C_2(R_g + R_1)}\} - \{1 - e^{-(t-\delta)/C_2(R_g + R_1)}\} u(t - \delta)]$$

and the exponential decays are much sharper which gives a closer approximation to the rectangular pulse. This can be achieved in practice by connecting a buffer amplifier of low output impedance (e.g. an 'emitter-follower') directly between the source, say a Geiger–Müller tube or a photo-multiplier, and the cable.

For longer lengths of cable and/or shorter pulse durations then the properties of the cable as a transmission line have to be reckoned with; it is usual to terminate the cable with an impedance equal to its characteristic impedance so that pulses are not reflected at the end of the cable.

In practical situations there are present always stray capacitances and inductances which cause distortion of pulses; indeed we have treated networks containing inductances only very briefly. The general effect of $L-C-R$ circuits is to cause 'ringing' if the damping is small enough. However, the characteristics of $L-C-R$ networks were treated in Chapter 4, albeit not by Laplace transform methods, so we will not deal further with them here apart from setting some pertinent exercises.

PROBLEMS

6.1. (a) Derive the expression in eqn (6.4) for the square wave shown in Fig. 6.1(c). (b) Derive the expression in eqn (6.3) for the sawtooth wave shown in Fig. 6.1(f).

6.2. Show that if the wave $v = \hat{V} \cos \omega t$ is full-wave rectified as depicted in Fig. 6.14 then the Fourier series $v(t)$

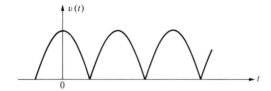

FIG. 6.14.

for this waveform is given by

$$v(t) = a_0 + \sum_{n=1}^{\infty} a_n \cos 2n\omega t$$

where

$$a_0 = \frac{2\hat{V}}{\pi}$$

$$a_n = -\frac{4\hat{V}}{\pi(4n^2 - 1)} \quad (n \text{ even})$$

$$a_n = \frac{4\hat{V}}{\pi(4n^2 - 1)} \quad (n \text{ odd}).$$

6.3. Show that the Fourier series which represent the square waves of Fig 6.1(b) and (d) and the sawtooth wave of Fig. 6.1(e) are given, respectively, by

$$v(t) = \frac{2\hat{V}}{\pi} \cos \omega t + \frac{2\hat{V}}{3\pi} \cos 3\omega t + \frac{2\hat{V}}{5\pi} \cos 5\omega t + \ldots$$

$$v(t) = \frac{\hat{V}}{2} + \frac{2\hat{V}}{\pi} \cos \omega t + \frac{2\hat{V}}{\pi} \cos 3\omega t + \frac{2\hat{V}}{\pi} \cos 5\omega t + \ldots$$

$$v(t) = \frac{\hat{V}}{\pi} \sin \omega t - \frac{\hat{V}}{2\pi} \sin 2\omega t + \frac{\hat{V}}{3\pi} \sin 3\omega t - \frac{\hat{V}}{4\pi} \sin 4\omega t + \ldots.$$

6.4. What are the r.m.s. values of the non-sinusoidal periodic waveforms shown in Figs. 6.1(a), (b), and (f)?

6.5. Show that the Fourier series for the train of triangular pulses shown in Fig. 6.15 is given by

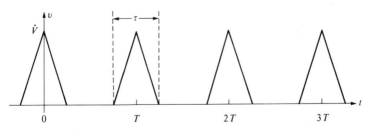

FIG. 6.15.

$$v = \hat{V}\left[\frac{\omega\tau}{4\pi} + \frac{4}{\pi\omega\tau}\left\{\sum_{n=1}^{\infty}\frac{(1-\cos\frac{1}{2}n\omega\tau)}{N^2}\cos n\omega t\right\}\right]$$

where $\omega \equiv 2\pi/T$.

6.6. Derive an expression for the time domain voltage across the inductor in the circuit shown in Fig. 6.16 if the excitation is a ramp voltage $e(t) = At$.

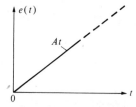

FIG. 6.16.

6.7. For the network and excitation $e(t)$ given in Fig. 6.17 show that the voltage $V(s)$ in the s-domain is given by

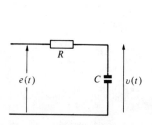
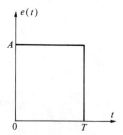

FIG. 6.17.

$$V(s) = \frac{A}{CR}\frac{1}{s\left(s+\frac{1}{CR}\right)} - \frac{A}{CR}\frac{1}{s\left(s+\frac{1}{CR}\right)}.$$

(You may assume that the capacitor is uncharged at time $t = 0$). Hence show that

$$v(t) = A\{1 - u(t-T) - e^{-t/CR} + e^{-(t-T)/CR}u(t-T)\}.$$

Sketch the general form of $v(t)$ and also the form of $v(t)$ for $CR \gg T$.

6.8. A non-periodic signal voltage $e(t)$ of the form shown in the diagram is applied to the high-pass C–R network shown in Fig. 6.18. Show that

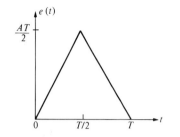
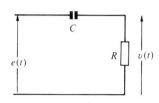

FIG. 6.18.

$$v(t) = ACR\{1 - e^{-t/CR}\} - 2ACR\{1 - e^{-(t-T/2)/CR}\}u(t - \tfrac{1}{2}T)$$
$$+ ACR\{1 - e^{-(t-T)/CR}\}u(t - T).$$

6.9. A pulse of the form shown in Fig. 6.19 is applied to the low-pass C–R network. Show that for the pulse

$$E(s) = \frac{A}{\tau}\left\{\frac{1}{s^2} - \frac{e^{-sT}}{s^2} - \frac{e^{-s(\tau+T)}}{s^2} + \frac{e^{-s(2\tau+T)}}{s^2}\right\}$$

and hence show that

$$v(t) = \frac{ACR}{\tau}\left\{\left[-1 + \frac{t}{CR} + e^{-t/CR}\right]\right.$$
$$-\left[-1 + \frac{(t-\tau)}{CR} + e^{-(t-\tau)/CR}\right]u(t-\tau)$$
$$-\left[-1 + \frac{(t-(\tau+T))}{CR} + e^{-(t-(\tau+T))/CR}\right]u[t-(\tau+T)]$$
$$+\left[-1 + \frac{(t-(2\tau+T))}{CR} + e^{-(t-(2\tau+T))/CR}\right] - [t-(2\tau+T)]\right\}.$$

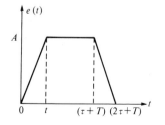
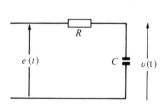

FIG. 6.19.

6.10. Show that the response $v(t)$ of the network to an excitation $e(t)$ of the form shown in Fig. 6.20 is given by

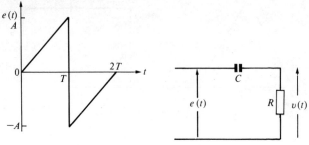

FIG. 6.20.

$$v(t) = A\left\{\frac{CR}{T}(1 - e^{-t/CR}) - 2e^{-(t-T)/CR}u(t-T)\right.$$

$$\left. - \frac{CR}{T}(1 - e^{-(t-2T)/CR})u(t-2T)\right\}.$$

7. Network analysis and synthesis

7.1. Introduction

THE way in which the currents and potential differences in the branches of a d.c. network can be determined in terms of the characteristics of the sources (voltage and current generators) and the resistances of the circuit elements by using the techniques of mesh and nodal analysis has been explained in Chapter 2. Further it was stated in Section 3.3. that Kirchhoff's laws could be taken over to linear a.c. netowrks providing that they were applied to the instantaneous values of current and potentials. It was also noted that Thévenin's and Norton's theorems could be adapted for a.c. networks.

In the consideration of a network the problem is frequently restricted simply to a determination of the response, or 'output', in one branch when a generator, or 'input' or 'excitation', is connected in another specified branch. In other words it is the external behaviour of the network which is of prime interest and the network can be replaced by a black box having two input and output terminals; the pairs of input and output terminals are usually referred to as the 'input port and output port', respectively. Such a network is called a two-port network and the system function H relates the response and the excitation, e.g. the output voltage (response) to input voltage (excitation), or output voltage (response) to input current (excitation) and so on.

The situation can be summarized as follows:

$$\text{Response} = (\text{system function}) \times (\text{excitation})$$

$$R = H \times E$$

Analysis
To determine the resonse given the excitation and the network.

Synthesis
To design the network given the excitation and the desired response.

In this chapter we will make the following assumptions about the networks and components:

(a) *Linear.* If the excitation is multiplied by a constant factor then the response is multiplied by the same factor.

(b) *Lumped.* A real circuit component is assumed to be resolvable into a combination of a pure resistor and/or a pure capacitor and/or a pure inductor. Furthermore the component is assumed to be of zero physical size which implies that, at any instant of time, the electrical field is constant along the length of the component. This in turn implies that the dimensions of the network are much less than the wavelength of an electromagnetic wave at the frequency in use so that there is negligible radiation of power; so energy is conserved in the network and Kirchhoff's laws can be applied. The situations where this assumption breaks down are discussed in Chapter 8, 'Transmission Lines'.

(c) *Passive.* Apart from the section devoted to linear amplifiers, it will be assumed that there are no energy sources within the networks considered.

As a specific example consider the network of Fig. 7.1(a) which is similar to that of Fig. 2.3 but with general impedances $Z_1 \ldots Z_6$ as the passive circuit elements instead of pure resistors and with the source of e.m.f. e_2 removed; e_1 is now a generator of a sinusoidal e.m.f. of course. In this example one of the input terminals and one of the output terminals are connected together internally and form a common connection; this often occurs in practice. The set of equations which are equivalent to eqns (2.10) are

$$i_1(Z_5 + Z_6) + (i_2 Z_6 - i_3 Z_5) = e_1 - i_1 Z_1 = v_1$$
$$i_1 Z_6 + i_2(Z_4 + Z_6) + i_3 Z_4 = -i_2 Z_2 = v_2 \quad (7.1)$$
$$-i_1 Z_5 + i_2 Z_4 + i_3(Z_3 + Z_4 + Z_5) = 0$$

and using Cramer's rule (eqn 2.12) it follows that

$$i_1 = v_1 D_{11}/D + v_2 D_{21}/D; \qquad i_2 = v_1 D_{12}/D + v_2 D_{22}/D \quad (7.2)$$

where $D_{12} = D_{21}$ in the case of a passive network such as this one.

Notice that we are continuing to employ the convention of using lower case symbols for currents and voltages in the time domain which we adopted in Chapter 3.

The determinant D and its cofactors may be calculated if the structure of the network is known but more important is the fact that eqns (7.2) may be written in the form

$$i_1 = y_{11} v_1 + y_{12} v_2; \qquad i_2 = y_{21} v_1 + y_{22} v_2 \quad (7.3)$$

where y_{11}, y_{21}, etc. are known as y-parameters, or admittance parameters since they have the dimensions of $(\text{ohm})^{-1}$. The sign convention

Network analysis and synthesis 171

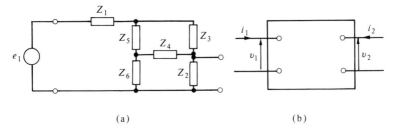

FIG. 7.1. (a) A general bridge network. (b) A representation of a two-port network.

for the currents and voltages at the input and output ports of a two-port network are shown in Fig. 7.1(b).

It follows from eqn (7.3) that

$$v_1 = z_{11}i_1 + z_{12}i_2; \qquad v_2 = z_{21}i_1 + z_{22}i_2 \qquad (7.4)$$

where z_{11}, z_{21}, etc. are known as impedance parameters. The derivation of the explicit relationships between the y- and z-parameters is left as an exercise; for instance

$$z_{12} = -y_{12}/(y_{11}y_{22} - y_{12}y_{21}).$$

Finally v_1, i_1, v_2, i_2 can be related by

$$v_1 = h_{11}i_1 + h_{12}v_2; \qquad i_2 = h_{21}i_1 + h_{22}v_2 \qquad (7.5)$$

where h_{11}, h_{21}, etc. are known as h-parameters or hybrid parameters since h_{12}, h_{21} are dimensionless, h_{11} has the dimensions of impedance, and h_{22} of admittance.

Remember that, although a specific network has been considered, eqns (7.3), (7.4), and (7.5) can be used to describe any four-terminal network.

Notice that at most four, and in the case of passive networks, three parameters are required to specify one pair of currents or voltages if the other is known.

From eqns (7.3) it can be seen that the y-parameters can be determined by short-circuiting the input and output terminals successively ($v_1 = 0$ and $v_2 = 0$, respectively) and *measuring* i_1, i_2, v_2 and i_1, i_2, v_1, respectively.

$$v_1 = 0: y_{12} = i_1/v_2 \qquad y_{22} = i_2/v_2 \qquad (7.6)$$

$$v_2 = 0: y_{11} = i_1/v_1 \qquad y_{21} = i_2/v_1. \qquad (7.7)$$

The z-parameters can be determined by open-circuiting the input and output terminals successively and measuring the appropriate

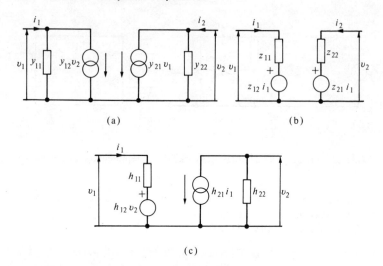

FIG. 7.2. Equivalent circuits for a two-port network in terms of (a) y-parameters, (b) z-parameters, (c) h-parameters.

voltages and currents:

$$i_1 = 0 : z_{12} = v_1/i_2 \qquad z_{22} = v_2/i_2 \qquad (7.8)$$

$$i_2 = 0 : z_{11} = v_1/i_1 \qquad z_{21} = v_2/i_1. \qquad (7.9)$$

Similarly, the h-parameters can be determined by successively open-circuiting the input terminals and short-circuiting the output terminals:

$$i_1 = 0 : h_{12} = v_1/v_2 \qquad h_{22} = i_2/v_2 \qquad (7.10)$$

$$v_2 = 0 : h_{11} = v_1/i_1 \qquad h_{21} = i_2/i_1. \qquad (7.11)$$

In the foregoing 'short circuit' does not necessarily mean a d.c. short circuit since a large enough capacitor may form a good enough a.c. short circuit. In any case we should proceed with caution when short-circuiting a pair of terminals in case damage is caused to the circuit components.

The equivalent circuits for a two-port network in terms of z-, y-, and h- parameters are shown in Fig. 7.2. The h-parameters as defined above are very useful as a description of transistors, at least at low frequencies.

8.2. T and Π networks

Since there are only three independent elements in any set of z-, y-, or h-parameters for a passive two-port network it can be represented

Network analysis and synthesis 173

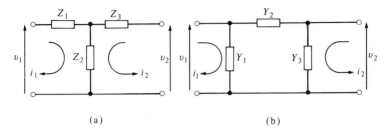

FIG. 7.3. (a) T-form and (b) Π-form of equivalent network for a two-port network.

by an equivalent network containing only three elements; the T- and Π-forms of such an equivalent network are shown in Fig. 7.3.

If we set up the mesh equations for the network of Fig. 7.3(a) then we obtain

$$i_1(Z_1 + Z_2) + i_2 Z_2 = v_1$$

So
$$i_1 Z_2 + i_2(Z_3 + Z_2) = v_2. \quad (7.12)$$

$$z_{11} = Z_1 + Z_2 \qquad z_{12} = Z_2$$

$$z_{21} = Z_2 \qquad z_{22} = Z_3 + Z_2. \quad (7.13)$$

From eqn (7.12) using Cramer's rule,

$$i_1 = \frac{v_1 \Delta_{11}}{\Delta_z} + \frac{v_2 \Delta_{21}}{\Delta_z}$$

$$i_2 = \frac{v_1 \Delta_{12}}{\Delta_z} + \frac{v_2 \Delta_{22}}{\Delta_z} \quad (7.14)$$

where
$$\Delta_{11} = (-1)^2 (Z_3 + Z_2) \qquad \Delta_{12} = \Delta_{21} = (-1)^3 Z_2$$

$$\Delta_{22} = (-1)^4 (Z_1 + Z_2)$$

$$\Delta_z = (Z_1 + Z_2)(Z_3 + Z_2) - Z_2^2 \quad \text{or} \quad \Delta_z = Z_1 Z_2 + Z_1 Z_3 + Z_2 Z_3$$

Now if we set up the nodal equations for the Π network of Fig. 7.3(b) we have

$$v_1 (Y_1 + Y_2) - v_2 Y_2 = i_1$$

So
$$-v_1 Y_2 + v_2 (Y_2 + Y_3) = i_2. \quad (7.15)$$

$$y_{11} = Y_1 + Y_2 \qquad y_{12} = y_{21} = -Y_2$$

$$y_{22} = Y_2 + Y_3. \quad (7.16)$$

From eqn (7.15), using Cramer's rule,

$$v_1 = \frac{i_1 \Delta_{11}}{\Delta_y} + \frac{i_2 \Delta_{21}}{\Delta_y}$$

$$v_2 = \frac{i_1 \Delta_{12}}{\Delta_y} + \frac{i_2 \Delta_{22}}{\Delta_y} \quad (7.17)$$

where
$$\Delta_{11} = (-1)^2(Y_2 + Y_3) \qquad \Delta_{12} = \Delta_{21} = (-1)^3(-Y_2)$$
$$\Delta_{22} = (-1)^4(Y_1 + Y_2)$$
$$\Delta_y = (Y_1 + Y_2)(Y_2 + Y_3) - Y_2^2 \text{ or } \Delta_y = Y_1 Y_2 + Y_1 Y_3 + Y_2 Y_3.$$

On comparing eqns (7.13) and (7.16) we see that the elements of the equivalent T- and Π-networks are related as follows;

$$Z_1 + Z_2 = (Y_2 + Y_3)/\Delta_y \quad Z_2 = Y_2/\Delta_y \quad Z_3 + Z_2 = (Y_1 + Y_2)/\Delta_y$$

whence
$$Z_1 = Y_3/\Delta_y \quad Z_3 = Y_1/\Delta_y \quad Z_2 = Y_2/\Delta_y. \quad (7.18)$$

Similarly the admittances Y_1, Y_2, Y_3 can be expressed in terms of the impedances Z_1, Z_2, Z_3 and Δ_z; you should be able to show that

$$Y_1 = Z_3/\Delta_z \quad Y_3 = Z_1/\Delta_z \quad Y_2 = Z_2/\Delta_z. \quad (7.19)$$

A two-port network is said to be symmetrical if open-, or short-circuit measurements, as the case may be, yield identical values for the z-, y-, or h-parameters irrespective of which pair of terminals is designated as the input port i.e. the network 'looks' identical when looked into from either port. In the case of an equivalent T-network, or an actual T-network, this implies that $Z_1 = Z_3$ or that $Y_1 = Y_3$. Actually it is common practice to designate each of the series elements in a symmetrical T-network by $Z_1/2$ and the shunt elements in a symmetrical Π-network by $Y_1/2$ as shown in Fig. 7.4.

For a symmetrical T-network with load impedance Z_L as shown in Fig. 7.4(a) we can see, by inspection, that the input impedance Z_{in} is given by

$$Z_{in} = \frac{1}{2}Z_1 + \frac{Z_2(Z_1/2 + 2Z_L)}{(Z_1/2 + Z_2 + Z_L)}$$

The characteristic impedance Z_{0_T} is defined to be such that if $Z_L = Z_{0_T}$ then $Z_{in} = Z_{0_T}$ also; whence we can show that

$$Z_{0_T} = \sqrt{(Z_1 Z_2 + Z_1^2/4)}. \quad (7.20)$$

Now the open-circuit input impedance ($Z_L = \infty$) is given by

$$Z_{oc} = Z_1/2 + Z_2 \quad (7.21)$$

and the short circuit input impedance ($Z_L = 0$) is given by

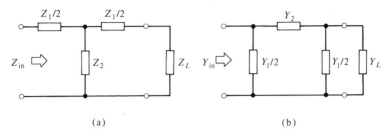

FIG. 7.4. (a), (b) T- and Π-forms respectively of a symmetrical two-port network.

$$Z_{sc} = Z_1/2 + \frac{\frac{1}{2}Z_1Z_2}{(\frac{1}{2}Z_1 + Z_2)}$$

or

$$Z_{sc} = \frac{Z_1Z_2 + \frac{1}{4}Z_1^2}{Z_2 + \frac{1}{2}Z_1} \quad (7.22)$$

Hence

$$Z_{oc}Z_{sc} = Z_1Z_2 + \frac{1}{4}Z_1^2$$

or

$$Z_{0_T} = \sqrt{(Z_{oc}Z_{sc})}. \quad (7.23)$$

For the Π-representation (see Fig. 7.4(b)) it follows from the duality principle (See Section 3.4) that

$$Y_{in} = \frac{1}{2}Y_1 + \frac{Y_2(\frac{1}{2}Y_1 + Y_L)}{(\frac{1}{2}Y_1 + Y_2 + Y_L)}$$

and that the characteristic admittance

$$Y_{0_\Pi} = \sqrt{(Y_1Y_2 + \tfrac{1}{4}Y_1^2)}. \quad (7.24)$$

The open-circuit admittance ($Y_L = 0$) is

$$Y_{oc} = \frac{Y_1Y_2 + \frac{1}{4}Y_1^2}{(Y_2 + \frac{1}{2}Y_1)} \quad (7.25)$$

and the short-circuit admittance ($Y_L = \infty$) is

$$Y_{sc} = \tfrac{1}{2}Y_1 + Y_2. \quad (7.26)$$

Thus,

$$Y_{0_\Pi} = \sqrt{(Y_{oc}Y_{sc})}. \quad (7.27)$$

The results of eqns (7.21) and (7.24) are true for any symmetrical two-port network.

As a simple example of the use of the foregoing analyses consider the problem of finding the T-network which is equivalent to the Π-network shown in Fig. 7.5(a). For simplicity we will perform the calculation for a frequency of $10^4/2\pi$ Hz so that $\omega = 10^4$ radians per

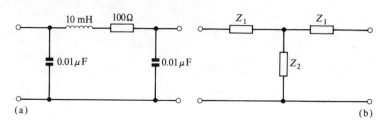

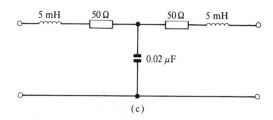

FIG 7.5. (a) A symmetrical Π-network, (b) A general symmetrical T-network, (c) The symmetrical T-network which is equivalent to the network of (a).

second (the values of the components of the equivalent network will be frequency dependent, of course).

$$Y_1 = Y_3 = j\,10^4 \cdot 10^{-8} = j\,10^{-4}\ \text{S}$$

$$Y_2 = (100 + j\,10^4 \cdot 10^{-2})^{-1} = \frac{(1-j)}{2 \cdot 10^2}\ \text{S}$$

$$\Delta_y = 2Y_1 Y_2 + Y_1^2 \approx 10^{-6}(1+j)\,\text{S}^2.$$

So using eqns (7.18) we find

$$Z_1 = \frac{j \cdot 10^{-4}}{10^{-6}(1+j)} \qquad Z_2 = \frac{(1-j)}{2 \cdot 10^2 \cdot 10^{-6}(1+j)}$$

$$= \frac{j \cdot 10^2 (1-j)}{2} \qquad = \frac{10^4 (1-j)(1-j)}{4}$$

or
$$Z_1 = 50(1+j)\,\Omega \qquad = -j\,5 \cdot 10^3\,\Omega.$$

Thus, at a frequency of $10^4/2\pi$ Hz,

$$Z_1 = 50 + j\,10^4(5 \cdot 10^{-3})\,\Omega \qquad Z_2 = \frac{-j}{10^4 \cdot 2 \cdot 10^{-8}}\,\Omega.$$

7.3. The interconnection of two-port networks

Many electrical and electronic systems are composed of interconnected two-port networks and it is important, as well as interesting,

Network analysis and synthesis 177

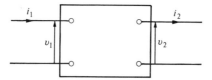

FIG. 7.6. The definitions, in relation to the transmission parameters, of the input variables v_1, i_1 and the output variables v_2, i_2 of a two-port network.

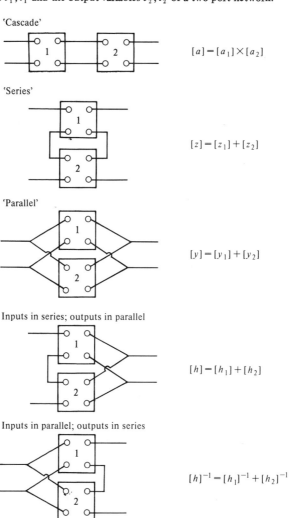

FIG. 7.7. Interconnections of two-port networks.

to consider the various ways in which the interconnections can be made. In Fig. 7.7 the five modes of interconnection are shown but before we discuss them further it is necessary to introduce another set of parameters in addition to the z-, y-, and h-parameters which we have already met. This additional set are called transmission parameters, or a-parameters, and are defined as follows (see Fig. 7.6)

$$v_1 = a_{11}v_2 + a_{12}i_2$$
$$i_1 = a_{21}v_2 + a_{22}i_2. \tag{7.28}$$

We can see that these parameters relate the output variables v_2, i_2 to the input variables v_1, i_1. It is necessary to take careful note of the fact that the sense of i_2 is opposite to that which was adopted in the context of the z-, y-, and h-parameters; the reason for this choice will be discussed below.

The definitive equations (eqns (7.3), (7.4), (7.5), and (7.26)) for the y-, z-, h-, and a-parameters can be written in the form;

$$\begin{bmatrix} i_1 \\ i_2 \end{bmatrix} = \begin{bmatrix} y_{11} & y_{12} \\ y_{21} & y_{22} \end{bmatrix} \begin{bmatrix} v_1 \\ v_2 \end{bmatrix} \qquad \begin{bmatrix} v_1 \\ v_2 \end{bmatrix} = \begin{bmatrix} z_{11} & z_{12} \\ z_{21} & z_{22} \end{bmatrix} \begin{bmatrix} i_1 \\ i_2 \end{bmatrix}$$

or

$$\begin{bmatrix} i_1 \\ i_2 \end{bmatrix} = [y] \begin{bmatrix} v_1 \\ v_2 \end{bmatrix} \qquad \begin{bmatrix} v_1 \\ v_2 \end{bmatrix} = [z] \begin{bmatrix} i_1 \\ i_2 \end{bmatrix}$$

$$\tag{7.29}$$

$$\begin{bmatrix} v_1 \\ i_2 \end{bmatrix} = \begin{bmatrix} h_{11} & h_{22} \\ h_{21} & h_{22} \end{bmatrix} \begin{bmatrix} i_1 \\ v_2 \end{bmatrix} \qquad \begin{bmatrix} v_1 \\ i_1 \end{bmatrix} = \begin{bmatrix} a_{11} & a_{12} \\ a_{21} & a_{22} \end{bmatrix} \begin{bmatrix} v_2 \\ i_2 \end{bmatrix}$$

or

$$\begin{bmatrix} v_1 \\ i_2 \end{bmatrix} = [h] \begin{bmatrix} i_1 \\ v_2 \end{bmatrix} \qquad \begin{bmatrix} v_1 \\ i_1 \end{bmatrix} = [a] \begin{bmatrix} v_2 \\ i_2 \end{bmatrix}.$$

Although we will not be concerned with the manipulation of matrices at a sophisticated level in this text it is worth making some comments on the last one of the above equations, say, which can be expressed in words as;

column 'vector' $\begin{bmatrix} v_1 \\ i_1 \end{bmatrix}$ equals column 'vector' $\begin{bmatrix} v_2 \\ i_2 \end{bmatrix}$ multiplied by

('operated on') by the square 2 × 2 matrix $\begin{bmatrix} a_{11} & a_{12} \\ a_{21} & a_{22} \end{bmatrix}$.

A comparison of this matrix equation and eqns (7.28) (or look ahead to

Network analysis and synthesis 179

eqn (7.37)) reveals the rule to be followed in multiplying together a 2 × 2 matrix and a column vector having 2 rows (a 2 × 1 matrix).

The utility of the matrix representation is that when networks are interconnected (as shown in Fig. 7.7) the matrices combine in a mathematically simple fashion to give a resultant matrix:

(a) Series connection $[z] = [z_1] + [z_2]$
(b) Parallel connection $[y] = [y_1] + [y_2]$
(c) Inputs in series; outputs in parallel $[h] = [h_1] + [h_2]$
(d) Inputs in parallel; outputs in series $[h]^{-1} = [h_1]^{-1} + [h_2]^{-1}$
(e) Cascade connection $[a] = [a_1] \times [a_2]$.

(7.30)

We will be concerned for the most part with the cascade, or tandem, connection of two-port networks in this text so the definition of the inverse $[h]^{-1}$ of a matrix $[h]$ will not be discussed here (consult a text dealing with matrices and determinants if you wish to pursue this particular matter). However one example of the parallel connection of two networks which is of interest is the 'twin-T' network (see Fig. 7.8) which consists of two symmetrical T-networks connected in parallel. On using the results (eqns 7.16) obtained in Section 7.2 for the y-parameters of a T-network, together with eqns (7.19) and the relevant equation from eqns (7.30), we have

$$[y] = \begin{bmatrix} \dfrac{\frac{1}{2}Z' + Z}{\Delta_1} & \dfrac{-\frac{1}{2}Z'}{\Delta_1} \\ \dfrac{-\frac{1}{2}Z'}{\Delta_1} & \dfrac{Z + \frac{1}{2}Z'}{\Delta_1} \end{bmatrix} + \begin{bmatrix} \dfrac{\frac{1}{2}Z + Z'}{\Delta_2} & \dfrac{-\frac{1}{2}Z}{\Delta_2} \\ \dfrac{-\frac{1}{2}Z}{\Delta_2} & \dfrac{Z' + \frac{1}{2}Z}{\Delta_2} \end{bmatrix}$$

where $\Delta_1 \equiv Z(Z + Z')$ and $\Delta_2 \equiv Z'(Z + Z')$.

In the addition of two matrices (which must be of the same dimensions) each matrix element of the resultant matrix is simply the sum of the corresponding elements of the two component matrices. Hence

$$[y] = \{ZZ'(Z + Z')\}^{-1} \begin{bmatrix} (\frac{1}{2}Z'^2 + 2ZZ' + \frac{1}{2}Z^2) & -\frac{1}{2}(Z'^2 + Z^2) \\ -\frac{1}{2}(Z'^2 + Z^2) & (\frac{1}{2}Z'^2 + 2ZZ' + \frac{1}{2}Z^2) \end{bmatrix}$$

(7.31)

The twin-T network is said to be 'balanced' (see Fig. 7.8(b)) when the current flowing through a load connected to the output terminals BC is zero. Since, under these circumstances, v_2 is also zero we have from eqn (7.3) that the value of y_{21} must be zero. For this to be so we see from eqn (7.31) that we must have $Z'^2 = -Z^2$. It is usual to

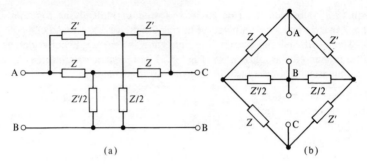

FIG. 7.8. Two ways of drawing the circuit diagram for a twin-T network.

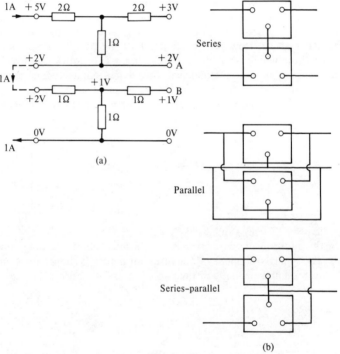

FIG. 7.9. (a) Two two-port networks with their input ports connected in series (see p. 179). (b) Interconnections of 'three-terminal' networks.

construct a twin-T network from resistors and capacitors since it is easy to make them variable; for $Z = R$, $Z' = -j/\omega C$, then $Z'^2 = -Z^2$ if $\omega RC = 1$. In this situation the network acts as a rejection filter at an angular frequency $\omega = (CR)^{-1}$. A useful practical feature of a twin-T network is that there is one terminal which is common to both the input and output ports so that one terminal of the source and load can be earthed. In contrast in a conventional bridge network the output port does not share a common terminal with the input port.

It must not be assumed automatically that two individual two-port networks retain their two port characteristics after being interconnected. For instance consider the two T-networks shown in Fig. 7.9(a) in which the input ports are connected in series; if the output ports were to be connected in series by connecting A and B, then the conditions at the input ports would be disturbed since the voltages at terminals A and B are not equal. This situation is exemplified by the fact that the sum of the z-matrices of the two-component T-networks does not equal the z-matrix for the combined network: Using eqns (7.13) we have

$$\begin{bmatrix} 3 & 1 \\ 1 & 3 \end{bmatrix} + \begin{bmatrix} 2 & 1 \\ 1 & 2 \end{bmatrix} \neq \begin{bmatrix} 9/2 & 5/2 \\ 5/2 & 9/2 \end{bmatrix} \text{ outputs connected in series.}$$

There are general rules, or validity tests, for interconnected two-port networks but it would not be very profitable to discuss them in detail here for two reasons:

(a) Many of the two-port networks in which we are interested have a common connection between an input and an output terminal (so-called 'three-terminal' networks) in which case the interconnections shown in Fig. 7.9(b) satisfy the validity tests; the twin-T network falls into this category.

(b) We will be concerned mainly with the cascade, or tandem, interconnection of two-port networks in which case this particular problem does not arise.

The relationships between the elements of the z-, y-, h-, and a-matrices of a two-port network can be derived easily, if somewhat tediously. For example consider the equations which define the z- and y-parameters (eqns (7.3) and (7.4)).

From eqns (7.3),

$$v_1 = \frac{i_2}{y_{21}} - \frac{y_{22} v_2}{y_{21}} \quad \text{and} \quad v_2 = \frac{i_1 - y_{11} v_1}{y_{12}}.$$

Hence,

$$v_1 \left(1 - \frac{y_{22} y_{11}}{y_{21} y_{12}} \right) = \frac{i_2}{y_{21}} - \frac{y_{22} i_1}{y_{21} y_{12}}$$

or

$$v_1 \frac{\{y_{12} y_{21} - y_{22} y_{11}\}}{y_{21} y_{12}} = \frac{-y_{22} i_1}{y_{21} y_{12}} + \frac{i_2}{y_{21}}$$

Thus
$$v_1 = \frac{y_{22}i_1}{\Delta_y} - \frac{y_{12}i_2}{\Delta_y} \tag{7.32}$$

where $\Delta_y \equiv y_{11}y_{22} - y_{12}y_{21}$.

On comparing eqn (7.32) with eqns (7.4) it can be seen that

$$z_{11} = \frac{y_{22}}{\Delta_y} \qquad z_{12} = \frac{-y_{12}}{\Delta_y}.$$

By using similar arguments the matrix parameter conversion table (Fig. 7.10) can be constructed.

The application of transmission parameters, in the case of the cascade connection of two-port networks, can be illustrated usefully by some simple examples (see Fig. 7.11). For the network of Fig. 7.11(a)

$$\begin{aligned} v_1 &= v_2 + i_2 Z \\ i_1 &= i_2 \end{aligned} \qquad [a] = \begin{bmatrix} 1 & Z \\ 0 & 1 \end{bmatrix}. \tag{7.33}$$

For the network of Fig. 7.11(b),
$$v_1 = v_2$$
$$\frac{(i_1 - i_2)}{Y} = v_2$$

or
$$\begin{aligned} v_1 &= v_2 \\ i_1 &= v_2 Y + i_2 \end{aligned} \qquad [a] = \begin{bmatrix} 1 & 0 \\ Y & 1 \end{bmatrix}. \tag{7.34}$$

	[z]		[y]		[h]		[a]	
[z]	z_{11}	z_{12}	$\frac{y_{22}}{\Delta_y}$	$\frac{-y_{12}}{\Delta_y}$	$\frac{\Delta_h}{h_{22}}$	$\frac{h_{12}}{h_{22}}$	$\frac{a_{11}}{a_{21}}$	$\frac{\Delta_a}{a_{21}}$
	z_{21}	z_{22}	$\frac{-y_{21}}{\Delta_y}$	$\frac{y_{11}}{\Delta_y}$	$\frac{-h_{21}}{h_{22}}$	$\frac{1}{h_{22}}$	$\frac{1}{a_{21}}$	$\frac{a_{22}}{a_{21}}$
[y]	$\frac{z_{22}}{\Delta_z}$	$\frac{-z_{12}}{\Delta_z}$	y_{11}	y_{12}	$\frac{1}{h_{11}}$	$\frac{-h_{12}}{h_{11}}$	$\frac{a_{22}}{a_{12}}$	$\frac{-\Delta_a}{a_{12}}$
	$\frac{-z_{21}}{\Delta_z}$	$\frac{z_{11}}{\Delta_z}$	y_{21}	y_{22}	$\frac{h_{21}}{h_{11}}$	$\frac{\Delta_h}{h_{11}}$	$\frac{-1}{a_{12}}$	$\frac{a_{11}}{a_{12}}$
[h]	$\frac{\Delta_z}{z_{22}}$	$\frac{z_{12}}{z_{22}}$	$\frac{1}{y_{11}}$	$\frac{-y_{12}}{y_{11}}$	h_{11}	h_{12}	$\frac{a_{12}}{a_{22}}$	$\frac{\Delta_a}{a_{22}}$
	$\frac{-z_{21}}{z_{22}}$	$\frac{1}{z_{22}}$	$\frac{y_{21}}{y_{11}}$	$\frac{\Delta_y}{y_{11}}$	h_{21}	h_{22}	$\frac{-1}{a_{22}}$	$\frac{a_{21}}{a_{21}}$
[a]	$\frac{z_{11}}{z_{21}}$	$\frac{\Delta_z}{z_{21}}$	$\frac{-y_{22}}{y_{21}}$	$\frac{-1}{y_{21}}$	$\frac{-\Delta_h}{h_{21}}$	$\frac{-h_{11}}{h_{21}}$	a_{11}	a_{12}
	$\frac{1}{z_{21}}$	$\frac{z_{22}}{z_{21}}$	$\frac{-\Delta_y}{y_{21}}$	$\frac{-y_{11}}{y_{21}}$	$\frac{-h_{22}}{h_{21}}$	$\frac{-1}{h_{21}}$	a_{21}	a_{22}

N.B. $\Delta_z = z_{11}z_{22} - z_{12}z_{21}$, etc.

FIG. 7.10. Matrix parameters: conversion table.

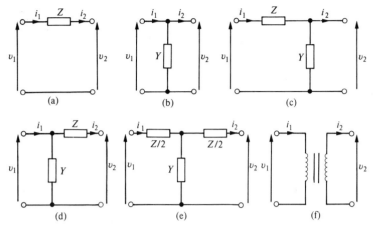

FIG. 7.11. Some two-port networks for which the transmission parameters are derived in the text.

For the network of Fig. 7.11(c), which is the resultant of the cascade connection of the previous two networks,

$$vv_1 = i_1 Z + \frac{(i_1 - i_2)}{Y}$$

$$v_2 = \frac{(i_1 - i_2)}{Y} \quad \text{or} \quad i_1 = v_2 Y + i_2.$$

So
$$v_1 = v_2(1 + YZ) + i_2 Z$$
$$i_1 = v_2 Y + i_2 \qquad [a] = \begin{bmatrix} (1 + YZ) & Z \\ Y & 1 \end{bmatrix}. \quad (7.35)$$

It should be noticed that this matrix is equal to the product of the previous two matrices, i.e.

$$\begin{bmatrix} 1 & Z \\ 0 & 1 \end{bmatrix} \times \begin{bmatrix} 1 & 0 \\ Y & 1 \end{bmatrix} = \begin{bmatrix} (1 + YZ) & Z \\ Y & 1 \end{bmatrix} \quad (7.36)$$

where use has been made of the matrix multiplication rule

$$\begin{bmatrix} a_{11} & a_{12} \\ a_{21} & a_{22} \end{bmatrix} \times \begin{bmatrix} b_{11} & b_{12} \\ b_{21} & b_{22} \end{bmatrix} = \begin{bmatrix} (a_{11}b_{11} + a_{12}b_{21}) & (a_{11}b_{12} + a_{12}b_{22}) \\ (a_{21}b_{11} + a_{22}b_{21}) & (a_{21}b_{12} + a_{22}b_{22}) \end{bmatrix}. \quad (7.37)$$

With regard to the convention which we have adopted for the sense of i_2 it can easily be checked, using the above type of procedure, that if i_2 is taken in the opposite sense to that which we have adopted then the matrices (7.33), (7.34), and (7.35) become

$$\begin{bmatrix} 1 & -Z \\ 0 & -1 \end{bmatrix}, \begin{bmatrix} 1 & 0 \\ Y & -1 \end{bmatrix} \quad \text{and} \quad \begin{bmatrix} (1 + YZ) & -Z \\ Y & -1 \end{bmatrix}, \text{respectively.}$$

Now

$$\begin{bmatrix} 1 & -Z \\ 0 & -1 \end{bmatrix} \times \begin{bmatrix} 1 & 0 \\ Y & -1 \end{bmatrix} = \begin{bmatrix} (1+YZ) & Z \\ -Y & 1 \end{bmatrix} \neq \begin{bmatrix} (1+YZ) & -Z \\ Y & -1 \end{bmatrix}.$$

However, if the matrix

$$\begin{bmatrix} 1 & 0 \\ 0 & -1 \end{bmatrix},$$

is interposed between the two matrices on the left-hand side of the equation then you can confirm that the correct resultant matrix is obtained.

It is a matter of taste as to which convention is adopted with regard to the sense of i_2 but we will continue here to use the convention specified in Fig. 7.6.

The transmission matrices for the networks of Fig. 7.11(d) and (e) can be derived by the methods used above and are

$$\begin{bmatrix} 1 & Z \\ Y & (1+YZ) \end{bmatrix} \tag{7.38}$$

and

$$\begin{bmatrix} (1+\tfrac{1}{2}YZ) & (Z+\tfrac{1}{4}YZ^2) \\ Y & (1+\tfrac{1}{2}YZ) \end{bmatrix}. \tag{7.39}$$

For the ideal transformer depicted in Fig. 7.11(f),

$$v_1 = \frac{n_1 v_2}{n_2}$$

$$i_1 = \frac{n_2 i_2}{n_1}$$

and so the transmission matrix is

$$\begin{bmatrix} \dfrac{n_1}{n_2} & 0 \\ 0 & \dfrac{n_2}{n_1} \end{bmatrix}. \tag{7.40}$$

As an example of the use of transmission parameters consider the circuit of Fig. 7.12(a) where a source of internal resistance R_0 is coupled to a load of resistance R_0 by a symmetrical T-network. The

Network analysis and synthesis   185

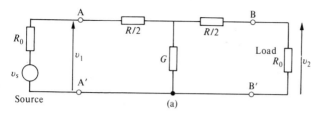

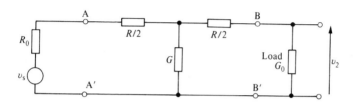

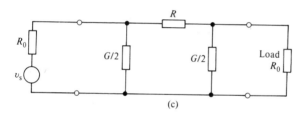

FIG. 7.12.   A matched attenuator network.

voltage at AA′, the accessible terminals of the source, is v_1 and the problem is to design an attenuator to give $v_2/v_1 = 1/n$ whilst arranging that the effective resistance presented to the source is still R_0. If the circuit is redrawn as in Fig. 7.12(b), where $G_0 \equiv 1/R_0$, then we can see that the two-port network with terminals AA′, BB′ has a transmission matrix $[a]$ given by

$$[a] = \begin{bmatrix} (1 + \tfrac{1}{2}RG) & (R + \tfrac{1}{4}GR^2) \\ G & (1 + \tfrac{1}{2}GR) \end{bmatrix} \times \begin{bmatrix} 1 & 0 \\ G_0 & 0 \end{bmatrix}$$

where we have applied eqns (7.39) and (7.34). Hence,

$$[a] = \begin{bmatrix} (1 + \tfrac{1}{2}RG + G_0R + \tfrac{1}{4}G_0GR^2) & 0 \\ G + G_0(1 + \tfrac{1}{2}GR) & 0 \end{bmatrix}. \quad (7.41)$$

Now if we consider eqn (7.28) in the light of this problem as specified by Fig. 7.12(b) we see that 'i_2' = 0 since the voltage v_2 across the load is imagined to be measured by a voltmeter of infinite resistance. Thus in eqn (7.28) $v_2/v_1 = 1/a_{11}$ and from eqn (7.41) this means that for $v_2/v_1 = 1/n$ we have

$$1 + \tfrac{1}{2}RG + G_0 R + \tfrac{1}{4}G_0 GR^2 = n$$

It is common practice to write such equations in terms of 'normalized' resistances (or impedances or admittances) and since $R_0 G_0 = 1$ it follows that

$$\frac{G}{G_0} \cdot \left(\frac{R}{R_0}\right)^2 + 2\frac{G}{G_0} \cdot \frac{R}{R_0} + 4\frac{R}{R_0} + 4(1-n) = 0. \quad (7.42)$$

Now we want a relationship between R and G such that the characteristic impedance of the symmetrical T-network is equal to the source and load resistance, namely R_0. Hence, we use eqn (7.20) in the form

$$R_0^2 = \frac{R}{G} + \frac{1}{4}R^2 \quad \text{or} \quad \frac{G}{G_0}\left(\frac{R}{R_0}\right)^2 + 4\frac{R}{R_0} - 4\frac{G}{G_0} = 0. \quad (7.43)$$

After substituting for (G/G_0) in eqn (7.42) from eqn (7.43) we find

$$\left(\frac{R}{R_0}\right)^2 (1+n) + 4\frac{R}{R_0} + 4(1-n) = 0.$$

On solving this quadratic equation for (R/R_0) we find

$$\frac{R}{R_0} = \frac{2(n-1)}{(n+1)} \quad \text{and thence} \quad \frac{G}{G_0} = \frac{(n^2-1)}{2n}. \quad (7.44\text{(a)})$$

For example, for $n = 2$ then $R/R_0 = 2/3$ and $G/G_0 = 3/4$.

An alternative way of realizing a matched symmetrical network to obtain the required attenuation factor $v_2/v_1 = 1/n$ is with a symmetrical Π-network as shown in Fig. 7.12(c); it is left as an exercise to show, by following an analysis analagous to that above, that the required relationships between G, R, and R_0 are

$$\frac{G}{G_0} = \frac{2(n-1)}{(n+1)} \quad \text{and} \quad \frac{R}{R_0} = \frac{(n^2-1)}{2n} \quad (7.44\text{(b)})$$

as we would expect from the principle of duality.

If identical matched T or Π-networks are cascaded then the combination remains matched to the characteristic impedance (and could be reduced to a single equivalent T- or Π-network). For the case of purely resistive networks such as we have just considered above the

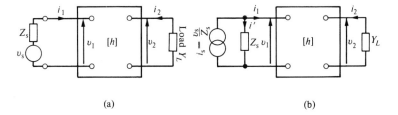

FIG. 7.13. A loaded two-port network with (a) a Thévenin equivalent of the source and (b) a Norton equivalent of the source.

over-all attenuation of the voltage, or current, would be n^m. A transmission line, usch as a coaxial cable, can be represented as a series of cascaded T- or Π-networks (see Chapter 8).

It is appropriate now to consider a loaded two-port network as depicted in Fig. 7.13; the network, which is represented by its h-parameters has its input port loaded by a source v_s of impedance Z_s and its output port loaded by an admittance Y_L. The defining equations for the h-parameters are

$$v_1 = h_{11}i_1 + h_{12}v_2 \tag{7.45}$$

$$i_2 = h_{22}i_1 + h_{22}v_2 \tag{7.46}$$

and also

$$v_1 = v_s - i_1 Z_s \tag{7.47}$$

$$v_2 = -i_2/Y_L. \tag{7.48}$$

From eqns (7.46) and (7.48)

$$i_1 = -v_2(h_{22} + Y_L)/h_{21} \tag{7.49}$$

and from eqns (7.45) and (7.47)

$$v_s = i_1(h_{11} + Z_s) + h_{12}v_2. \tag{7.50}$$

If we substitute for i_1 in eqn (7.50) from eqn (7.49) we then obtain

$$\frac{v_2}{v_s} = \frac{-h_{21}}{(h_{11} + Z_s)(h_{22} + Y_L) - h_{12}h_{21}} \tag{7.51}$$

or

$$\frac{i_2}{v_s} = \frac{h_{21}Y_L}{(h_{11} + Z_s)(h_{22} + Y_L) - h_{12}h_{21}}. \tag{7.52}$$

To determine the input impedance Z_{in} ($\equiv v_1/i_1$) of the loaded two-port network we substitute for v_2 in eqn (7.45) from eqn (7.49) and find

$$Z_{\text{in}} = h_{11} - \frac{h_{22}h_{21}}{(h_{22} + Y_L)} \qquad (7.53)$$

The output admittance $Y_0 \ (\equiv i_2/v_2)$ can be derived by setting $v_s = 0$ and using a similar method of analysis. However, it is useful to consider another approach. From eqns (7.45) and (7.47),

$$v_2 = \{v_s - (h_{11} + Z_s)i_1\}/h_{12}$$

and using eqn (7.46) to substitute for i_1 in terms of i_2, v_2 we find

$$v_2 = \frac{-h_{21}v_s}{\{h_{22}(h_{11} + Z_s) - h_{12}h_{21}\}} - \frac{(h_{11} + Z_s)(-i_2)}{\{h_{22}(h_{11} + Z_s) - h_{12}h_{21}\}}. \qquad (7.54)$$

We notice, by comparison with eqn (7.51), that the coefficient of v_s is the open-circuit voltage gain (i.e. the voltage gain with $Y_L = 0$), i.e.

$$\text{Open-circuit voltage gain} = \left(\frac{v_2}{v_s}\right)_{oc} = \frac{-h_{21}}{\{h_{22}(h_{11} + Z_s) - h_{12}h_{21}\}}. \qquad (7.55)$$

Under ideal voltage coupling conditions ($Z_s = 0$; $Y_L = 0$) then

$$\left(\frac{v_2}{v_1}\right)_{\text{ideal}} = \frac{-h_{21}}{\{h_{22}h_{11} - h_{12}h_{21}\}}. \qquad (7.56)$$

Also, by comparing eqn (7.54) with eqn (2.3), and bearing in mind the sense of i_2, we can see that the Thévenin equivalent of the output port has an impedance

$$Z_0 = \frac{(h_{11} + Z_s)}{\{h_{22}(h_{11} + Z_s) - h_{12}h_{21}\}} \qquad (7.57)$$

or an output admittance

$$Y_0 = h_{22} - \frac{h_{12}h_{21}}{(h_{11} + Z_s)}. \qquad (7.58)$$

Similar analyses can be used in an obvious way to obtain the equivalent expressions for the z-, y-, and a-parameter models.

From the Norton equivalent of the signal source (see Fig. 7.13(b))

$$i_1 = \frac{v_s}{Z_s} - i' \quad \text{or} \quad i_1 = i_s - i'$$

where $i_s \equiv v_s/Z_s$. Hence on using eqns (7.45) and (7.48),

$$i_1 \left(1 + \frac{h_{11}}{Z_s}\right) = i_s + \frac{h_{12}i_2}{Z_s Y_L}.$$

Network analysis and synthesis 189

If we now substitute for i_1 in eqn (7.46) from this last equation, and again use eqn (7.48) for v_2 then it follows that

$$\frac{i_2}{i_s} = \frac{h_{21}}{\left(1 + \frac{h_{11}}{Z_s}\right)\left\{1 - \frac{h_{12}h_{21}}{Z_s Y_L \left(1 + \frac{h_{11}}{Z_s}\right)} + \frac{h_{22}}{Y_L}\right\}}. \quad (7.59)$$

The short-circuit current gain ($Y_L = \infty$) is equal to

$$\left(\frac{i_2}{i_s}\right)_{s/c} = \frac{h_{21}}{\left(1 + \frac{h_{11}}{Z_s}\right)} \quad (7.60)$$

and under ideal current coupling conditions ($Z_s = \infty$; $Y_L = \infty$)

$$\left(\frac{i_2}{i_1}\right)_{ideal} = h_{21}. \quad (7.61)$$

7.4. Equivalent circuits for active devices

Active devices are circuit elements across the terminals of which there is a potential difference even if there is no current flowing (in the case of a voltage generator) or elements between whose terminals a current flows even when there is no potential difference (current generator). This is in contrast to a passive element between whose terminals there is a potential difference only when a current flows or vice versa (resistors, inductors, capacitors). The active devices of particular interest are transistors and thermionic valves.

Although it is not an aim of this text to give a detailed description of the equivalent circuits of active devices in general it is appropriate to pay some brief attention to the bipolar transistor. The latter is a device with three terminals which are called the base, emitter, and collector (see Fig. 7.14(a)); the relationships between the collector current I_C, collector–emitter voltage V_{CE}, base current I_B, and base–emitter voltage V_{BE} are shown schematically in Fig. 7.14(b)) and Fig. 7.14(c) for a so-called n–p–n bipolar transistor (N.B. these are the d.c. or static characteristics). A vital point to notice is that over a large part of the graph of I_C against V_{CE}, I_C is virtually independent of V_{CE} and, in the conventional notation, if V_{CE} is held constant then $\Delta I_c = \beta \Delta I_B$, or $\beta = \partial I_C / \partial I_B |_{V_{CE}}$. Also from the input characteristic the input resistance is given by the reciprocal of $\partial I_B / \partial V_{BE} |_{V_{CE}}$. The weak

190   Network analysis and synthesis

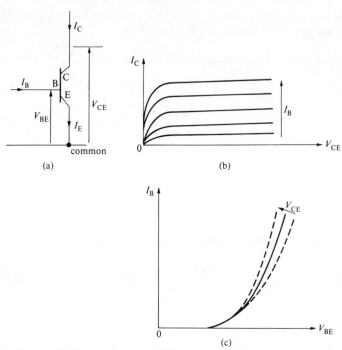

FIG. 7.14.   (a) The conventional symbol for a bipolar transistor (n–p–n type); B-base, C-collector, E-emitter. (b) The I_C versus V_{CE} characteristic (schematic) with I_B as parameter. (c) The I_B versus V_{BE} characteristic (schematic); note the weak dependence on the parameter V_{CE}.

dependence of I_C on V_{CE} for a given value of I_B can be represented by a shunt conductance of value $\partial I_C / \partial V_{CE}|_{I_B}$ connected between the output terminals. Finally the dependence of the input characteristic on the parameter V_{CE} can be represented by the value of $\partial V_{BE} / \partial V_{CE}|_{I_B}$; this means that for a change ΔV_{CE} in the output voltage a voltage $(\partial V_{BE}/V_{CE})_{\Delta V_{CE}}$ appears in the input circuit.

In summary we have

$$\text{input resistance} = \left.\frac{\partial V_{BE}}{\partial I_B}\right|_{V_{CE}} \qquad \text{reverse voltage transfer ratio} = \left.\frac{\partial V_{BE}}{\partial V_{CE}}\right|_{I_B}$$

$$\text{forward current gain} = \left.\frac{\partial I_C}{\partial I_B}\right|_{V_{CE}} \qquad \text{output conductance} = \left.\frac{\partial I_C}{\partial V_{CE}}\right|_{I_B}.$$

(7.62)

Network analysis and synthesis

For 'small' signal conditions, in which the voltage and current variations represented by ΔV_{BE}, ΔI_C, etc., are small enough so that the properties defined in eqn (7.62) above can be assumed to be constant, then it is possible and useful to define h-parameters for the transistor by making a direct comparison with eqns (7.5). If we identify small changes in currents and voltages such as ΔV_{BE}, ΔI_C... with the amplitudes of sinusoidally varying quantities v_{BE}, i_c... then eqns (7.5) can be rewritten as;

$$v_{BE} = \left(\frac{\partial V_{BE}}{\partial I_B}\right) i_B + \left(\frac{\partial V_{BE}}{\partial V_{CE}}\right) v_{CE} \quad \text{or} \quad v_{BE} = h_{ie} i_B + h_{re} v_{CE} \quad (7.63)$$

$$i_C = \left(\frac{\partial I_C}{\partial I_B}\right) i_B + \left(\frac{\partial I_C}{\partial V_{CE}}\right) v_{CE} \quad \text{or} \quad i_C = h_{fe} i_B + h_{oe} v_{CE} \quad (7.64)$$

The suffix notation 'ie', 're', etc. is used to denote the fact that these h-parameters refer to a transistor connected in the so-called common-emitter configuration where the emitter terminal is common to the input and output ports. For the common-collector and common-base configurations, whose names should be self-explanatory now, the corresponding h-parameters have suffixes 'ic', 'rc', and 'ib', 'rb' respectively; it is important to note that the h-parameters have different values, in general, in the different configurations.

Typical values for the common-emitter h-parameters for a transistor for use at audio and low radio-frequencies are:

$$h_{ie} \approx 5\,k\Omega \qquad h_{re} \approx 2 \times 10^{-4}$$
$$h_{fe} \approx 400 \qquad h_{oe} \approx 5 \times 10^{-5}\,S. \qquad (7.65)$$

The h-parameters of a transistor could be obtained from the graphical static characteristics, but not with any great accuracy; it is usual to determine the parameters from measurements made under open- and short-circuit conditions as specified in Section 7.1 using alternating voltages and currents of small enough amplitudes.

In a typical working situation such as in the single-stage common-emitter amplifier shown in Fig. 7.15(a) the transistor is coupled to a source and to a load by coupling capacitors C_1, say, and also there will be a biassing network (R_1, R_2). If the operating frequency is high enough so that the reactances of the coupling capacitors, and of the emitter resistor by-pass capacitor C_2, are negligible, as also is the impedance of the power supply, then the a.c. equivalent circuit is as shown in Fig. 7.15(b) where R_B is equal to the parallel combination of R_1 and R_2. Using eqn (7.63) together with

$$v_{BE} = (i_1 - i_B) R_B \qquad (7.66)$$

192   Network analysis and synthesis

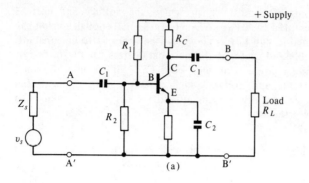

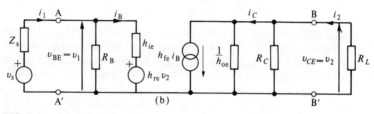

FIG. 7.15. (a) A single-stage common-emitter transistor amplifier. (b) The a.c. equivalent circuit.

it follows that

$$v_{BE} = \frac{h_{ie} i_1}{\left(1 + \dfrac{h_{ie}}{R_B}\right)} + \frac{h_{re}}{\left(1 + \dfrac{h_{ie}}{R_B}\right)} v_{CE} \qquad (7.67)$$

So for the two-port network having ports AA', BB',

$$h_{11} = \frac{h_{ie}}{\left(1 + \dfrac{h_{ie}}{R_B}\right)} \quad \text{and} \quad h_{12} = \frac{h_{re}}{\left(1 + \dfrac{h_{ie}}{R_B}\right)} \qquad (7.68)$$

Now

$$v_{CE} = (i_2 - i_c)R_C$$

and so after substituting for i_c from eqn (7.64) and then for i_B from eqn (7.66) we have

$$i_2 = h_{fe} i_1 - \frac{h_{fe}}{R_B} v_1 + \left(h_{oe} + \frac{1}{R_C}\right) v_2$$

where now we have put $v_{BE} = v_1$ and $v_{CE} = v_2$ to fall in with our notation for two-port networks. If we now substitute for v_1 in this equation from eqns (7.5) and then substitute for h_{11}, h_{12} from eqns (7.68) in the equation so obtained then

$$i_2 = \left\{ h_{fe} - \frac{h_{fe}h_{ie}}{R_B\left(1 + \frac{h_{ie}}{R_B}\right)} \right\} i_1 + \left\{ h_{oe} - \frac{h_{fe}h_{re}}{R_B\left(1 + \frac{h_{ie}}{R_B}\right)} + \frac{1}{R_C} \right\} v_2.$$
(7.69)

Thus

$$h_{21} = h_{fe} - \frac{h_{fe}h_{ie}}{R_B\left(1 + \frac{h_{ie}}{R_B}\right)} \qquad h_{22} = h_{oe} - \frac{h_{fe}h_{re}}{R_B\left(1 + \frac{h_{ie}}{R_B}\right)} + \frac{1}{R_C}.$$
(7.70)

The general expressions of eqns (7.68) and (7.70) are rather cumbersome and so it is useful to examine a practical situation. For a supply voltage of 10 V, R_1 and R_2 could have the values $100\,k\Omega$ and $10\,k\Omega$ respectively so that $R_B \approx 10\,k\Omega$; also we will take R_C to be $1\,k\Omega$. If we then use the transistor h-parameters of eqns (7.65) we find;

$$h_{11} \approx \tfrac{2}{3}h_{ie} \qquad h_{12} \approx \tfrac{2}{3}h_{fe}$$
$$h_{21} \approx \tfrac{2}{3}h_{fe} \qquad h_{22} \approx 1/R_C.$$

This analysis serves as a reminder that we should always be aware of the loading effect of one network, or circuit element, upon another network. (see Section 3.8.2.).

It must be emphasized that the h-parameters, as we have described them, have been derived in a relatively crude way; improved equivalent circuit can be derived which are based on more sophisticated physical models. Also the equivalent circuits which are appropriate to the operation of transistors at high frequencies assume radically different forms to those which are used at audio-frequencies.

Readers will find that as practical experience in the field is gained they will acquire quickly the facility of making sensible approximations to equations such as eqns (7.68) and (7.70).

7.5. Basic features of linear amplifiers

For the purposes of this discussion an amplifier will be described as a two-port network, containing one or more active elements, which will produce in a load connected to the output port an amplified version of a signal applied to the input port (see Fig. 7.16(a)). Any change in a voltage and/or current constitutes a signal but in this context it will be assumed that the input signal is a sinusoidal voltage or current.

The gain of an amplifier can be expressed as v_2/v_1, i_2/i_1, or as P_2/P_1 which is equal to $(v_2 i_2)/(v_1 i_1)$, i.e. as voltage, current, or power gains respectively; for a linear amplifier $v_2 \propto v_1$, $i_2 \propto i_1$, $P_2 \propto P_1$. The

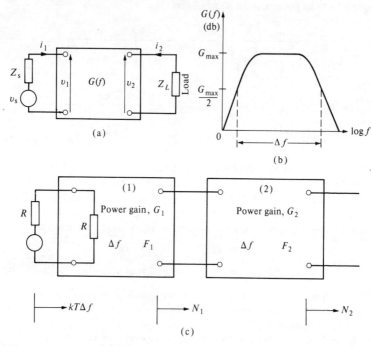

FIG. 7.16. (a) A block diagram of an amplifier having a frequency-dependent power gain $G(f)$ (see (b)). (c) The model for calculating the noise figure of two cascaded amplifier stages.

frequency dependence of the above ratios is of great practical importance and usually, for convenience, the logarithm of the ratio is expressed as a function of the logarithm of the frequency. In this event the gain is commonly expressed in 'decibels' (db) where

$$\text{Power gain (db)} \equiv 10 \log P_2/P_1. \tag{7.71}$$

Notice that if the input and output signals are developed across resistances R_{in} and R_L respectively then

$$\text{Power gain} = 10 \log \{(v_2^2/R_L)/(v_1^2/R_{in})\}$$
$$= 20 \log (v_2/v_1) + 10 \log (R_{in}/R_L). \tag{7.72}$$

It is common practice to refer to *voltage* gains (and *current* gains) in decibels, that is ignoring the second term on the right-hand side of eqn (7.72); obviously in this case a decibel value does not correspond to a power ratio. However, if a voltage or current gain is expressed in relation to some reference value then the term $10 \log (R_{in}/R_L)$

disappears. For example,

$$\frac{\text{power gain at frequency } f_0}{\text{power gain at frequency } f} = \left(\frac{v_2^2 R_{\text{in}}}{R_L v_1^2}\right)_{f_0} \bigg/ \left(\frac{v_2^2 R_{\text{in}}}{R_L v_1^2}\right)_{f} = \left(\frac{v_2^2}{v_1^2}\right)_{f_0} \bigg/ \left(\frac{v_2^2}{v_1^2}\right)_{f}$$

and, assuming that the gain at f is lower than that at f_0, it would be said that the power (or voltage or current) gain at frequency f is 'down' by N db compared with that at f_0 where

$$N = 20 \log \{(v_2/v_1)_{f_0}/(v_2/v_1)_f\} = 20 \log \{(i_2/i_1)_{f_0}/(i_2/i_1)_f\}. \quad (7.73)$$

The gain of an amplifier is never independent of frequency (see Fig. 7.16(b)); so an important feature of an amplifier is its (frequency) 'bandwidth' which is usually defined as that frequency range lying between the upper and lower 'cut-off' frequencies at which the voltage (or current) gain is $1/\sqrt{2}$ of the maximum gain, or midband gain, whichever term is appropriate. The power gain is $1/2$ of the maximum power gain at these cut-off frequencies. Notice that at the cut-off frequencies, as just defined, the amplifier gain (voltage, or current, or power) is three decibels (3 db) 'down' on the maximum gain.

The values of the input and output impedances of an amplifier, (v_1/i_1) and (v_2/i_2) respectively, are of great practical significance and, broadly speaking, two situations can be distinguished.

Firstly, one or both of the input and output impedances can be matched for maximum power transfer from a source or to a load respectively. In telephony a repeater amplifier has to be matched to the circuits carrying the incoming and outgoing signals because of the limited available signal power; similarly in radio receivers it is usual to arrange for maximum power transfer between the aerial and the input of the receiver. As another example, in audio-amplifier systems the output of the amplifier is matched to the impedance of the loudspeaker.

Secondly, there are situations where voltage or current amplification is important (i.e. 'voltage coupling' and 'current coupling' as discussed in Section 3.8.2), and matching for maximum power transfer is unnecessary; for example a C.R.O. is used essentially as a voltmeter and so should have a very high input impedance in order to be versatile. Also since the outputs of its amplifiers feed into high impedance loads (the deflector plates of the cathode ray tube) power matching is again unnecessary.

So far no mention has been made of the coexistence with the signal of unwanted signals or 'noise'. Strictly speaking any unwanted signal could be classified as noise but, in practice, two broad categories of undesired signals may be distinguished. On the one hand there are such phenomena as 'hum' and 'pick-up' which arise from the amplifier's power supply and from external sources respectively. On the other hand all amplifiers introduce noise themselves and it is in this second category that attention will be focused; hum and pick-up signals can be largely eliminated by suitable filter circuits and by screening the amplifier and the connecting leads (see Section 10.5.1).

A detailed account of the many sources of noise in amplifiers is out of place here but it can be pointed out that the random thermal motion of the conduction electrons in a resistor is one well known example of noise arising from the statistical fluctuations which are occurring all the time in physical systems. The noise power generated in a resistor is distributed uniformly across the spectrum of frequencies (so-called 'white' noise) with a mean spectral density of $4\,kT$ watts per unit of frequency bandwidth. Another source of white noise which is important in thermionic valves is 'shot' noise which is associated with the random nature of the process of electron emission from heated filaments. Flicker noise is important in transistors at frequencies of a few kilohertz and below.[†]

The noise characteristics of an amplifier can be expressed in terms of a noise figure F which is defined through

$$F \equiv \frac{\text{(available input signal power)/(available input noise power)}}{\text{(available output signal power)/(available output noise power)}}$$

(7.74)

where it is usual to assume that the only source of input noise is thermal noise in the resistance of the signal source. The available noise power from a resistor R is equal to the power that would be delivered to a matched load, also of resistance R, and is equal to $kT\Delta f$ for a bandwidth of Δf Hz. Hence from eqn (7.74)

$$F = \text{(available output noise power)}/(GkT\Delta f) \qquad (7.75)$$

where G is the power gain of the amplifier. The form of this expression for F means that the available output noise power is converted to an equivalent noise power at the input (by dividing by the power gain) and is then compared with the available thermal noise power. Obviously the minimum possible value for F is unity and in practice F is greater than this. The noise figure is often expressed in decibels.

It is of interest to consider the output noise power from two stages of amplification as is shown in Fig. 7.16(c) and to determine the overall noise figure F_{12}. From eqn (7.75)

$$N_1 = kT\Delta f G_1 F_1 \qquad (7.76)$$

[†] It is assumed that uncorrelated noise signals combine as the sum of the powers associated with the various sources of noise.

and
$$N_{12} = kT\Delta f G_1 G_2 F_{12}. \qquad (7.77)$$

Eqn (7.76) can for the first stage be rewritten as

$$N_1 = kT\Delta f G_1 + (F_1 - 1)kT\Delta f G_1$$

which emphasizes (for any stage) that the total noise power at the output is equal to the input noise power multiplied by the power gain of the stage together with the noise added by the stage itself. On applying this principle to the second stage we obtain an alternative expression for N_{12} viz.

$$N_{12} = kT\Delta f G_1 F_1 G_2 + (F_2 - 1)kT\Delta f G_2. \qquad (7.78)$$

$$\left\{ \begin{matrix} \text{input} \\ \text{noise} \end{matrix} \times G_2 \right\} \qquad \qquad \left\{ \begin{matrix} \text{noise} \\ \text{added} \end{matrix} \right\}$$

On comparing eqns (7.77) and (7.78) it follows that for two cascaded stages
$$F_{12} = F_1 + (F_2 - 1)/G_1. \qquad (7.79)$$

For further amplification stages in cascade $F = F_1 + (F_2 - 1)/G_1 + (F_3 - 1)/G_1 G_2 + \ldots$ and the most important feature is that, providing $G_1, G_1 G_2$, etc. are large enough, the noise figure of the first stage (or 'pre-amplifier') is dominant. A great deal of effort has gone into devising active devices and associated circuitry to provide low-noise amplifiers.

The signal power to a typical radio receiver might be $\sim 10^{-12}$ watt and the output power, to a loudspeaker say, will be ~ 1 watt; thus the over-all power gain is in the region of 120 db. The basic source of noise is the thermal noise in the aerial and a typical noise figure for a receiver not employing, such ultra-sophisticated low-noise devices as masers and parametric amplifiers, is $F = 5$ or 7 db (10 log 5 = 6.990). At low radio- and audio-frequencies externally generated unwanted signals, such as hum and pick-up of various forms, are usually much more serious than 'statistical' noise.

7.6. Excitations, system functions, and responses

One of the most fascinating features of the theory of electrical networks is the fact that it is possible, generally speaking, to synthesize a network which will have a desired response to a specified excitation. Network synthesis, particularly with regard to the design of filters (although all networks may be thought of as filters in a general sense), has become an exceedingly important aspect of electronics in general and in telecommunications and radar in particular; computer-aided

198 Network analysis and synthesis

design is now a well established technique which is available to the designer. The aim of this section, and of those immediately following, is to describe the basic principles on which the sophisticated procedures of network synthesis rely.

Before discussing the synthesis of networks it is necessary to say something about the general properties of the mathematical functions which describe the excitations and responses of networks. Although in the familiar context in which filters are used a steady-state situation is assumed, in which transient currents and voltages have decayed to negligible proportions, let us represent the excitation of a network in the general form Ae^{st} where, as before, s is the complex frequency. As we saw in Section 6.5 the most important excitations, namely exponential decay and growth, step function, damped and growing sinusoids, and constant amplitude sinusoids, can be represented by making s real, zero, complex, and imaginary, respectively. Let us therefore represent the system function H of a network as a function of s also, i.e. $H(s)$. Now although we have not used the term as such, we have, in fact, determined and/or used many system functions in the earlier sections of this text. For instance consider the networks shown in Fig. 7.17. For the network of Fig. 7.17(a) we have, using the rules and notation developed in Section 6.5, that

$$V(s) = I(s)\frac{R/sC}{(R + 1/sC)}$$

or the system function $H(s)$ is given by

$$H(s) \equiv \frac{V(s)}{I(s)} = \frac{1}{C\left(s + \dfrac{1}{CR}\right)} \quad (7.80(a))$$

It is left as an exercise to show that for the situations (b), (c), and (d) of Fig. 7.17 the system functions $H(s)$ are

$$H(s) \equiv \frac{V(s)}{E(s)} = \frac{1}{CR(s + 1/CR)} \quad (7.80(b))$$

$$H(s) \equiv \frac{I(s)}{E(s)} = \frac{1}{L(s + R/L)} \quad (7.80(c))$$

$$H(s) \equiv \frac{V(s)}{E(s)} = \frac{1}{LC\left(s^2 + \dfrac{Rs}{L} + \dfrac{1}{LC}\right)} \quad (7.80(d))$$

or

$$H(s) = \frac{1}{LC(s - \alpha_1)(s - \alpha_2)}$$

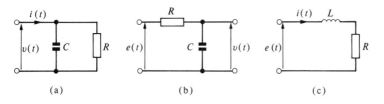

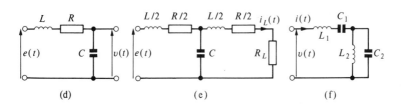

FIG. 7.17. Some simple networks illustrating a variety of excitations and responses.

where

$$\begin{matrix}\alpha_1\\ \alpha_2\end{matrix} \equiv -\frac{R}{2L} \pm j\sqrt{\left(\frac{1}{LC} - \frac{R^2}{4L^2}\right)}.$$

For the two-port network of Fig. 7.17(e) it can be shown, rather tediously, that

$$H(s) \equiv \frac{I_L(s)}{E(s)}$$

$$= \frac{4}{CL^2 s^3 + 2LC(R + R_L)s^2 + (4L + R^2 C + RR_L C)s + 4(R + R_L)}$$
(7.80(e))

and, for the network of Fig. 7.17(f),

$$H(s) \equiv \frac{V(s)}{I(s)}$$

$$= [L_1 C_2 s^4 + (C_2/C_1 + L_1/L_2 + 1)s^2 + 1/L_2 C_1]/(C_2 s^3 + s/L_2).$$
(7.80(f))

The reader will have noticed that system functions can be defined in a number of different ways depending on whether the specified responses and excitations are voltages or currents.

If the excitation is a current source and the response is a voltage, with both the excitation and the response being measured between the same pair of terminals of the network, then the system function is a *driving point impedance* (Fig. 7.17(a)); if the current and voltage are interchanged then we have a *driving point admittance* (Fig. 7.17(c)). Since it turns out that the mathematical descriptions for dual networks are identical, with voltages and currents interchanged, it is common practice to use the embracing term driving point *immittance*.

A system function in which the excitation and response are measured at different ports is called a transmittance; it could be a transfer admittance (Fig. 7.17(e)) or a transfer impedance, a voltage ratio transfer function (Fig. 7.17(d)) or a current ratio transfer function.

Now the reason for giving the cumbersome expressions of eqns (7.80(e)) and (7.80(f)) was to illustrate the point, which can be proved quite generally, that all system functions for physically realizable networks are rational functions which can be written as the quotient of two polynomials in s. So we can write $H(s)$ generally as

$$H(s) = \frac{a_p s^p + a_{p-1} s^{p-1} + \ldots a_1 s + a_0}{b_q s^q + b_{q-1} s^{q-1} + \ldots b_1 s + b_0}$$

$$= \frac{N_H(s)}{b_q s^q + b_{q-1} s^{q-1} + \ldots b_1 s + b_0}$$

where $N_H(s)$ represents the polynomial of the numerator. Now such polynomials can be factorized, although the process may be tedious and the availability of a programmed computer welcomed if p and $q \geqslant 4$; in particular the polynomial in the denominator of the last equation can be written in term of its factors and

$$H(s) = \frac{N_H(s)}{(s - s_{H_1})(s - s_{H_2})(s - s_{H_3}) \ldots (s - s_{H_q})} \quad (7.81)$$

The quantities $s_{H_1}, s_{H_2}, \ldots$, etc. are called the roots of the polynomial and in this context are known as the poles of $H(s)$ since $H(s) \to \infty$ for $s \to s_{H_1}, s_{H_2}, \ldots$, etc. Further it can be seen from Table 6.1 (and from more extensive tables of Laplace transformations[†]) that the Laplace transformations of excitation functions are also ratios of polynomials in s. Hence, for reasons identical to those given just above we can write

$$E(s) = \frac{N_E(s)}{(s - s_{E_1})(s - s_{E_2})(s - s_{E_3}) \ldots}$$

and so

[†] Holbrook, J. G. *Laplace transforms for electronic engineers* (2nd edn). Pergamon Press, Oxford (1960).

Network analysis and synthesis

$$R(s) = H(s)E(s) = \frac{N(s)}{(s - s_{E_1})(s - s_{E_2}) \dots (s - s_{H_1})(s - s_{H_2}) \dots}.$$
(7.82)

A quotient such as the right-hand side of eqn (7.82) can be expanded by the method of partial fractions (see Appendix 2) so that $R(s)$ can be expressed as a *sum* of terms

$$R(s) = \frac{K_{E_1}}{(s - s_{E_1})} + \frac{K_{E_2}}{(s - s_{E_2})} + \dots + \frac{K_{H_1}}{(s - s_{H_1})} + \frac{K_{H_2}}{(s - s_{H_2})} + \dots$$
(7.83)

where the $K_{E_1}, \dots K_{H_1} \dots$ are constants for a particular situation. By using the inverse Laplace transformation of each term of the right-hand side of eqn (7.83) the time-domain response $r(t)$ can be obtained. It can be seen therefore that the response contains a term for each pole of the excitation and a term for each pole of the system function. As the two examples given below will indicate, the terms arising from the system function represent the natural response of the system and the terms arising from the excitation function represent the forced response, i.e. these two sets of terms arise from the 'complementary function' and the 'particular integral', respectively, in the solution of the integro-differential equation which describes the excited network (see Section 3.1). Although the foregoing general discussion has been kept brief it has been rather formal and it is more interesting if the general point is illustrated by examining some relatively simple situations.

Consider the circuit shown in Fig. 7.18(a) where the network is excited by a step function current source of unit amplitude. In the s-domain the response, which is the indicated voltage, can be shown easily to be given by

$$V(s) = \frac{1}{s} \cdot \frac{(s - 2s_1)}{(s - s_1)} \cdot \frac{R}{2}$$
(7.84)

where $s_1 \equiv -(2RC)^{-1}$; i.e. the response is given by the product of the excitation $(I(s) = 1/s)$ and the system function (in this case a transfer impedance).

On using the partial fraction technique to convert the right-hand side of eqn (7.84) into a *sum* of functions of s which are given in tables of Laplace transformations we have

$$\frac{(s - 2s_1)}{s(s - s_1)} = \frac{A}{s} + \frac{B}{(s - s_1)}$$

or

$$(s - 2s_1) = A(s - s_1) + B_s$$
$$= s(A + B) - As_1.$$

Whence $(A + B) = 1$ and $As_1 = 2s_1$ and so $A = 2$; $B = -1$. Thus,

$$V(s) = \frac{R}{2}\left\{\frac{2}{s} - \frac{1}{(s-s_1)}\right\}.$$

On making the inverse Laplace transformation,

$$v(t) = \underset{\substack{\text{step}\\ \text{(forced response)}}}{Ru(t)} - \underset{\substack{\text{damped exponential}\\ \text{(natural response)}}}{\frac{R}{2} \cdot e^{-t/2CR}}. \qquad (7.85)$$

This result shows that the total response in the time domain consists of a term arising from the excitation (the forced response) plus a term which is characteristic of the natural response of the network alone.

As another example which is worth examining in detail consider the circuit shown in Fig. 7.19 which can be used to examine the response of an $L-C-R$ network to shock excitation by a square wave

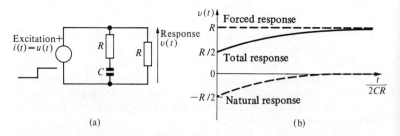

FIG. 7.18. The response (b) to a step function excitation applied to the $C-R$ network shown in (a).

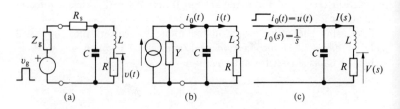

FIG. 7.19. A representation of a shock excited $L-C-R$ network.

signal; in practice the signal source will be usually a voltage generator v_g of low output impedance Z_g which must be decoupled from the L–C–R network by a large resistor R_s in order not to damp out the resonance. So, to a reasonable approximation, the generator can be transformed to a current generator having a very small admittance Y, i.e. to an ideal current generator for practical purposes. Further, the square wave form, of unit amplitude say, can be decomposed into a positive-going step plus a delayed negative-going step; so, in determining the response $v(t)$ we need concern ourselves only with a unit step function excitation. Now,

$$I(s) = I_0(s) \frac{1/sC}{(R + sL + 1/sC)}$$

where $I_0(s) = 1/s$ and so, since $V(s) = RI(s)$ we find that

$$V(s) = \frac{R}{LC} \cdot \frac{1}{s\left(s^2 + s\frac{R}{L} + \frac{1}{LC}\right)}$$

or

$$V(s) = \frac{R}{LC} \cdot \frac{1}{s(s - s_1)(s - s_2)} \tag{7.86}$$

where

$$\begin{matrix} s_1 \\ s_2 \end{matrix} = -\frac{R}{2L} \pm j\sqrt{\left(\frac{1}{LC} - \frac{R^2}{4L^2}\right)} \equiv (\alpha \pm j\beta), \text{ say,} \tag{7.87}$$

are the roots of the polynomial (a quadratic) in the demoninator of eqn (7.86).

Using the partial fraction technique again,

$$\frac{1}{s(s - s_1)(s - s_2)} = \frac{A}{s} + \frac{B}{(s - s_1)} + \frac{D}{(s - s_2)}$$

whence

$$A = \frac{1}{s_1 s_2}; \qquad B = \frac{1}{s_1(s_1 - s_2)}; \qquad D = \frac{1}{s_2(s_2 - s_1)}$$

so that

$$V(s) = \frac{R}{LC}\left\{\frac{1}{(s_1 s_2)s} + \frac{1}{s_1(s_1 - s_2)(s - s_1)} - \frac{1}{s_2(s_1 - s_2)(s - s_2)}\right\}. \tag{7.88}$$

Now

$$s_1 s_2 = \alpha^2 + \beta^2 \qquad s_1(s_1 - s_2) = j\,2\beta(\alpha + j\beta)$$
$$(s_1 - s_2) = j\,2\beta \qquad s_2(s_1 - s_2) = j\,2\beta(\alpha - j\beta).$$

So on making the inverse Laplace transformations

204 Network analysis and synthesis

$$v(t) = \frac{R}{LC}\left\{\frac{u(t)}{(\alpha^2+\beta^2)} + \frac{e^{s_1 t}}{j2\beta(\alpha+j\beta)} - \frac{e^{s_2 t}}{j2\beta(\alpha-j\beta)}\right\}$$

$$= \frac{R}{LC}\left\{\frac{u(t)}{(\alpha^2+\beta^2)} - \frac{(\beta+j\alpha)e^{s_1 t}}{2\beta(\alpha^2+\beta^2)} - \frac{(\beta-j\alpha)e^{s_2 t}}{2\beta(\alpha^2+\beta^2)}\right\}$$

Now since $(\alpha^2+\beta^2) = 1/LC$ we have

$$v(t) = R\left\{u(t) - \frac{(\beta+j\alpha)e^{(\alpha+j\beta)t}}{2\beta} - \frac{(\beta-j\alpha)e^{(\alpha-j\beta)t}}{2\beta}\right\}. \quad (7.89)$$

Earlier we defined β as

$$\beta \equiv \sqrt{\left(\frac{1}{LC} - \frac{R^2}{4L^2}\right)} = \frac{1}{\sqrt{LC}}\sqrt{\left(1 - \frac{R^2 C}{4L}\right)}.$$

Now in Section 4.1 we defined the resonance angular frequency ω_0 and the quality factor Q for an L–C–R circuit through $\omega_0 = (LC)^{-1/2}$ and $Q = \omega_0 L/R$ respectively. Thus

$$\beta = \omega_0\sqrt{\left(1 - \frac{1}{4Q^2}\right)} \qquad \alpha = -\frac{R}{2L}. \quad (7.90)$$

If $1/(4Q^2) < 1$ (i.e. $Q > 1/2$), then β is real (or, alternatively, the roots s_1, s_2 are complex) and the expression of eqn (7.89) for $v(t)$ takes the form of a step function plus damped oscillatory terms. For instance if $Q \geqslant 5$ then $1/(4Q^2) \leqslant 10^{-2}$ and so $\beta \approx \omega_0$ and $\alpha/\beta \approx -1/2Q$ so that we have

$$v(t) = R\left[u(t) - e^{\alpha t}\left\{\frac{e^{j\beta t} + e^{-j\beta t}}{2}\right\} + \frac{j\alpha}{2\beta}\{e^{j\beta t} - e^{-j\beta t}\}\right]$$

$$= R\left[u(t) - e^{\alpha t}\left\{\cos\beta t - \frac{\alpha}{\beta}\sin\beta t\right\}\right]$$

and finally, on substituting for α, β, we obtain

$$v(t) = Ru(t) - e^{-Rt/2L}\left\{\cos\omega_0 t + \frac{1}{2Q}\sin\omega_0 t\right\}. \quad (7.91)$$

For $Q \geqslant 10$ say then the term in $\sin\omega_0 t$ will have a small amplitude compared to the $\cos\omega_0 t$ term and the over-all response will be of the form of Curve (i) in Fig. 7.20. It should be noted again that eqns (7.89) and (7.91) for the response $v(t)$ are in the form of the sum of the forced response and the natural response.

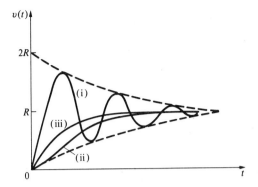

FIG. 7.20. The response of the $L-C-R$ network of Fig. 7.19(c), to a step function excitation: (i) underdamped; (ii) overdamped, (iii) critically damped.

If $1/(4Q^2) > 1$, then β is imaginary and s_1, s_2 are real, not complex. So if we now write $\beta = j\gamma$ where $\gamma \equiv \omega_0 \sqrt{\{1/(4Q^2) - 1\}}$ then eqn (7.89) becomes

$$v(t) = R\left[u(t) - e^{\alpha t}\left\{\cosh \gamma t - \frac{\alpha}{\gamma} \sin h\gamma t\right\}\right]. \quad (7.92)$$

In this situation the response is non-oscillatory and is depicted as Curve (ii) in Fig. 7.20; the network is said to be 'overdamped'. In an analogous mechanical system, for example the suspension system of a galvanometer or a moving-coil meter, the term 'dead-beat' is often used to describe this condition.

Finally if $1/(4Q^2) = 1$, then $s_1 = s_2 = s$, say, and eqn (7.86) becomes

$$v(s) = \frac{R}{LC} \cdot \frac{1}{s(s-\alpha)^2}. \quad (7.93)$$

It is left as an exercise to express the right-hand side of this equation as partial fractions and hence to show that

$$v(t) = R\left\{u(t) - e^{-Rt/2L} - \frac{Rt}{2L} \cdot e^{-Rt/2L}\right\}. \quad (7.94)$$

Again the response is non-oscillatory and is depicted as Curve (iii) in Fig. 7.20; the network is said to be 'critically damped'. The suspension systems of galvanometers and moving-coil meters are usually arranged to be slightly less than critically damped; the consequent small degree of 'over shoot' is an aid to estimating what the final equilibrium position will be.

This detailed analysis of the response of an $L-C-R$ network has

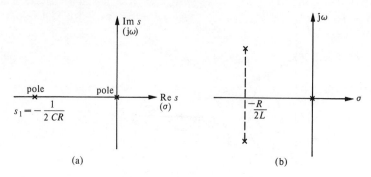

FIG. 7.21. (a) The poles of the function $V(s)$ of eqn (7.84). (b) The poles of the function $V(s)$ of eqn (7.86).

revealed the great power of the technique of using Laplace transformations; with practice the technique becomes purely routine. The precise response of a system from $t = 0$ to infinity can be obtained, if necessary, although in particular cases it may be possible to make simplifying approximations; for instance exponential terms can be replaced by their series expansions and the significant terms, only, retained.

The poles of response and system functions can be displayed conveniently in the s-domain; for example the poles of the function of eqns (7.84) and of (7.86) are displayed in Fig. 7.21(a) and (b) respectively. There are two very important general points which are demonstrated in these two diagrams. Firstly, complex poles always occur in pairs, one pole being the conjugate complex of the other. Secondly, for passive networks the poles lie on the left-hand half-plane, including the imaginary axis; remember (see Fig. 7.6) that points in the right-hand half-plane in the s-domain correspond to functions whose amplitudes are increasing with time and this is impossible, physically. in a passive network. Notice, also that the farther the poles are away from the imaginary axis the more rapidly damped is the natural response.

The relationship between a system function and the response of the network to an excitation in the form of an impulse $\delta(t)$ is of considerable interest (see Fig. 6.5 for the definition of the 'delta' function $\delta(t)$). Since $R(s) = \mathsf{E}(s)H(s)$ and $\mathsf{E}(s) = \mathscr{L}[\delta(t)] = 1$ it follows that in this case $R(s) = H(s)$. Hence the impulse response in the time domain $\{\mathscr{L}^{-1}[R(s)]\}$ is given by the inverse Laplace transformation of the system function. Further, since the 'step' response is the integral of the impulse response we can see, by using Table 7.1 that

$$\text{Unit step response} = \mathscr{L}^{-1} \frac{H(s)}{s}$$

and similarly, since the 'ramp' response is the integral of the 'step' response, that
$$\text{Unit ramp response} = \mathscr{L}^{-1}\, \frac{H(s)}{s^2}$$

Let us return our attention now to the general form of the system function $H(s)$ given in eqn (7.81); the numerator can be factorized in a like manner to the denominator and the roots so obtained are called the zeros of the system function since for these values of s, $H(s) = 0$. Incidentally it can be shown quite generally that for a stable network, to which class passive networks belong automatically, the degree of the numerator cannot exceed the degree of the denominator by more than unity.[†] The properties of a system function are represented in a compact way by its pole–zero diagram in the s-domain. In fact a system function for a realizable network is specified completely by its poles and zeros (apart from a scale factor). For example, for the network shown in Fig. 7.18 the transfer impedance $H(s)$ $(= -V(s)/E(s)$, where $E(s)$ is the current step of unit magnitude) can be derived from eqn (7.84);

$$H(s) = \frac{R}{2} \cdot \frac{(s - 2s_1)}{(s - s_1)}$$

So $H(s)$ has a zero at $s = 2s_1$ and a pole at $s = s_1$, where $s_1 = -(2CR)^{-1}$, and the pole–zero diagram is as shown in Fig. 7.22.

If we are concerned with sinusoidal excitations only then $s = j\omega$ and

$$H(j\omega) = \frac{R}{2} \cdot \frac{\left(j\omega + \dfrac{1}{CR}\right)}{\left(j\omega + \dfrac{1}{2CR}\right)}.$$

Now, if we use the polar form of expression for complex numbers, then

$$\frac{(a + jb)}{(c + jd)} = \frac{|(a + jb)| e^{j\phi_N}}{|(c + jd)| e^{j\phi_D}}$$
$$= \frac{|(a + jb)| e^{j(\phi_N - \phi_D)}}{|(c + jd)|}$$

where ϕ_N, ϕ_D are the phase angles of $(a + jb)$, $(c + jd)$ respectively.

[†]Weinberg, L. *'Network analysis and synthesis'*, Chapter 6, p. 246. McGraw-Hill, New York (1962).

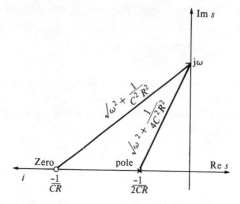

FIG. 7.22. The pole–zero diagram for the network of Fig. 7.18(a).

Hence we can say that

$$|H(j\omega)| = \frac{R}{2}\sqrt{\left\{\frac{\omega^2 + \dfrac{1}{C^2 R^2}}{\omega^2 + \dfrac{1}{4C^2 R^2}}\right\}}$$

and

$$\angle H(j\omega) = \tan^{-1}\left(\frac{\omega}{CR}\right) - \tan^{-1}\left(\frac{\omega}{2CR}\right).$$

An inspection of Fig. 7.22, will reveal that, apart from a scale factor $R/2$, $|H(j\omega)|$ is given by the ratio of the magnitudes of the vectors in the s-domain from the zero and pole, respectively, to the point $(j\omega)$ on the imaginary axis. Further the phase angle, or argument, of $H(j\omega)$ is equal to (angle made by the zero) − (angle made by the pole).

For a system function of the general form

$$H(s) = K\left\{\frac{(s-z_1)(s-z_2)(s-z_3)\ldots}{(s-p_1)(s-p_2)(s-p_3)\ldots}\right\}$$

it follows, by extending the previous arguments, that

$$|H(j\omega)| =$$

$$\frac{K(\text{Product of the magnitudes of the vectors from the zeros to }(j\omega))}{(\text{Product of the magnitudes of the vectors from the poles to }(j\omega))}$$

and

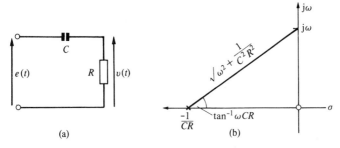

FIG. 7.23. (a) A simple C–R network and (b) the pole–zero diagram for the system functions $H(s) = V(s)/E(s)$.

$$\angle H(j\omega) =$$

(Sum of the angles made by the vectors from the zeros to $(j\omega)$) −

(Sum of the angles made by the vectors from the poles to $(j\omega)$).

As a very simple example consider the high-pass C–R network shown in Fig. 7.23(a): The voltage transfer function $H(s)\{= V(s)/E(s)\}$ is given by

$$H(s) = \frac{s}{s + \dfrac{1}{CR}}.$$

Hence,

$$|H(j\omega)| = \frac{\omega}{\sqrt{\left\{\omega^2 + \dfrac{1}{C^2 R^2}\right\}}}$$

$$\angle H(j\omega) = 90° - \tan^{-1}(\omega CR).$$

As we saw earlier the transfer impedance $H(s)$ for the L–C–R network shown in Fig. 7.19(c) has a pair of conjugate complex poles (see Fig. 7.21(b)). From eqn (7.86) we can see that

$$\frac{V(s)}{I_0(s)} = H(s) = \frac{R}{LC} \cdot \frac{1}{(s - \alpha - j\beta)(s - \alpha + j\beta)} \quad (7.95)$$

and so we have

$$|H(j\omega)| = \frac{R}{LC} \cdot \frac{1}{\sqrt{\{\alpha^2 + (\beta - \omega)^2\}}\sqrt{\{\alpha^2 + (\beta + \omega)^2\}}}$$

$$\angle H(j\omega) = \tan^{-1}\left(\frac{\beta - \omega}{\alpha}\right) - \tan^{-1}\left(\frac{\beta + \omega}{\alpha}\right).$$

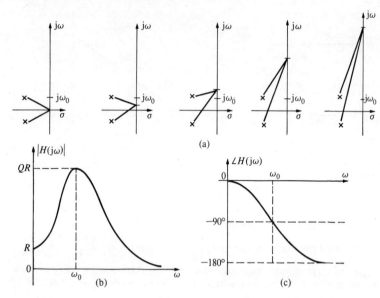

FIG. 7.24. (a) The variation with frequency of the vectors from the poles of an L–C–R network giving the resonance curve in (b). (c) The phase characteristic.

If it is assumed that $Q \geqslant 5$, in which case $\beta \approx \omega_0$, then the dependence of $|H(j\omega)|$ and $\angle H(j\omega)$ on ω is as shown in Fig. 7.24.

Eqn (7.95) for $H(s)$ can be expressed as

$$H(s) \cdot \frac{LC}{R} = (s^2 + 2\alpha s + \alpha^2 + \beta^2)^{-1}$$

and using the definitions of α and β (see eqn (7.87)) together with $\omega_0^2 \equiv (LC)^{-1}$ it follows that

$$H(s) \cdot \frac{\omega_0^2}{R} = \left(s^2 + \frac{Rs}{L} + \omega_0^2\right)^{-1}. \qquad (7.96)$$

If we introduce a normalized (complex) frequency $s = s/\omega_0$ and remember that $Q \equiv \omega_0 L/R$ then

$$H(s) \cdot \frac{1}{R} = \left(s^2 + \frac{s}{Q} + 1\right)^{-1}. \qquad (7.97)$$

The function of the right-hand side of this equation is often called the 'universal resonance curve'.

It is common practice to express $|H(j\omega)|$ in decibels and so if $|H(s)| = |N(s)|/|D(s)|$ we have

Network analysis and synthesis   211

$20 \log |H(j\omega)| = 20 \times$ [Sum of the logs of the magnitudes of the factors in the numerator]

$- 20 \times$ [Sum of the logs of the magnitudes of the factors in the denominator]

where, as we have seen, the factors referred to can have the following forms:

K (a constant) s $1/s$

$(s - s_1)$ $\dfrac{1}{(s - s_1)}$ $\dfrac{1}{(s - s_1)(s - s_2)}$

For sinusoidal excitations the frequency dependence of the magnitude and phase of these functions is usually described as a function of a normalized frequency; for instance for the network illustrated in Fig. 7.7(b) the system function is of the form $(s + 1/CR)^{-1}$ where the pole is at $s = -1/CR$. Apart from a scale factor this function can be written as $(s + 1)^{-1}$ where now s is normalized through $s \equiv sCR$. A pair of graphical plots of the magnitude and phase angle of this system function for sinusoidal excitations are shown in Fig. 7.25 and the straight-line approximations to the curves are commonly termed Bode plots. It is left as an exercise to show that the straight-line approximation differs from the function $(\omega^2 + 1)^{-1/2}$ by about 1 db at an octave above and below the 'corner' frequency ($\omega = 1/CR$). For frequencies higher than the octave of the corner frequency then, to a very good approximation, $|H(j\omega)| \propto 1/\omega$ and $20 \log |H(j\omega)| =$ (constant) $- 20 \log \omega$; hence the slope (or 'roll-off') is 6 db per octave, very closely, or 20 db per decade of frequency. The straight-line approximation to the phase angle variation has a slope of $45°$ per decade and its maximum deviation from the true curve is $6°$ approximately. The Bode plots for the six kinds of system function listed above are shown in Fig. 7.26.

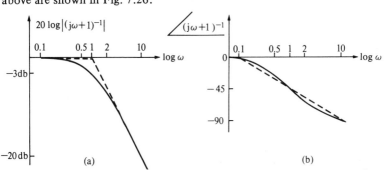

FIG. 7.25.   A pair of Bode plots for a normalized system function $(s + 1)^{-1}$: (a) the amplitude plot; (b) the phase plot.

212   Network analysis and synthesis

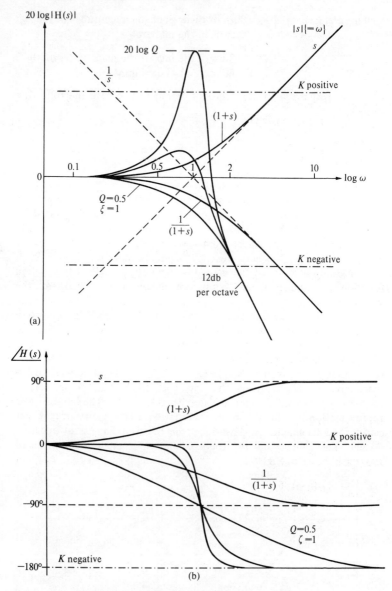

FIG. 7.26. Bode plots for the six types of system function listed on p. 211: (a) amplitude plots; (b) phase plots.

The notion of a normalized frequency, and of a normalized impedance (and of a normalized system function in general) are of great practical importance since they facilitate the presentation of

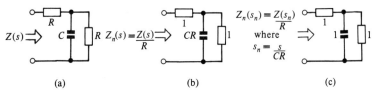

FIG. 7.27. An illustration of the normalization of frequency and impedance for a network.

information about networks in a compact form. It is common practice to specify system functions in terms of a normalized frequency (i.e. normalized to a critical frequency such as a 'corner' frequency, or a resonance frequency) and in relation to a terminating impedance of 1 ohm. This procedure is illustrated for a simple one-port $C-R$ network in Fig. 7.27. For this network the driving point impedance $Z(s)$ is given by

$$Z(s) = \frac{R\left(R + \dfrac{2}{sC}\right)}{\left(R + \dfrac{1}{sC}\right)} = \frac{R(sCR + 2)}{(scR + 1)}.$$

If we divide through by R then the value of the resistors of the network is normalized to $1\,\Omega$ (Fig. 7.27(b)). Further if a normalized frequency s_n is defined through $s_n \equiv sCR$ then the normalized driving point impedance is given by

$$Z_n(s_n) = \frac{(s_n + 2)}{(s_n + 1)} \quad \text{(see Fig. 7.27(c))}$$

which is of the general form

$$Z(s) = \frac{(s + 2)}{(s + 1)}. \tag{7.98}$$

Again, for the low-pass $L-R$ network shown in Fig. 7.28(a), the voltage transfer ratio $H(s) \equiv V(s)/E(s)$ is given by

$$H(s) = \frac{R}{(R + sL)} = \frac{1}{\left(1 + \dfrac{sL}{R}\right)}.$$

Here we notice that if we divide the value of each element of the network by R then $H(s)$ is unchanged; this is true for voltage and current transfer functions in general. If in this case a normalized frequency is defined in terms of the corner frequency $(L/R)^{-1}$ (i.e. $s_n \equiv sL/R$) then

$$H(s) = \frac{1}{(s + 1)}$$

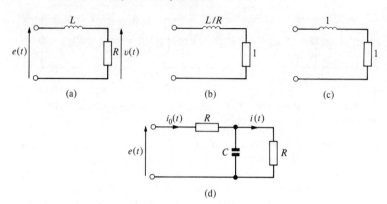

FIG. 7.28. (a), (b), (c) The normalization of an $L-R$ network. (d) (See text).

which is the voltage transfer ratio for the generalized network of Fig. 7.28(c).

It should be apparent, by inspection almost, that for the network of Fig. 7.28(d), the normalized current transfer ratio $I(s)/I_0(s) = (s+1)^{-1}$. Also for this same network it is left as a simple exercise to show that the transfer admittance $H(s)$ [$\equiv I(s)/E(s)$] is given by

$$H(s) = \frac{1}{R} \cdot \frac{1}{(sCR+2)}.$$

If we divide through by $1/R$ then the conductances of the network are normalized to $1\,\Omega^{-1}$ (and the capacitor value is again multiplied by the factor R) and if the normalized frequency is again defined in terms of the corner frequency $(CR)^{-1}$ we have that the normalized transfer admittance is given by

$$H(s) = (s+2)^{-1}.$$

From eqn (7.97) we can see that for the network of Fig. 7.19(c) the normalized transfer impedance is

$$H(s) = (s^2 + s/Q + 1)^{-1}.$$

In summary, it can be said that to increase the impedance level of a generalized network by a factor η and to increase the frequency by a factor γ:

Each network resistor should be multiplied by η.
Each network inductor should be multiplied by $(\gamma\eta)$.
Each network capacitor should be multiplied by $(\gamma\eta)^{-1}$.
These rules can be extended in an obvious way to the dual networks and elements.

7.7. General aspects of filters

For our purposes filters can be divided into four classes namely low-pass, high-pass, band-pass, and band-stop for which the ideal characteristics are shown in Fig. 7.29. The reader will suspect, and rightly, that such ideal characteristics, having perfectly sharp cut-offs, infinite attenuation in the 'stop' band, and 'zero' attenuation in the 'pass' band cannot be realized in a practical circuit which is constructed from a finite number of elements. We would say that the system function $H(j\omega)$ did not belong to that class of mathematical functions (so-called positive real, or p.r., functions)[†] which can be realised through networks composed of real, physical, circuit elements. So, given a desired filter characteristic, the problem is to synthesize a practical network whose characteristic is a reasonable approximation to the desired one; a serious aspect of the over-all problem is to choose the criteria for deciding what constitutes a reasonable approximation. As mentioned at the beginning of Section 7.6, very powerful mathematical and computational techniques have been developed in order to facilitate the design of filters but a detailed account of these would lie outside the scope of this text. However it will be useful to make brief mention of some of the salient features of some of the important ways of approaching the approximation problem. Let us concentrate on the low-pass filter characteristic (it turns out that the other forms of characteristic can be obtained from this by means of suitable transformations).

Since such an ideal characteristic could not be realized through a network having a finite number of elements the characteristic is relaxed in a number of ways. Thus the attenuation in the stop-band is allowed to have finite (but acceptable) value which may be allowed to decrease significantly for frequencies which are far enough above the critical frequency ω_0. Similarly the attenutation in the pass-band will not be zero and furthermore a certain degree of 'ripple' will be allowed in the characteristic in both the pass- and stop-bands. Finally, the cut-off cannot be perfectly sharp and so an acceptable value for the slope of the attenuation versus frequency curve in the cut-off region must be specified. Tolerances have to be specified in a similar fashion for the phase characteristic of the network.

In the Butterworth approximation

$$|H(j\omega)|^2 = (1 + \omega^{2n})^{-1}$$

and the characteristics obtained for first-order ($n = 1$) and fifth-order ($n = 5$) filters are sketched in Fig. 7.30(a). The rate of change of

[†]Weinberg, L. *'Network analysis and synthesis'*, Chapter 6. McGraw-Hill, New York (1962).

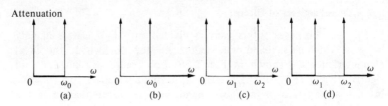

FIG. 7.29. Ideal filter characteristics: (a) low-pass; (b) high-pass; (c) band-pass; (d) band-stop.

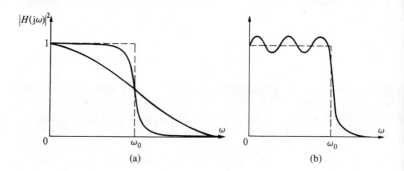

FIG. 7.30. Sketches of (a) Butterworth filter characteristics and (b) a Chebyshev filter characteristic.

$H(j\omega)$ with ω in the stop-band is $-20n \log \omega$ (for $\omega^{2n} \gg 1$) and so the roll-off is $6n$ db per octave, or $20n$ per decade, very closely.

The Chebyshev approximation is specified in terms of Chebyshev polynomials $C_n(\omega)$ and has the form

$$|H(j\omega)|^2 = [1 + \epsilon C_n^2(\omega)]^{-1}$$

where $\epsilon < 1$. The general form of the filter characteristic is sketched in Fig. 7.30(b) and it should be noted that for a given maximum tolerable value of attenuation in the pass-band the Chebyshev approximation gives a sharper cut-off. An even sharper cut-off is obtained if ripple can be tolterated in the stop-band also, the filters arising in this approximation are usually called elliptic filters.

In practice trade-offs have to be made; the characteristics of a filter improve as the order of the filter is increased but, unfortunately, so does the complexity (and hence cost) since more circuit elements are required. Another aspect of the various approximations which can be very significant is their sensitivity to the tolerances on the values of the components used.

A low-pass filter can be transformed to a high-pass filter simply by replacing capacitors by inductors having the same reactance at the normalized critical frequency and vice versa. Band-pass filters can be obtained from low-pass filters by making $\sqrt{\omega_1 \omega_2} = \omega_0$ where ω_1, ω_2 are the limits of the pass-band and analogously for band-stop filters. A very useful exposition of a 'users' approach to filter design has been given by Lubkin.[†]

7.8. The synthesis of driving point immittances

Although a detailed consideration of the synthesis of system functions would be out of place, an examination of some simple examples of network synthesis should be instructive and impart some of the flavour of the more general approach.

For the lossless one-port $L-C$ network of Fig. 7.31 it can be shown, using mesh analysis, that, if $X_1 = X_3 = X_5 = sL$ and $X_2 = X_4 = X_6 = (sC)^{-1}$, then

$$Z(s) = \frac{1}{sC} \cdot \frac{(s^6 L^3 C^3 + 4s^4 L^2 C^2 + 2s^2 LC - 2)}{(s^4 L^2 C^2 + 4s^2 LC + 3)}.$$

The form of this expression illustrates the general point that for such networks the driving point impedance can be written as the ratio of two polynomials in s. In general

$$Z(s) = (s)^{\pm 1} \cdot \frac{(a_{2n} s^{2n} + a_{2n-2} s^{2n-2} + \ldots a_2 s^2 + a_0)}{(b_{2n} s^{2n} + b_{2n-3} s^{2n-2} + \ldots b_2 s^2 + b_0)}$$

and it should be clear that an admittance function can be obtained through an analogous argument. Now we are interested in the steady-state response ($s = j\omega$) and so

$$Z(j\omega) = jX(\omega) = (j\omega)^{\pm 1} \cdot \left(\frac{a_{2n}}{b_{2n}}\right) \times$$

$$\frac{(\omega^{2n} - A_{2n-2} \omega^{2n-2} + A_{2n-4} \omega^{2n-4} - \ldots + (-1)^n A_0)}{(\omega^{2n} - B_{2n-2} \omega^{2n-2} + B_{2n-4} \omega^{2n-4} - \ldots + (-1)^n B_0)}$$

where $A_{2n-2} \equiv a_{2n-2}/a_{2n}$, etc. and similarly for B_{2n-2}. Now the polynomials in the last expression are factorizable to give the zeros and poles of $X(\omega)$.

$$jX(\omega) = (j\omega)^{\pm 1} \left(\frac{a_{2n}}{b_{2n}}\right) \frac{(\omega^2 - \omega_{z_1}^2)(\omega^2 - \omega_{z_2}^2) \ldots (\omega^2 - \omega_{z_n}^2)}{(\omega^2 - \omega_{p_1}^2)(\omega^2 - \omega_{p_2}^2) \ldots (\omega^2 - \omega_{p_n}^2)}. \quad (7.99)$$

The most important property of functions such as that given in eqn 7.99

[†] Lubkin, Y. I. *Filter systems and design; electrical, microwave, and digital.* Addison-Wesley, Englewood Cliffs, New Jersey (1970).

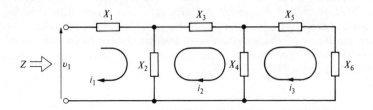

FIG. 7.31. An example of a lossless one-port $L-C$ network.

is that the function is established over the whole of the range $0 \leqslant \omega \leqslant \infty$ if the poles and zeros of $X(\omega)$ are specified and if the value of $X(\omega)$ is known at one frequency where $X(\omega) \neq 0$ (so that the scale factor (a_{2n}/b_{2n}) can be evaluated); these features are illustrated by the following example.

The problem is to synthesize an $L-C$ network having poles at ω_1 and ω_3 and zeros at $\omega = 0$, $\omega = \omega_2$, and $\omega = \infty$ where $\omega_1 < \omega_2 < \omega_3$; depending how the network is used, the poles and zeros could correspond to the mid-points of pass- and stop-bands, for example. It should be noted that the zeros at $\omega = 0$ and $\omega = \infty$ are called 'external' zeros and that $\omega_1, \omega_2, \omega_3$ are 'internal' poles and zeros and also that the zeros and poles alternate along the ω-axis. Since $X(\omega)$ is always an increasing function of ω (except at discontinuities, of course) it follows that $X(\omega)$ must have the general form shown in Fig. 7.32(a) which can be expressed mathematically as

$$jX(\omega) = j\omega H \frac{(\omega^2 - \omega_2^2)}{(\omega^2 - \omega_1^2)(\omega^2 - \omega_3^2)} \qquad (7.100)$$

where H is a scale factor which has not yet been specified; the factor $(j\omega)^{+1}$ has been chosen, as opposed to $(j\omega)^{-1}$, to accord with the specified external zero at $\omega = 0$. The right-hand side of this equation can be expressed in partial fractions

$$\frac{(\omega^2 - \omega_2^2)}{(\omega^2 - \omega_1^2)(\omega^2 - \omega_3^2)} = \frac{A_1\omega + D_1}{(\omega^2 - \omega_1^2)} + \frac{A_2\omega + D_2}{(\omega^2 - \omega_3^2)}$$

and you can show that

$$A_1 = -A_2 = 0; \qquad D_1 = \frac{(\omega_2^2 - \omega_1^2)}{(\omega_3^2 - \omega_1^2)}; \qquad D_2 = \frac{(\omega_3^2 - \omega_2^2)}{(\omega_3^2 - \omega_1^2)}.$$

Note that if $X(\omega) = X_0$ at $\omega = \omega_0$, where $0 < \omega_0 < \omega_1$, say, then

$$H = \frac{X_0(\omega_0^2 - \omega_1^2)(\omega_0^2 - \omega_3^2)}{\omega_0(\omega_0^2 - \omega_2^2)}.$$

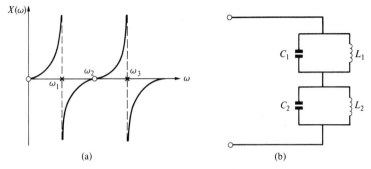

FIG. 7.32. (a) A reactance function $X(\omega)$ and (b) a network which will realize this function.

(Note that H is negative in this case.) So

$$jX(\omega) = -j\omega|H|\frac{(\omega_2^2-\omega_1^2)}{(\omega_3^2-\omega_1^2)(\omega^2-\omega_1^2)} - j\omega|H|\frac{(\omega_3^2-\omega_2^2)}{(\omega_3^2-\omega_1^2)(\omega^2-\omega_3^2)}. \quad (7.101)$$

Now, as readers will be able to check for themselves, the reactance of a parallel (or 'anti-resonant') $L-C$ circuit having an inductor L_1 and a capacitor C_1 is given by

$$X(\omega) = -\frac{\omega}{C_1}\frac{1}{(\omega^2-\omega_1^2)}$$

where $\omega_1^2 \equiv (L_1 C_1)^{-1}$. Hence the reactance function of eqn (7.101) can be realized physically by a network consisting of the series connection of two parallel $L-C$ networks (see Fig. 7.32(b)) where we have

$$\omega_1^2 = \frac{1}{L_1 C_1} \qquad \omega_3^2 = \frac{1}{L_2 C_2}$$

$$C_1 = \frac{(\omega_3^2-\omega_1^2)}{|H|(\omega_2^2-\omega_1^2)} \qquad C_2 = \frac{(\omega_3^2-\omega_2^2)}{|H|(\omega_3^2-\omega_2^2)}.$$

The reactances of the basic elements of one-port lossless $L-C$ networks are displayed in Fig. 7.33; the properties of these networks and their duals are summarized in Table 7.1.

As another example: Find the network to realize the susceptance function $B(\omega)$ given by

$$jB(\omega) = j\omega\frac{(\omega^2-2)(\omega^2-4)}{(\omega^2-1)(\omega^2-3)}.$$

The internal zeros of $B(\omega)$ occur at $\omega = \sqrt{2}$, $\omega = 2$ and the internal poles occur at $\omega = 1$, $\omega = \sqrt{3}$, and further there is an external zero at

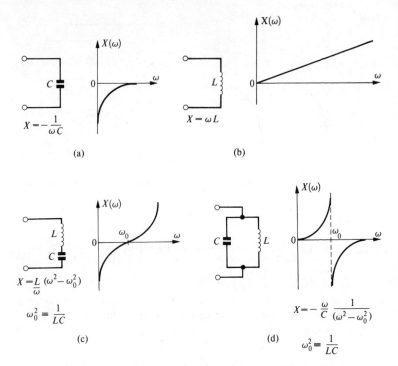

FIG. 7.33. The basic elements of one-port lossless networks.

$\omega = 0$ and an external pole at $\omega = \infty$; this susceptance function is sketched in Fig. 7.34(a). If we expand the expression for $B(\omega)$ in the form of partial fractions then

$$\frac{\omega(\omega^2 - 2)(\omega^2 - 4)}{(\omega^2 - 1)(\omega^2 - 3)} = A + \frac{D}{(\omega^2 - 1)} + \frac{F}{(\omega^2 - 3)}$$

whence
$$A = 1; \quad D = -3/2; \quad F = -1/2$$

and
$$B(\omega) = \omega - \frac{\omega}{\frac{2}{3}(\omega^2 - 1)} - \frac{\omega}{2(\omega^2 - 3)}.$$

By referring to Table 7.1 readers will be able to see that for this case $B(\omega)$ can be realized by the network shown in Fig. 7.34(b). The networks of Figs. 7.32 and 7.34 are examples of so-called 'Foster' networks; in general the 'first' Foster form is found from a partial fraction expansion of $X(s)$ and the 'second' form from a partial fraction expansion of $B(s)$.

Another synthesis technique which is used in the context of lossless

Network analysis and synthesis 221

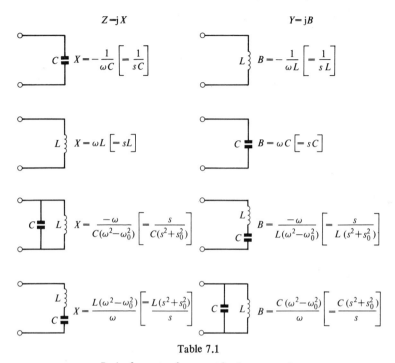

Table 7.1

Basic elements of one-port lossless networks

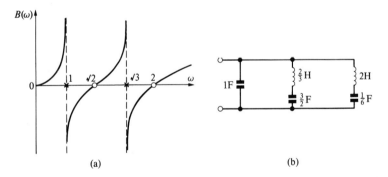

FIG. 7.34. (a) A susceptance function $B(\omega)$ and (b) a network which will realize this function.

one-port networks, although it can be used in wider contexts, is that of the 'continuous fraction expansion' which yields the so-called 'Cauer' networks. The physical basis for this technique can be illustrated by considering a ladder network such as that shown in Fig. 7.35.

222   Network analysis and synthesis

At plane P_6 the reactance 'looking to the right' is

$$X_6 = \frac{1}{B_6}$$

At plane P_5 the susceptance 'looking to the right' is

$$B_5 = \cfrac{1}{X_5 + \cfrac{1}{B_6}}$$

At plane P_4 the reactance 'looking to the right' is

$$X_4 = \cfrac{1}{B_4 + \cfrac{1}{X_5 + \cfrac{1}{B_6}}}$$

At plane P_3 the susceptance 'looking to the right' is

$$B_3 = \cfrac{1}{X_3 + \cfrac{1}{B_4 + \cfrac{1}{X_5 + \cfrac{1}{B_6}}}}$$

At plane P_2 the reactance 'looking to the right' is

$$X_2 = \cfrac{1}{B_2 + \cfrac{1}{X_3 + \cfrac{1}{B_4 + \cfrac{1}{X_5 + \cfrac{1}{B_6}}}}}$$

Finally we have

$$X = X_1 + \cfrac{1}{B_2 + \cfrac{1}{X_3 + \cfrac{1}{B_4 + \cfrac{1}{X_5 + \cfrac{1}{B_6}}}}}$$

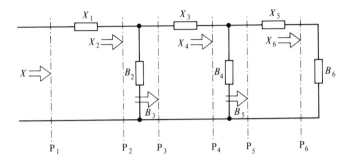

FIG. 7.35. The basis for the continuous fraction expansion which yields the Cauer networks.

As an example consider the problem of synthesizing an $L-C$ network to realize the reactance function

$$X(s) = (2s^4 + 20s^2 + 18)/s(s^2 + 4) \quad (7.102)$$

which, on factorizing, can be written as

$$X(s) = \frac{2(s^2 + 1)(s^2 + 9)}{s(s^2 + 4)}$$

and which has internal zeros at $s = \pm 1, \pm 3$, internal poles at $s = \pm 2$, and external poles at $s = 0, s = \infty$.

The partial fraction expansion of $X(s)$ yields

$$X(s) = 2s + \frac{9}{2s} + \frac{15s}{2(s^2 + 4)}. \quad (7.103)$$

$$X_1 \qquad X_2$$

Thus $X_2 = X(s) - X_1(s)$ and we see that we have 'removed a pole at infinity', namely the term $X_1(s) = 2s$. Now,

$$X_2(s) = \frac{12s^2 + 18}{s(s^2 + 4)}$$

and, on inverting,

$$\frac{1}{X_2(s)} = \frac{s(s^2 + 4)}{12(s^2 + 3/2)}$$

which expression has a pole at infinity. Using a partial fraction expansion again we find

$$\frac{1}{X_2(s)} = \frac{s}{12} + \frac{5s}{24(s^2 + 3/2)}.$$

$$B_2 \qquad B_3$$

Now $B_3 = \left(\dfrac{1}{X_2} - \dfrac{s}{12}\right)$ and we have removed another pole at infinity. Further,
$$B_3 = \frac{5s}{24(s^2 + 3/2)}$$
and on inverting we find
$$\frac{1}{B_3} = \frac{24}{5s}(s^2 + 3/2)$$
$$= \frac{24s}{5} + \frac{36}{5s}$$
$$\phantom{X_4 = \frac{1}{B_3} - \frac{24s}{5}}X_3 \quad X_4(=1/B_4)$$
and
$$X_4 = \frac{1}{B_3} - \frac{24s}{5}$$

and we have removed yet another pole at infinity. For this example we have reached the end of the process of 'successively removing poles at infinity'. In this case the continuous fraction expansion is

$$X = X_1 + \cfrac{1}{B_2 + \cfrac{1}{X_3 + \cfrac{1}{B_4}}}$$

where $X_1 = 2s$, $B_2 = s/12$, $X_3 = 24s/5$, $B_4 = 5s/36$ and the physical realization is shown in Fig. 7.36(a); it can be seen by inspection of eqn (7.103) and Table 7.1. that a Foster network to realize the same reactance function is as shown in Fig. 7.36(b). The mechanics of a continuous fraction expansion can be illustrated by returning to the

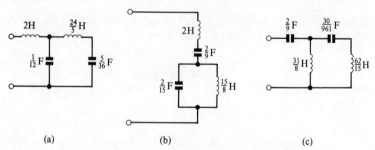

FIG. 7.36.   (a) The 'first' Cauer network realization of the reactance function
$$X(s) = (2s^4 + 20s^2 + 18)/s(s^2 + 4).$$
(b) A Foster series network to realize this same reactance. (c) The 'second' Cauer network realization of this reactance function.

reactance function of eqn (7.102); we arrange the numerator and demoninator in descending powers of s and

$$s^3 + 4s \overline{\smash{\big)}\,2s^4 + 20s^2 + 18} \quad \frac{2s(X_1)}{}$$

$$\underline{2s^4 + 8s^2}$$

$$12s^2 + 18 \overline{\smash{\big)}\,s^3 + 4s} \quad \frac{s/12(B_2)}{}$$

$$\underline{s^3 + \frac{3s}{2}} \quad \frac{24s}{5}(X_3)$$

$$\frac{5s}{2} \overline{\smash{\big)}\,12s^2 + 18}$$

$$12s^2 \quad \frac{5s}{36}(B_4)$$

$$18 \overline{\smash{\big)}\,\frac{5s}{2}} \, .$$

Now eqn (7.103) can be written also as

$$X(s) = \underbrace{\frac{9}{2s}}_{X_1} + \underbrace{2s + \frac{15s}{2(s^2 + 4)}}_{X_2}$$

and $X_2 = X(s) - 9/2s$, i.e. we have 'removed a pole at the origin'. Since $X_2 = \dfrac{s(4s^2 + 31)}{2(s^2 + 4)}$ and so has a pole at infinity, we invert and use partial fractions to find

$$\frac{1}{X_2} = \underbrace{\frac{8}{31s}}_{B_2} + \underbrace{\frac{30s}{31(4s^2 + 31)}}_{B_3}.$$

So $B_3 = \dfrac{1}{X_2} - \dfrac{8}{31s}$ and we have removed another pole at the origin.
Finally, if we invert B_3 we find

$$\frac{1}{B_3} = \underbrace{\frac{961}{30s}}_{X_3} + \underbrace{\frac{62s}{15}}_{X_4(=1/B_4)}$$

This realization of $X(s)$ (see Fig. 7.36(c)) can be obtained by the continuous fraction expansion in which the numerator and denominator are arranged in *ascending* powers of s; this is left as an exercise.

226 Network analysis and synthesis

A continuous fraction expansion with the polynomials arranged in descending powers of s, in which poles at infinity are removed successively, synthesizes a first Cauer form of ladder network (e.g. Fig. 7.36(a)); an expansion with the polynomials arranged in ascending powers of s, and with poles at the origin removed successively, synthesizes the 'second' Cauer form of ladder network (e.g. Fig. 7.36(c)).

The rules to be followed in synthesizing the Cauer networks can be summarized as follows:

Cauer I: Descending powers of s.

$$X(\infty) = \infty \qquad\qquad B(\infty) = \infty$$

Expansion on $X(s)$ Expansion on $B(s)$
Leading element; series L   Leading element; shunt C

Cauer II: Ascending powers of s.

$$X(0) = \infty \qquad\qquad B(0) = \infty$$

Expansion on $X(s)$ Expansion on $B(s)$
Leading element; series C   Leading element; shunt L

Number of elements equals the number of poles

If the expansion has been started correctly then the resultant ladder network will be terminated by the appropriate element but, in any case, the requirement on the terminating element can be checked easily from an inspection of the required behaviour of the immittance function at $s = 0$ or $s = \infty$.

PROBLEMS

7.1. Show that the z- and y- parameters of a two-port network are related by:

$$z_{11} = \frac{y_{22}}{\Delta_y} \qquad z_{12} = \frac{y_{12}}{\Delta_y}$$

$$z_{21} = -\frac{y_{21}}{\Delta_y} \qquad z_{22} = \frac{y_{11}}{\Delta_y}.$$

7.2. Derive eqns (7.19) relating the elements of equivalent T- and Π-networks as shown in Fig. 7.3.

7.3. Derive the h-parameters of the Π-network shown in Fig. 7.37.

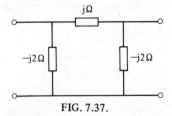

FIG. 7.37.

7.4. Obtain the expressions (eqns (7.20) and (7.24)) for the characteristic impedance and admittance of symmetrical T- and Π-networks.

7.5. Obtain the expressions given in eqns (7.38) and (7.39) for the transmission matrices of the two networks shown in Fig. 7.11(d) and (e).

Network analysis and synthesis 227

7.6. Derive the relationships of eqns (7.44(b)) for the elements of a symmetrical Π-network designed to match a source having resistance R_0 to a load resistance R_0 with a voltage attenuation ratio of n.

7.7. Design a matched symmetrical T-network to give an attenuation of 14 db between a generator of 50 Ω internal resistance and a 50 Ω load. Note that the ratio (power in the load before the coupling network is inserted) ÷ (power in the load with the coupling network inserted) is called the insertion loss; in the example of this question the insertion loss is 14 db.

7.8. Obtain expressions for the y-parameters of the bridged-T network shown in Fig. 7.38.

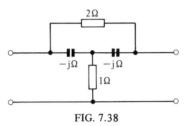

FIG. 7.38

7.9. Write the mesh equations in matrix form for the network shown in Fig. 7.39 and then give the equation for determining the current through Z_2.

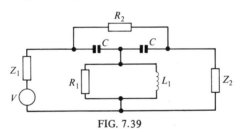

FIG. 7.39

7.10. Write the nodal equations in matrix form for the network shown in Fig. 7.40.

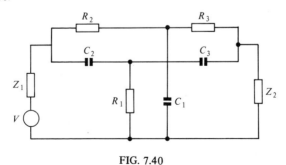

FIG. 7.40

7.11. Calculate the r.m.s. value of the noise voltage developed across a 10 k resistor at a temperature of 300 K in a frequency bandwidth of 10 MHz.

7.12. Obtain the transmission parameters in matrix form for the Π-network shown in Fig. 7.41.

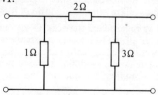

FIG. 7.41

7.13. Obtain expressions (see eqns (7.80)) for the system functions of the networks of Fig. 7.17(b), (c), and (d), where the respective system functions are defined by $V(s)/E(s)$, $I(s)/E(s)$, and $V(s)/E(s)$.

7.14. Obtain the expressions of eqns (7.80(e)) and (7.80(f)) for the specified system functions of the networks of Fig. 7.17(e) and (f).

7.15. Derive eqn (7.94) from eqn (7.93).

7.16. Show for the network of Fig. 7.31 that the input impedance $Z(s)$ is given by

$$Z(s) = \frac{1}{sC} \frac{(s^6 L^3 C^3 + 4s^4 L^2 C^2 + 2s^2 LC - 2)}{(s^4 L^2 C^2 + 4s^2 LC + 3)}.$$

7.17. Show that the reactance of the parallel L–C network shown in Fig. 7.42 is equal to $-(\omega/C)(\omega^2 - \omega_0^2)^{-1}$ where $\omega_0^2 \equiv (LC)^{-1}$.

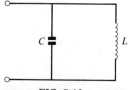

FIG. 7.42

7.18. Derive an expression for the susceptance $B(s)$ of an inductance L in series with a capacitance C.

7.19. Sketch the form of the susceptance function

$$B(\omega) = \frac{-2(1-\omega^2)}{\omega(2-\omega^2)}$$

and synthesize a parallel Foster network to realize this function.

7.20. Show that the amplitude characteristic of Fig. 7.25(a) differs from the Bode plot by ± 0.97 db at frequencies an octave below and an octave above the 'corner' frequency.

7.21. Synthesize a parallel Foster network to realize the susceptance function

$$B(s) = \frac{s^4 + 34s^2 + 225}{s(s^2 + 16)}.$$

7.22. Synthesize a series Foster network to realize the reactance function

$$X(\omega) = \frac{(\omega^2 - 10^{12})(\omega^2 - 9 \times 10^{12})10^{-3}}{\omega(\omega^2 - 4 \times 10^{12})} \, \Omega$$

7.23. Factorize the reactance function

$$X(\omega) = \frac{2(\omega^4 - 4\omega^2 + 3)}{\omega(\omega^2 - 2)}$$

and sketch the function. Synthesize a series Foster network to realize this reactance.

7.24. Sketch the following reactance function as a function of ω

$$X(s) = \frac{(s^3 + 9s)}{(s^4 + 20s^2 + 64)}.$$

Synthesize the series Foster network which realizes this reactance function.

7.25. Synthesize a series Foster network to realize the following reactance function:

Internal pole at $(2/2\pi)$ MHz
Internal zero at $(1/2\pi)$ MHz
External pole at zero Hz
External zero at infinity
Magnitude of reactance function at $(3/2\pi)$ MHz equals $80/15$ Ω.

7.26. Find the poles and zeros of the network shown in Fig. 7.41.

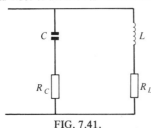

FIG. 7.41.

Hence find the condition that the circuit be non-oscillatory. What is the limiting value of R_C to meet the non-oscillatory requirement if $L = 100$ mH; $R_L = 10$ Ω; $C = 0.1 \mu$F?

7.27. Obtain the second Cauer network realization of the reactance function

$$X(s) = \frac{(2s^4 + 20s^2 + 18)}{(s^2 + 4s)}$$

by a continuous fraction expansion in which the numerator and denominator are arranged in ascending powers of s.

7.28. Express the numerator and denominator of the reactance function of Problem 27 in descending powers of s and so obtain the first Cauer network realization of the function.

7.29. Sketch the following reactance function as a function of ω,

$$X(s) = \frac{2s^4 + 8s^2 + 6}{(s^3 + 2s)}.$$

Synthesize the two types of Cauer network which realize this reactance function.

8. Transmission lines and waveguides

8.1. Introduction

AN implicit assumption which was made in the discussions of those of the foregoing chapters which are concerned directly with a.c. circuit theory and network analysis was that the dimensions of circuit elements, and of networks, were much smaller than the wavelength of electromagnetic waves at the operating frequency of interest. This being so then it followed that, at any instant of time, the strength of the electric field was constant throughout a circuit element, to an acceptable degree of approximation. For instance, if the electric field E in a resistor is written as $E = \hat{E} \sin(\omega t - (2\pi x)/\lambda)$, that is a wave of angular frequency ω and wavelength λ travelling in the $+x$-direction, then if

$$\delta E = \frac{\partial E}{\partial x} \delta x = -\frac{2\pi \hat{E}}{\lambda} \cos\left(\omega t - \frac{2\pi x}{\lambda}\right) \delta x,$$

we have

$$\frac{|\delta E|_{\max}}{\hat{E}} = \frac{2\pi \delta x}{\lambda}.$$

If we take as δx the length of a typical resistor say 1 cm, and assume a frequency of $3/2\pi$ MHz (i.e. about 500 kHz) then $|\delta E|_{\max}/\hat{E} = 10^{-4}$; for practical purposes this represents a negligible variation in electric field strength and we can regard the resistor (or inductor or capacitor) as a *'lumped'*, or localized, circuit element. However if the frequency is 10^4 times higher (5 GHz approx.) then $|\delta E|_{\max}/\hat{E} \approx 0.2$ which is a very significant variation indeed and we have to use the notion of *distributed* resistance (and inductance and capacitance). The same argument can be applied with even more force to connecting wires and cables since their linear dimensions are generally greater than those of circuit elements.

Another physical factor which has increasing significance as the operating frequency is increased is the radiation of energy as electromagnetic waves; in simple terms the power radiated depends on the ratio $(l/\lambda)^2$ where l is the length of the circuit element. There are two major consequences of such radiation: firstly it constitutes an additional significant source of power dissipation in a circuit and, as such, sets a limit to the quality factor of $L-C-R$ networks and, secondly, it provides coupling between the components of a circuit which is usually very undesirable. This latter feature is the reason that it is necessary to provide screening in and around high-frequency circuits and to pay great attention to the lay-out of the circuit components.

There are also problems arising from the point of view of 'hardware'. Consider, for instance, an $L-C-R$ network made of lumped elements which is resonant at about 10 MHz; the value of the capacitor will be $\sim$ 10 pF. For a reasonant frequency of 1000 MHz the dimensions of the lumped elements would have to be 100 times smaller; their small size would limit severely the amount of energy which could be stored in the reactive elements because of their small volumes and, in addition, there would be considerable technical problems in manufacturing elements to the required precision. Also, as we saw in Chapter 5, the radio-frequency resistance of coils is proportional to (frequency)$^{1/2}$ and so the quality factor of the circuit will be ten times smaller, roughly speaking. In summary, the concept of 'lumped' circuit elements can be used when the functions of energy storage and dissipation can be associated uniquely, for practical purposes, with identifiable, localized, circuit elements, i.e. reactive and resistive elements respectively.

Hence, if the ratio (l/λ) is not small, circuits are described in terms of distributed parameters; the following list will serve as a reminder of the wavelengths of electromagnetic waves in free space at a variety of frequencies:

Frequency	*Wavelength*	
50 Hz	6000 km	a.c. Mains
1 kHz	300 km	Audio
1 MHz	300 m	
100 MHz	3 m	U.H.F.
1 GHz	30 cm	Microwaves
10 GHz	3 cm	

So even at the mains frequency of 50 Hz the wavelength is comparable with transcontinental distances. At the other end of this list, in the realm of solid-state microwave *integrated* circuits, a lumped element description can be used again since, even though the wavelength is only 3 cm, the dimensions of the circuit components are much smaller.

The most common forms of elements in distributed circuits are coaxial cables and waveguides, the latter being hollow metal pipes which are rectangular in form most commonly. Strip lines (see Section 8.5) are becoming increasingly important in integrated circuits. Coaxial

232 Transmission lines and waveguides

cables are generally used for the interconnection of pieces of electronic equipment, even at low frequencies where the ratio (l/λ) is very small, because of their convenience and also because there is negligible radiation loss since electromagnetic fields are confined almost exclusively to the region between the inner and outer conductors (see Section 1.1.2.). Coaxial cables can be used as distributed circuit elements at microwave frequencies but have greater attenuation per unit length than waveguides and their lack of rigidity also poses problems.

8.2. Transmission line equations

Before we discuss the properties of real lines and waveguides it is necessary to introduce some general theoretical considerations — this will be done as simply and briefly as possible.

A portion of a transmission line which consists of two conductors, and which is uniform (i.e. its physical properties do not vary with distance along the line), can be represented in a general way as shown in Fig. 8.1(a) where R is the total of the distributed series resistance per

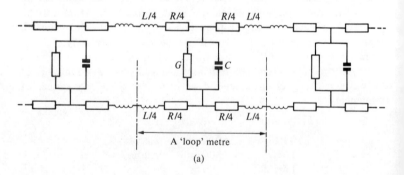

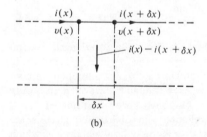

FIG. 8.1. (a) A general schematic representation of a transmission line. (b) The basis for a description of the variation of voltage and current along a transmission line.

Transmission lines and waveguides

loop meter, and L, G, C relate in a similar way to the distributed series inductance, shunt conductance, and shunt capacitance respectively.

If the line is imagined to extend in the x-direction then the variation in the voltage across the line and the current along it at a particular instant of time can be represented as in Fig. 8.1(b). Furthermore, if it is assumed that $\dfrac{\partial i}{\partial x} \delta x$ can be neglected compared to $i(x)$ then we can write

$$R\delta x\, i(x) + L\delta x\, \frac{\partial i(x)}{\partial t} = v(x) - v(x + \delta x).$$

Now,

$$v(x + \delta x) = v(x) + \frac{\partial v(x)}{\partial x} \delta x$$

and so we have

$$R\delta x\, i(x) + L\delta x\, \frac{\partial i(x)}{\partial t} = -\frac{\partial v(x)}{\partial x} \delta x. \tag{8.1}$$

Also, assuming that $(\partial v(x)/\partial x)\delta x$ can be neglected compared to $v(x)$,

$$G\delta x\, v(x) + C\delta x\, \frac{\partial v(x)}{\partial t} = i(x) - i(x + \delta x)$$

or

$$G\delta x\, v(x) + C\delta x\, \frac{\partial v(x)}{\partial t} = -\frac{\partial i(x)}{\partial x} \delta x. \tag{8.2}$$

If $v(x)$, $i(x)$ are sinusoidal then we can represent them by the phasors

$$v(x, t) = \hat{V}(x) e^{j\omega t} \qquad i(x, t) = \hat{I}(x) e^{j\omega t}. \tag{8.3}$$

On substituting for the currents and voltages in eqns (8.1) and (8.2) using eqns (8.3) we find

$$-\frac{\partial \hat{V}(x)}{\partial x} = (R + j\omega L)\hat{I}(x) \qquad -\frac{\partial \hat{I}(x)}{\partial x} = (G + j\omega C)\hat{V}(x). \tag{8.4}$$

Eqns (8.4) are known, commonly, as the 'Telegrapher's' equations.

If eqns (8.4) are differentiated then it is found that

$$\frac{\partial^2 \hat{V}(x)}{\partial x^2} = -\gamma^2 \hat{V}(x) \qquad \frac{\partial^2 \hat{I}(x)}{\partial x^2} = -\gamma^2 \hat{I}(x) \tag{8.5}$$

where $\gamma \equiv \sqrt{\{(R + j\omega L)(G + j\omega C)\}}$ is usually called the propagation coefficient. Since γ is complex it can be written as $\gamma = (\alpha + j\beta)$ where α is the attenuation coefficient and β is the phase coefficient (phase change per unit length),

$$\alpha^2 - \beta^2 = RG - \omega^2 LC$$
$$2\alpha\beta = \omega(LG + RC). \tag{8.6}$$

The solutions of the second-order differential eqns (8.5) are

$$\hat{V}(x) = A e^{-\gamma x} + B e^{\gamma x} \tag{8.7}$$

$$\hat{I}(x) = \frac{A}{Z_0} e^{-\gamma x} - \frac{B}{Z_0} e^{\gamma x} \tag{8.8}$$

where

$$Z_0 \equiv \sqrt{\left(\frac{R + j\omega L}{G + j\omega C}\right)} \tag{8.9}$$

is the so-called *characteristic impedance* of the line. Now,

$$v(x, t) = A e^{j\omega t - \gamma x} + B e^{j\omega t + \gamma x}$$

or

$$v(x, t) = (A e^{-\alpha x}) e^{j(\omega t - \beta x)} + (B e^{-\alpha(-x)}) e^{j(\omega t + \beta x)}. \tag{8.10}$$

Here the first term on the right-hand side represents a voltage wave travelling in the positive x-direction with phase velocity $u = \omega/\beta$, the amplitude of the wave being exponentially damped by the factor $A e^{-\alpha x}$. The second term describes a similar wave travelling in the negative x-direction; obviously there are concomitant current waves.

The coefficients A, B are determined by the boundary conditions on the line: For instance imagine that a generator is connected at the 'sending end' of the line ($x = 0$) where the voltage and current are denoted by V_s, I_s respectively (see Fig. 8.2). On putting $x = 0$ in eqns (8.7), (8.8) we find

$$V_s = A + B \qquad I_s Z_0 = A - B$$

and so

$$A = \frac{V_s + I_s Z_0}{2} \qquad B = \frac{V_s - I_s Z_0}{2}. \tag{8.11}$$

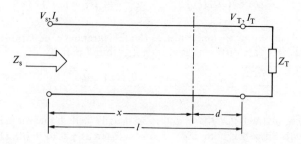

FIG. 8.2. The specification of voltages, currents, and impedances on a general terminated line.

On substituting for A and B in eqns (8.7), (8.8) and using the definitions of the hyperbolic functions, namely,

$$\cosh\theta = (e^\theta + e^{-\theta})/2$$
$$\sinh\theta = (e^\theta - e^{-\theta})/2$$
$$\tanh\theta = \frac{\sinh\theta}{\cosh\theta}$$

it follows that

$$V(x) = V_s \cosh\gamma x - I_s Z_0 \sinh\gamma x \qquad (8.12.\text{a})$$

$$I(x) = I_s \cosh\gamma x - \frac{V_s}{Z_0} \sinh\gamma x \qquad (8.12.\text{b})$$

where $\hat{V}(x)$, $\hat{I}(x)$ have been replaced by $V(x)$, $I(x)$ following the convention adopted in Section 3.3.

At $x = l$, $Z_T = V_T/I_T$. Now we have

$$V_T = V_s \cosh\gamma l - I_s Z_0 \sinh\gamma l$$

$$I_T = I_s \cosh\gamma l - \frac{V_s}{Z_0} \sinh\gamma l$$

or

$$Z_T = \frac{V_s \cosh\gamma l - I_s Z_0 \sinh\gamma l}{I_s \cosh\gamma l - \frac{V_s}{Z_0} \sinh\gamma l}.$$

Further, $Z_s = V_s/I_s$ and so

$$Z_T = \frac{Z_s \cosh\gamma l - Z_0 \sinh\gamma l}{\cosh\gamma l - \frac{Z_s}{Z_0} \sinh\gamma l}$$

or

$$\frac{Z_s}{Z_0} = \frac{Z_T/Z_0 + \tanh\gamma l}{1 + \frac{Z_T}{Z_0} \tanh\gamma l}. \qquad (8.13)$$

For the special case of a lossless line, for which $\gamma = j\beta$

$$\frac{Z_s}{Z_0} = \frac{Z_T/Z_0 + j\tan\beta l}{1 + \frac{Z_T}{Z_0} j\tan\beta l}$$

since $\tanh j\beta l = j\tan\beta l$.

Eqn (8.13) will turn out to be very important in practice but, for the moment, note that if $Z_T = Z_0$ then $Z_s = Z_0$. Furthermore, if $Z_T = Z_0$, then from eqn (8.11), $B = 0$ and there is no wave travelling in the negative x-direction, i.e. there is no reflected wave from the termination; we say that the termination is matched to the line. It is common practice to work in terms of impedances/resistances/reactances, or admittances/conductances susceptances which are normalized with respect to the characteristic impedance of the line in question, e.g. the normalized impedance of the termination is Z_T/Z_0 and a normalized terminating impedance/admittance of value unity represents a matched termination.

If a line is open-circuit then $Z_T = \infty$ and $Z_s = Z_{oc}$ where

$$Z_{oc} = Z_0 \coth \gamma l. \tag{8.14}$$

For a short-circuited line $Z_T = 0$ and $Z_s = Z_{sc}$ where

$$Z_{sc} = Z_0 \tanh \gamma l. \tag{8.15}$$

From eqns (8.13) and (8.14) it follows that

$$\tanh \gamma l = \sqrt{\frac{Z_{sc}}{Z_{oc}}} \tag{8.16}$$

and

$$Z_0 = \sqrt{(Z_{sc} Z_{oc})}. \tag{8.17}$$

This latter relationship has been found already for T- and π-networks (see Section 7.2); indeed the uniform transmission line can be thought of as a ladder network composed of identical symmetrical T- or π-sections. We see that the values of Z_0 and γ for a line can be determined from measurements of Z_{sc} and Z_{oc}.

It will be useful, at this stage, to introduce the neper as a unit specifying the ratio of two voltages or two currents; if the two voltages, say, are V_1, V_2 then their ratio is expressed in nepers as $\ln(V_1/V_2)$. For instance, consider the voltage wave travelling in the positive x-direction which is specified by the first term on the right-hand side of eqn (8.10):

$$\frac{V(x_1)}{V(x_2)} = \frac{A e^{-\alpha x_1}}{A e^{-\alpha x_2}} = e^{\alpha(x_2 - x_1)}$$

and

$$\ln \left| \frac{V(x_1)}{V(x_2)} \right| = \alpha(x_2 - x_1).$$

So we see the convenience of the definition of nepers; the attentuation factor α is in nepers per meter.[†] Now for the important types of practical line $\alpha \sim 10^{-3}$ nepers per meter, or less, and the attenuation per wavelength at a frequency of 100 MHz, say, has the very small value of 3×10^{-3} nepers. Hence to simplify the description of many of the important features of transmission lines, and their uses, we will assume

---

[†] A neper value N is related to a decibel value D through $N = 0.115 D$.

the lines to be lossless; the effects of losses in typical lines will be discussed later as the occasion arises.

For a lossless line ($\alpha = 0$) eqns (8.7) and (8.8) become

$$V(x) = A e^{-j\beta x} + B e^{j\beta x} \tag{8.18}$$

$$I(x) = \frac{A}{Z_0} e^{-j\beta x} - \frac{B}{Z_0} e^{j\beta x}$$

and the reflection coefficient Γ_T of the termination to the line is defined through

$$\Gamma_T \equiv \frac{B e^{j\beta l}}{A e^{-j\beta l}} \quad \text{or} \quad \Gamma_T = \frac{B}{A} e^{j2\beta l}. \tag{8.19}$$

Obviously Γ_T is a complex quantity in general and can be written in polar form as

$$\Gamma_T = |\Gamma_T| e^{j\phi_T}. \tag{8.20}$$

So, using eqns (8.18), it follows that

$$\frac{Z_T}{Z_0} = \frac{1 + \Gamma_T}{1 - \Gamma_T} \tag{8.21}$$

and

$$\Gamma_T = \frac{Z_T/Z_0 - 1}{Z_T/Z_0 + 1}. \tag{8.22}$$

If we substitute for B in eqn (8.18) from eqn (8.19) then

$$V(x) = A \{ e^{-j\beta x} + \Gamma_T e^{j(\beta x - 2\beta l)} \}$$
$$= A e^{-j\beta l} \{ e^{j\beta(l-x)} + \Gamma_T e^{-j\beta(l-x)} \}. \tag{8.23}$$

Now $(l - x) = d$ (see Fig. 8.2) and so we can write

$$V(x) = A e^{-j\beta l} \{ e^{j\beta d} + \Gamma_T e^{-j\beta d} + \Gamma_T e^{j\beta d} - \Gamma_T e^{j\beta d} \}$$
$$= A e^{-j\beta x}(1 - \Gamma_T) + 2 A \Gamma_T e^{-j\beta l} \cos \beta d. \tag{8.24}$$

Bearing in mind that $v(x, t) = V(x) e^{j\omega t}$ we can see that the first term of eqn (8.24) will give rise to a factor $e^{j(\omega t - \beta x)}$ which represents a travelling wave and the second term, which does not contain any x-dependence, gives rise to a 'standing' wave.

For the present let us consider the two extreme possibilities for the termination namely short-circuit and open-circuit: if the termination is a short-circuit then $|\Gamma_T| = 1$ and $\phi_T = \pi$ since there is a node of $V(x)$, i.e. the incident and reflected waves cancel out each other at the termination. In this situation eqn (8.24), with $e^{j\omega t}$ reintroduced, becomes

$$v(x, t) = 2A\{e^{j(\omega t - \beta x)} - \cos \beta d \cdot e^{j(\omega t - \beta l)}\}$$

since $e^{j\pi} = -1$. Also $x = (l - d)$ and so it follows that

$$v(x, t) = 2A e^{j(\omega t - \beta l)}\{e^{j\beta d} - \cos \beta d\}$$
$$= j2A \sin \beta d \{\cos(\omega t - \beta l) + j \sin(\omega t - \beta l)\}.$$

Now the voltage $v(x, t)$ must be real and so taking the real part of the last expression we obtain finally a pure standing wave

$$v(x, t) = -2A \sin(\omega t - \beta l) \sin \beta d$$

$$= (2A \sin \beta d) \cos \left\{(\omega t - \beta l) + \frac{\pi}{2}\right\}.$$

So, for a short-circuit termination, the voltage standing wave pattern has nodes at positions given by $\sin \beta d = 0$, i.e. for $\beta d = n\pi$ or

$$d = \frac{n\lambda}{2} \quad (n = 0, 1, 2, \ldots). \tag{8.25.a}$$

If the termination is an open-circuit then in eqn (8.24), $|\Gamma| = 1$ and $\phi_T = 0$ since there is an anti-node of $V(x)$ at the termination. Thus, since $e^{j\phi_T} = 1$ in this case, we have that the travelling wave term in eqn (8.23) disappears and again a pure standing wave remains, i.e.

$$v(x, t) = 2A[\cos(\omega t - \beta l) + j \sin \omega t] \cos \beta d$$

or

$$v(x, t) = (2A \cos \beta d) \cos(\omega t - \beta l)$$

since the voltage must be real. This standing wave pattern has nodes where $\cos \beta d = 0$, i.e. for

$$\beta d = (2n + 1)\frac{\pi}{2} \quad \text{or} \quad d = (2n + 1)\frac{\lambda}{4}. \tag{8.25.b}$$

The voltage standing wave patterns for short- and open-circuit terminations on a lossless line are sketched in Fig. 8.3.

You may well have wondered what happens to the reflected wave travelling in the negative x-direction when it reaches the sending end of the line. If the source impedance Z_s is equal to Z_0 then the reflected wave will be completely absorbed but otherwise there is at least a partial reflection at the sending end which causes a reflected wave to travel back in the positive x-direction. Suppose that the reflection coefficient at the sending end is Γ_s in which case the wave travelling in the positive x-direction on the line has amplitude

$$A + \Gamma_s \Gamma_T A + \Gamma_s^2 \Gamma_T^2 A + \Gamma_s^3 \Gamma_T^3 A \ldots = A',$$

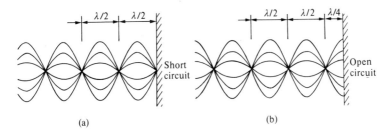

FIG. 8.3. The voltage standing wave pattern for (a) a short-circuited line, and (b) for an open-circuited line.

say. The wave travelling in the negative x-direction has an amplitude

$$\Gamma_T A + \Gamma_s \Gamma_T^2 A + \Gamma_s^2 \Gamma_T^3 A + \ldots = \Gamma_T A'.$$

Hence, the effect of a reflection at the sending end is to change the amplitude of the wave travelling in the positive x-direction from A to A' but, otherwise, the analysis is unchanged.

So far we have considered two extreme examples of termination, that is short-circuit and open-circuit, and also the intermediate example of a matched termination in which situation there is no reflected wave and hence no standing wave pattern. Let us now consider the case of a general termination having a reflection coefficient Γ_T as defined in eqn (8.20). For convenience we will write Γ_T in the form

$$\Gamma_T = e^{-2(a+jb)}$$

which means that $\quad a = \ln \dfrac{1}{\sqrt{|\Gamma_T|}}$ and $b = -\dfrac{1}{2}\phi_T.$

On substituting this new form of Γ_T in eqn. (8.23) and re-introducing the $e^{j\omega t}$ term we have

$$v(x,t) = A e^{j(\omega t - \beta l)}\{e^{j\beta d} + e^{-2a - j2b - j\beta d}\}$$
$$= A e^{j(\omega t - \beta l)}\sqrt{\Gamma_T}\{e^{a + j(b + \beta d)} + e^{-a - j(b + \beta d)}\}$$

or
$$v(x,t) = 2A e^{j(\omega t - \beta l)}\sqrt{\Gamma_T}\cosh\{a + j(b + \beta d)\}.$$

Using the standard expansion for $\cosh[a + j(b + \beta d)]$ and noting that $\cosh j\theta = \cos\theta$ and $\sinh j\theta = j\sin\theta$ we find

$$|\cosh\{a + j(b + \gamma d)\}|$$
$$= \{\cosh^2 a \cos^2(b + \beta d) + \sinh^2 a \sin^2(b + \beta d)\}^{1/2}$$
$$= \{\sinh^2 a + \cos^2(b + \beta d)\}^{1/2}$$

since $\cosh^2 a = (1 + \sinh^2 a)$. Thus we have that

$$|v(x, t)| \propto \{\sinh^2 a + \cos^2 (b + \beta d)\}^{1/2}. \qquad (8.26)$$

For given values of Z_0 and Z_T, a is constant and so $|V(x)|$ is a minimum for $\cos^2 (b + \beta d) = 0$, i.e. for $(b + \beta d) = \pm (2n + 1)\pi/2$. This yields

$$\frac{d}{\lambda} = \frac{1}{4}\left(\pm 1 + \frac{\phi_T}{\pi}\right) \pm \frac{n}{2} \qquad n = 0, 1, 2, 3, \ldots.$$

but a consideration of the extreme cases of $\phi_T = 0$ (open-circuit) and $\phi_T = \pi$ (short-circuit) for $n = 0$ (giving the minimum nearest to the termination) shows that for d to be positive, or zero, then the permissible choice of sign gives

$$\frac{d}{\lambda} = \frac{1}{4}\left(1 + \frac{\phi_T}{\pi}\right) \pm \frac{n}{2}. \qquad (8.27)$$

For the minimum in the voltage standing wave pattern which is nearest to the termination

$$\frac{d}{\lambda} = \frac{1}{4}\left(1 + \frac{\phi_T}{\pi}\right). \qquad (8.28)$$

For a given line and termination $\{2A\sqrt{\Gamma_T}e^{j(\omega t - \beta l)}\}$ is just a scale factor and so the variation of the modulus of the voltage along the line is given by $\{\sinh^2 a + \cos^2 (b + \beta d)\}^{1/2}$; the general form of $|V(x)|$ is sketched in Fig. 8.4.

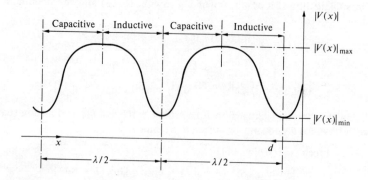

Fig. 8.4. A schematic illustration of the spatial variation along a line of the modulus of the voltage in a standing wave pattern. The changes in sign of the reactance on the line are indicated also.

From eqn (8.26)

$$\frac{|V(x)|_{max}}{|V(x)|_{min}} = \left\{\frac{\sinh^2 a + 1}{\sinh^2 a}\right\}^{1/2}$$

$$= \left\{\frac{\cosh^2 a}{\sinh^2 a}\right\}^{1/2}$$

$$= \frac{(1 + e^{-2a})}{(1 - e^{-2a})}.$$

Now the ratio $|V(x)|_{max}/|V(x)|_{min}$ is called the *voltage standing wave ratio (vswr)* which we will denote by r. Since $e^{-2a} = |\Gamma_T|$ we have

$$r = \frac{1 + |\Gamma_T|}{1 - |\Gamma_T|}$$

and (8.29)

$$|\Gamma_T| = \frac{(r-1)}{(r+1)}.$$

If we use eqn (8.21) and write $Z_T = R_T + jX_T$ and $\Gamma_T = |\Gamma_T|e^{j\phi_T}$, then we can show easily that

$$\frac{R_T}{Z_0} = \frac{(1 - |\Gamma_T|^2)}{[1 + |\Gamma_T|^2 - 2|\Gamma_T|\cos\phi_T]} \quad (8.30)$$

and

$$\frac{X_T}{Z_0} = \frac{2|\Gamma_T|\sin\phi_T}{[1 + |\Gamma_T|^2 - 2|\Gamma_T|\cos\phi_T]}. \quad (8.31)$$

So we now have the very important result that if the value of the vswr and the position of the nearest minimum to the termination can be measured then $|\Gamma_T|$ and ϕ_T can be determined and thence Z_T can be found.

Since the vswr pattern repeats itself along the line with a spatial periodicty of $\lambda/2$, measurements made at any known position d can be used to determine the characteristics of the termination.

Consider a general reflection coefficient $\Gamma(d)$ which is defined through

$$\Gamma(d) = \frac{Be^{j\beta x}}{Ae^{-j\beta x}} = \frac{B}{A}e^{j2\beta(l-d)}$$

or

$$\Gamma(d) = \left(\frac{B}{A}e^{j2\beta l}\right)e^{-j2\beta d}.$$

Thus,

$$\Gamma(d) = \Gamma_T e^{-j2\beta d}$$

242 Transmission lines and waveguides

and using eqn (8.20) we have

$$\Gamma(d) = |\Gamma_T|e^{j(\phi_T - 2\beta d)}. \tag{8.32}$$

Thus by comparison with eqns (8.30) and (8.31) we have

$$\frac{R(d)}{Z_0} = \frac{(1 - |\Gamma_T|^2)}{\{1 + |\Gamma_T|^2 - 2|\Gamma_T|\cos(\phi_T - 2\beta d)\}}$$

$$\frac{X(d)}{Z_0} = \frac{2|\Gamma_T|\sin(\phi_T - 2\beta d)}{\{1 + |\Gamma_T|^2 - 2|\Gamma_T|\cos(\phi_T - 2\beta d)\}}.$$

Using eqn (8.27) we can show that at the position of any minimum in the vswr pattern

$$2\beta d_{min} = \phi_T \pm (2n+1)\pi$$

so that $X(d_{min}) = 0$ ($X = 0$ at the position of a maximum in the vswr pattern also), i.e. the line impedance $Z_d(= V(d)/I(d))$ is *real* at the position of the minima and maxima in the pattern. Furthermore, since $\cos(\phi_T - 2\beta d_{min}) = \cos[\mp(2n+1)\pi] = -1$, we have that

$$\frac{R(d_{min})}{Z_0} = \frac{(1 - |\Gamma_T|^2)}{(1 + |\Gamma_T|^2)}$$

Thus, using eqns (8.29) we have

$$Z(d_{min}) = Z_0/r \tag{8.33}$$

and it follows that

$$Z(d_{max}) = rZ_0.$$

We can illustrate an important general feature of voltage standing wave patterns by considering, for simplicity, a short-circuited lossless line. For this situation eqn (8.15) becomes

$$\frac{Z(d)}{Z_0} = j\tan\beta d \tag{8.34}$$

and it can be seen that $Z(d)$ is purely reactive and that the sign of the reactance changes every quater-wavelength. This variation of the reactance occurs also in general for lines which are not short-circuited as shown schematically in Fig. 8.4.

8.3. Classes of transmission line

We had earlier that the propagation constant γ is given by

$$\gamma = (\alpha + j\beta) = \sqrt{(R + j\omega L)(G + j\omega C)}$$

or
$$(\alpha + j\beta) = j\omega\sqrt{(LC)}\left(1 + \frac{R}{j\omega L}\right)^{1/2}\left(1 + \frac{G}{j\omega C}\right)^{1/2}.$$

If we expand each of the last two factors by using the binomial theorem† and retain terms to order $R^2/\omega^2 L^2$, $G^2/\omega^2 C^2$, then

$$\alpha = \left(\frac{R}{2}\sqrt{\frac{C}{L}} + \frac{G}{2}\sqrt{\frac{L}{C}}\right)\left\{1 - \frac{1}{2}\left(\frac{R}{2\omega L} - \frac{G}{2\omega C}\right)^2\right\}$$
$$+ \frac{R}{2}\sqrt{\frac{C}{L}}\left(\frac{R^2}{16\omega^2 L^2}\right) + \frac{G}{2}\sqrt{\frac{L}{C}}\left(\frac{G^2}{16\omega^2 C^2}\right)$$

$$\beta = \omega\sqrt{(LC)}\left\{1 + \frac{1}{2}\left(\frac{R}{2\omega L} - \frac{G}{2\omega C}\right)^2\right\}.$$

Now if $\omega L/R$, $\omega C/G$ are sufficiently large (the so-called 'high-frequency' condition) then α and ω/β are effectively independent of frequency.

Also we had
$$Z_0 = \sqrt{(R + j\omega L)}/\sqrt{(G + j\omega C)}$$
so that
$$Z_0 \equiv R_0 + jX_0 = \sqrt{\frac{L}{C}}\left(1 + \frac{R}{j\omega L}\right)^{1/2}\left(1 + \frac{G}{j\omega C}\right)^{1/2}$$

and on expanding and approximating as above we find

$$R_0 = \sqrt{\frac{L}{C}}\left\{1 + \frac{1}{2}\left(\frac{R}{2\omega L} - \frac{G}{2\omega C}\right)\left(\frac{R}{2\omega L} + \frac{3G}{2\omega C}\right)\right\}$$
and
$$X_0 = -\sqrt{\frac{L}{C}}\left(\frac{R}{2\omega L} - \frac{G}{2\omega C}\right) \times$$
$$\left[1 - \frac{1}{2}\left\{\left(\frac{R}{2\omega L}\right)^2 + 2\left(\frac{R}{2\omega L}\right)\left(\frac{G}{2\omega C}\right) + 5\left(\frac{G}{2\omega C}\right)^2\right\}\right].$$

Again if $\omega L/R$ and $\omega C/G$ are sufficiently large R_0 is effectively independent of frequency and $X_0 \to 0$.

† Using the binomial theorem
$$(1 + x)^{1/2} = 1 + \frac{x}{2} - \frac{x^2}{8} + \frac{x^3}{16} - \cdots.$$

So a 'high frequency' line has

$$\alpha_{hf} = R/2Z_0 + GZ_0/2$$

(since $X_0 = 0$ which means that $Z_0 = R_0 = \sqrt{(L/C)}$) and

$$u = \frac{\omega}{\beta_{hf}} = \frac{1}{\sqrt{(LC)}}.$$

The approximations which we have made are good enough for most practical purposes if $\omega L/R$ and $\omega C/G$ are greater than 10. Since $\omega C/G \gg \omega L/R$, usually, because G is so small in practical lines, it is the value of $\omega L/R$ which is most critical. L and C (see Sections 1.1.2 and 5.4.2 for examples of calculations of L and C) are independent of frequency, essentially, for practical purposes, from frequencies ~100 kHz up to microwave frequencies; so Z_0 and the phase velocity u are constants of the line, effectively. However R and G are frequency-dependent and hence α is frequency-dependent also. Now $R \propto$ (frequency)$^{1/2}$ at high enough frequencies (see Section 5.4.1), and G is very small and so X_0 is related to (frequency)$^{-1/2}$. Thus X_0 is always very small and the phase angle of Z_0 is very small, and decreases with increasing frequency.

A low loss line can be specified by

$$R \ll \omega L; \qquad G \approx 0$$

in which case

$$\gamma = \sqrt{\{(R + j\omega L)j\omega C\}} = j\omega\sqrt{(LC)}\left(1 - \frac{jR}{\omega L}\right)^{1/2}.$$

On expanding binomially and retaining terms to order $R/\omega L$ only

$$\gamma \approx j\omega\sqrt{(LC)}\left(1 - \frac{jR}{2\omega L}\right) = \frac{R}{2}\sqrt{\frac{C}{L}} + j\omega\sqrt{(LC)}$$

Thus $\alpha = \frac{1}{2}R\sqrt{(C/L)}$ and $u = \omega/\beta = 1/\sqrt{(LC)}$. Also Z_0 can be found in a similar way to be given by

$$Z_0 \approx \sqrt{\frac{L}{C}}\left(1 - \frac{jR}{2\omega L}\right)$$

or,

$$Z_0 \approx \underbrace{\sqrt{\frac{L}{C}}}_{'R_0'} - \underbrace{\frac{jR}{2\omega} \cdot \frac{1}{\sqrt{(LC)}}}_{'jX_0'}.$$

The conditions G, L small and $\omega L \ll R$, $G \ll \omega C$ are met on some lines at low (audio) frequencies and in this situation

$$\gamma = \sqrt{(j\omega RC)} = (\omega RC)^{1/2} \underline{/45°} \text{ since } (\sqrt{j})^4 = 180°.$$

Thus we can write

and
$$\gamma = \sqrt{(\omega RC)} \cos 45° + j\sqrt{(\omega RC)} \cdot \sin 45°$$

Also
$$\alpha = \beta = \sqrt{(\omega RC/2)}.$$

i.e.
$$Z_0 = \sqrt{R/j\omega C} \quad \text{or} \quad Z_0 = \sqrt{R/\omega C} \cdot \underline{/45°}$$

$$R_0 = X_0 = \sqrt{(R/2\omega C)}.$$

It is desirable that transmission lines should be non-dispersive i.e. sinusoidal waves of different frequencies should travel at the same speed. Since the phase velocity $u = \omega/\beta$, the non-dispersive requirement demands that β should be proportional to ω, a linear phase characteristic. We have already seen that this requirement is met for the 'high frequency' and 'low-loss' lines which we have just discussed but not for the 'low frequency' line. Fortunately the ear is not very sensitive to phase differences and so the dispersive character of the 'low frequency' line is not too troublesome.

Signals can be thought of as wave 'packets' which propagate with a group velocity $u_g \equiv d\omega/d\beta$. Hence in order that a 'packet' (e.g. a pulse) should not be distorted in its transmission along a line it is necessary that u_g should be independent of frequency — this requirement is met again if the line has a linear phase characteristic.

For many types of line of practical importance the attenuation is low (see Table 8.1) and the foregoing analyses of lossless lines can be applied with reasonable accuracy. For a line with significant losses eqn (8.26) becomes

and
$$|V(d)| \propto [\sinh^2(\alpha d + a) + \cos^2(\beta d + b)]^{1/2}$$

$$\frac{|V(d)|_{\max}}{|V(d)|_{\min}} = \frac{[\sinh^2(\alpha d + a) + 1]^{1/2}}{\sinh(\alpha d + a)}$$

$$= \coth(\alpha d + a)$$

$$= \frac{1 + e^{-2(\alpha d + a)}}{1 - e^{-2(\alpha d + a)}}$$

$$= \frac{1 + |\Gamma_T| e^{-2\alpha d}}{1 - |\Gamma_T| e^{-2\alpha d}}$$

If $(\alpha d + a) = 3$, then $e^{-2(\alpha d + a)} = 2.5 \times 10^{-3}$ and $|V(d)|_{\max}/|V(d)|_{\min} = 1$

Table 8.1

Type of line	(nepers/metre)	(pF/metre)	Z_0 (ohms)
Parallel cylindrical conductors: 1.4 mm dia. copper wires, spacing 22.5 cm, frequency 1 kHz	1.4×10^{-5}	4.4	900
Screened cable: 1.4 mm dia. copper wires, spacing 4.3 mm, polythene dielectric, cylindrical metal sheath, frequency 1 kHz	4.7×10^{-5}	44	270
Coaxial screened cable: polythene dielectric, outside dia. 5.1 mm.			
10 MHz	5×10^{-3}	56	75 (nominal)
100 MHz	1.3×10^{-2}	56	75
900 MHz	5×10^{-2}	56	75
Balanced twin feeder: polythene dielectric			
10 MHz	1.4×10^{-3}	13.2	300 (nominal)
1000 MHz	1.9×10^{-2}	13.2	300
Coaxial screened cable: polythene dielectric Frequency 10 GHz	1.2×10^{-1}		
Rectangular copper waveguide: Frequency 10 GHz	1.4×10^{-2}		

very closely; that is, under these conditions the input impedance of a line will be equal, closely, to the characteristic impedance irrespective of the value of Z_T. For example, if $\alpha = 0.5$ nepers per metre and $|\Gamma_T| = 0.5$ then, to meet the condition $(\alpha d + a) = 3$, $d = 5.3$ metres. So for lengths of this line greater than 5.3 metres the input impedance is equal to the characteristic impedance irrespective of the load.

8.4. Examples of calculations on transmission lines

Consider a coaxial line which is to operate at a frequency of 1 MHz and which has the following distributed parameters: $R = 10\,\Omega\,\text{km}^{-1}$; $L = 200\,\mu\text{H}\,\text{km}^{-1}$; $G = 10\,\mu\text{S}\,\text{km}^{-1}$; $C = 0.08\,\mu\text{F}\,\text{km}^{-1}$.

(a) Is the 'high frequency' approximation valid at a frequency of 1 MHz?
(b) What are the values of α, β, u?
In answer to question (a)

$$\omega L/R = 126$$
$$\omega C/G = 5.03 \times 10^4.$$

Hence the 'high frequency' formulae can be used with accuracy.
In answer to part (b)

$$Z_0 = \sqrt{\frac{L}{C}} = 50\,\Omega$$

$$\alpha = \frac{R}{2Z_0} + \frac{GZ_0}{2} = 10^{-4} \text{ nepers m}^{-1}$$

$$\beta = \omega\sqrt{(LC)} = 2.5 \times 10^{-2} \text{ rad m}^{-1}$$

$$u = \frac{\omega}{\beta} = \frac{1}{\sqrt{(LC)}} = 2.5 \times 10^8 \text{ m s}^{-1}.$$

It is of interest to note that a section of transmission line can be represented as a two-port network. In this case the 'transmission' parameters can be found easily from eqns (8.12); from eqn (8.12a),

$$V_s = \frac{V(x)}{\cosh \gamma x} + I_s Z_0 \tanh \gamma x$$

and on substituting for I_s from eqn (8.12b)

$$V_s(1 - \tanh^2 \gamma x) = \frac{V(x)}{\cosh \gamma x} + \frac{Z_0 I(x) \sinh \gamma x}{\cosh^2 \gamma x}$$

Since $(1 - \tanh^2 \gamma x) = (\cosh^2 \gamma x - \sinh^2 \gamma x)/\cosh^2 \gamma x = 1/\cosh^2 \gamma x$ we obtain

$$V(s) = V(x) \cosh \gamma x + I(x) Z_0 \sinh \gamma x.$$

In a similar fashion we can show that

$$I(s) = \frac{V(x)}{Z_0} \sinh \gamma x + I(x) \cosh \gamma x.$$

Hence the transmission matrix is

$$\begin{bmatrix} \cosh \gamma x & Z_0 \sinh \gamma x \\ \dfrac{\sinh \gamma x}{Z_0} & \cosh \gamma x \end{bmatrix}.$$

Impedance-, admittance-, and hybrid-parameter matrices can be derived also.

A line of length $\lambda/4$ has useful transforming properties between lines connected at its input and output ports; if $l = \lambda/4$ then $\beta l = \pi/2$ and $\tan \beta l = \infty$. So for a lossless section of line we have, from eqn (8.13)

$$\frac{Z_s}{Z_0} = \frac{Z_0}{Z_T} \quad \text{or} \quad Z_s = \frac{Z_0^2}{Z_T}.$$

Hence a quarter wavelength line of characteristic impedance $Z_0 = \sqrt{(Z_s Z_T)}$ will form a (maximum power) transformer between lines having characteristic impedances Z_s and Z_T (or between a source of impedance Z_s and a line of impedance Z_T).

Calculations on lines which can be treated as lossless are greatly facilitated by the use of impedance charts (or transmission lines charts) of which the 'Smith' chart is the most commonly encountered example. The theoretical derivation of this type of chart is given in many places[†] and so we will give what is largely an operational description.

For a lossless line we have from eqn (8.23), on reintroducing the $e^{j\omega t}$ term,

$$\frac{v(x, t)}{A} = e^{j(\omega t - \beta x)} \{1 + \Gamma_T e^{j2\beta(x-l)}\}$$

and using the polar form for Γ_T we have

$$\frac{v(x, t)}{A} = e^{j(\omega t - \beta x)} \{1 + |\Gamma_T| e^{j(\phi_T - 2\beta d)}\}.$$

The two phasors on the right-hand side of this equation can be represented in a vectorial phasor diagram as shown in Fig. 8.5(a). For a given value of $|\Gamma_T|$ then, as d varies the point T moves around a circle, centred on O. If d is increasing, that is if we are moving towards the 'sending end' at which the generator is connected, then the point T moves in a clockwise sense around the circle. Conversely, if d is decreasing we move towards the termination (or 'load') and T moves in an anticlockwise sense around the circle. $V(x)$ is proportional to MT and varies cyclically with d as T moves around its circle, the minimum value occuring at S where $(\phi_T - 2\beta d) = (2n + 1)\pi$ and the maximum value occurring at N where $(\phi_T - 2\beta d) = 2n\pi$. The circle on which T lies is a 'circle of constant $|\Gamma_T|$', and hence a 'circle of constant vswr'

[†] For a detailed discussion of the theory of impedance charts see, for example, Ramo, S., Whinnery, J.R., and van Duzer, T. *Fields and waves in communication electronics.* Wiley, New York (1965).

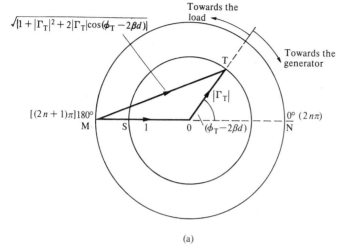

(a)

FIG. 8.5. The construction of a Smith chart: (a) The construction of a circle of constant reflection coefficient (and hence of constant vswr). (b) Circles of constant normalized resistance. (c) Circles of constant normalized reactance. (d) This figure shows how the circles of constant resistance and constant reactance are superimposed in a Smith chart.

also; $r = V_{max}/V_{min} = $ MN/MS. For a circle of zero radius $|\Gamma_T| = 0$ and $r = 1$ whereas for a circle of radius ON $|\Gamma_T| = 1$ and $r = \infty$.

It can be shown (see the cited texts) that families of circles having normalized resistance R/Z_0 and normalized reactance X/Z_0 as parameters, respectively, as depicted in Fig. 8.5(b) and (c), can be superimposed on the diagram of Fig. 8.5(a). So, finally, we have a pair of sets of orthogonal curvilinear coordinates superimposed on the voltage reflection coefficient diagram: this is the Smith chart (Fig. 8.5(d)). A transparent rotatable radial cursor, which is calibrated along its length in $|\Gamma_T|$ and vswr, is placed on the diagram with its central point ($|\Gamma_T| = 0$ or $r = 1$) coincident with O. The best way of continuing the description of the chart is by means of some illustrative examples:

Example 1. A line is terminated by a (normalized) impedance $(2 - j3)$. What is the vswr on the line?
The intersection of the circles $R/Z_0 = 2$ and $X/Z_0 = -3$ lie on the vswr circle $r = 7$: (see Fig. 8.6)

Example 2. What is the normalized admittance corresponding to a normalized impedance $(1.5 + j2.5)$?
From eqn (8.13) we have that for a lossless line ($\gamma = j\beta d$) that

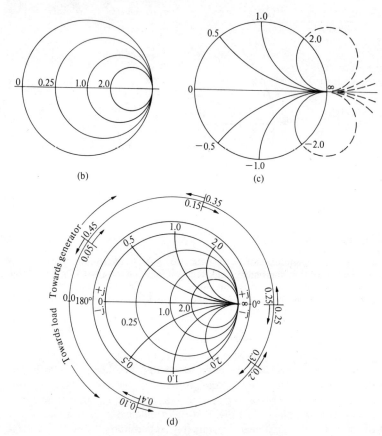

FIG. 8.5 (b–d)

$$\frac{Z(d)}{Z_0} = \frac{Z_T/Z_0 + j \tan \beta d}{1 + j\frac{Z_T}{Z_0} \tan \beta d}. \tag{8.35}$$

Thus,

$$\frac{Y(d)}{Y_0} = \frac{1 + j\frac{Z_T}{Z_0} \tan \beta d}{Z_T/Z_0 + j \tan \beta d}. \tag{8.36}$$

Now since $\beta\lambda/4 = \pi/2$ we have

$$Z\left(d + \frac{\lambda}{4}\right) = \frac{Z_T/Z_0 + j \tan (\beta d + \pi/2)}{1 + j\frac{Z_T}{Z_0} \tan (\beta d + \pi/2)}$$

Transmission lines and waveguides 251

If we now use the identity $\tan(\theta + \pi/2) = -(\tan\theta)^{-1}$ we can show very easily that

$$Z(d + \lambda/4) = Y(d). \qquad (8.37.a)$$

It is obvious that the converse will be true also, namely,

$$Y(d + \lambda/4) = Z(d). \qquad (8.37b)$$

For the problem posed find the intersection of the $R/Z_0 = 1.5$ and $X/Z_0 = 2.5$ circles and note the value of the vswr indicated by the cursor ($r \approx 6.5$)(see Fig. 8.7). If we now rotate through $180°$ on the circle of constant vswr, which operation corresponds to travelling a distance $\lambda/4$ along the line, we arrive at the point having coordinates $R/Z_0 \approx 0.18$, $X/Z_0 \approx -0.3$, so the normalized admittance $Y/Y_0 = 0.18 - j0.3$ corresponds to the normalized impedance $Z/Z_0 = 1.5 + j2.5$.

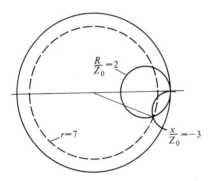

FIG. 8.6.

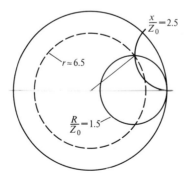

FIG. 8.7.

Example 3. The vswr on a line has the value 1.5. Find the normalized impedances at the positions of a voltage maximum and of a voltage minimum.

The positions at which the voltage maximum and minima occur are specified by the points at which the vswr = 1.5 circle intersects the horizontal diameter of the Smith chart. In this case the points corresponding to the maxima and minima have coordinates (1.5 + j0.0) and (0.66 + j0.0) respectively, which give the normalized impedances; notice that 0.66 = 1/1.5 in accordance with eqns (8.33)(see Fig. 8.8).

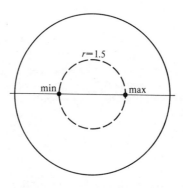

FIG. 8.8

Example 4. A line is terminated by a normalized impedance (0.8 + j0.5). Find

(a) The angle of the reflection coefficient.
(b) The vswr.
(c) The distance from the termination of the first minimum in the voltage standing wave pattern.

(a) Find the point with coordinates (0.8 + j0.5) and then use the cursor to read the angle (96°) on the circular scale; this is the angle of the reflection coefficient (the modulus of the reflection coefficient is about 0.3)(see Fig. 8.9).

(b) At the point (0.8 + j0.5) the cursor scale gives $r \approx 1.8$.

(c) The reading on the wavelength scale at the angle of 96° is 0.116λ. If we move from the 'position' (0.116λ) of the termination 'toward the generator' (i.e. if we move around the circle of constant vswr in a clockwise sense) we travel a distance of (0.500λ − 0.116λ), or 0.384λ, in order to reach the first minimum (at its angular position of −180°).

Transmission lines and waveguides 253

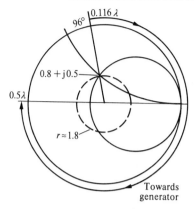

FIG. 8.9.

Example 5. The minima of a voltage standing wave pattern shift 0.1λ towards the termination of a line when the original terminating load is replaced by a short-circuit. If the original vswr was given by $r = 2.5$ find the normalized impedance and admittance of the original terminating load.

This problem illustrates a very commonly-encountered practical situation. We note that the measurement of r and of the position of a minimum, or of two adjacent minima, can be made on a portion of the line which is many wavelengths distant from the termination since the standing wave pattern repeats itself with a spatial periodicity of $\lambda/2$. If λ is known as well as the distance of a minimum from the termination then the distance from the termination to the first minimum can be calculated. However in this particular problem the information regarding the shift of position of the standing wave pattern through a distance of 0.1λ, on short-circuiting the line, tells us that, originally, the first minimum was distant 0.1λ from the terminating load.

Starting at the position of a vswr minimum rotate the cursor through an angle corresponding to 0.1λ towards the load and then note the values of R/Z_0 and X/Z_0 for the two circles which intersect at the point corresponding to $r = 2.5$; the result is $(0.56 - j0.56)$, very closely (see Fig. 8.10).

By rotating through $180°$ at constant vswr the normalized admittance of the termination is found to be $(0.88 + j0.88)$ very closely.

This example, and Example 2, illustrate the fact that these types of problems can be discussed in terms of admittances instead of impedances if the Smith chart is rotated, as a whole, through $180°$; see Fig. 8.11. In this case the resistance and reactance circles become circles of constant conductance and constant susceptance, of course.

254 Transmission lines and waveguides

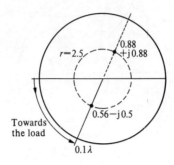

FIG. 8.10

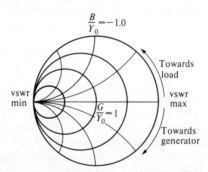

FIG. 8.11. The Smith chart as an admittance chart.

Example 6. A line is terminated such that there is a voltage standing wave pattern with $r = 3.0$ and the first minimum is at a distance $0.286\,\lambda$ from the termination. Find the normalized admittance of the load and then find the position and admittance of a matching 'stub' which will match the termination to the line. What is the length of the 'stub' line?

Starting at the position of a vswr minimum rotate the cursor through an angle corresponding to $0.286\,\lambda$ towards the load; this gives the normalized admittance of the load $(0.35 - j0.2)$ (see Fig. 8.12).

Now rotate the cursor towards the generator at constant $r = 3.0$ to meet the $G/Y_0 = 1$ circle after traversing an angle corresponding to a distance along the line of $0.203\,\lambda$. At this position the normalized

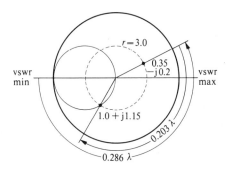

FIG. 8.12.

admittance on the line is $(1.0 + j1.15)$. If at this position we connect a stub line of normalized susceptance (-1.15) in parallel (or 'shunt') with the main line then the normalized admittance of the portion of the main line from the stub to the termination appears to be $(1.0 + j0.0)$, i.e. as far as the generator is concerned the line appears to be terminated by a matched load.

A stub line which has a purely reactive admittance of variable magnitude and sign can be realized conveniently by a short-circuited line of variable length. So in this example we require a short-circuited stub line for which

$$\frac{Y}{Y_0} = -j1.15 = -j \cot \beta l. \qquad \text{(see eqn 8.34)}$$

Thus we require $\beta l = 0.716$ rad and so the required length of the stub line is 0.114λ.

The attainment of a matched condition is usually the goal in transmission line circuits for three important reasons:
 (a) Maximum power is coupled into the terminating load.
 (b) The existence of a standing wave in a length of line due to a mismatch increases the total of the losses in the line; in high power systems there could even be electrical breakdown of insulators due to the high electric fields which could be set up.
 (c) The frequency of the generator can be affected by a reflected wave reaching it; we would say that the frequency of the generator was being 'pulled' by the effect of the mismatched load.

We have seen in Section 3.5 that a source of internal impedance $Z_g (= R_g + jX_g)$ is said to be conjugately matched for maximum power transfer to a load of impedance $Z_L (= R_L + jX_L)$ if $Z_g = Z_g^*$ i.e. if $R_g = R_L$ and $X_g = -X_L$. A conjugately matched situation is unsuitable for broadband operation since the condition $X_g = -X_L$ will be true, even approximately, only over a narrow band of frequencies. A reflectionless match (where $R_g = R_L$ and $X_g = X_L$) is better suited to broadband requirements; the power delivered to the load is less than for the conjugate match, by the factor R_L^2/Z_L^2 or $10 \log (R_L^2/Z_L^2)$ db, but a higher average power will be delivered over a wider band of frequencies.

The situation is a little more complex for the case of a source (e.g. a receiving aerial) which is connected to a load (a receiver) via a transmission line; the source impedance Z_g, the characteristic impedance Z_0 of the line, and the impedance of the load Z_L may all be different in general. In a so-called 'flat', or untuned, feeder both the load and the source are matched to the characteristic impedance of the line by means of suitable matching devices.

For the so-called tuned feeder the load is not matched to the line but the source is conjugately matched to the input impedance of the line at the sending end. This method of matching is applicable only where aggregate of losses in the line is small (~ 1 db or less) so that the simplicity of the system outweighs the potentially higher efficiency of the more complex tuned feeder.

8.5. Waveguides considered as transmission lines

Electromagnetic waves, carrying energy and signals, can be guided conveniently along hollow metal pipes[†] although such simple structures become a practical proposition only at frequencies of about 3 GHz and above (i.e. in the 'microwave' region) since their lateral dimensions are of the same order of magnitude as a wavelength (10 cm at 3 GHz). In this frequency region the values of the attenuation constants of waveguides are of the order of 0.01 nepers/metre which is very considerably better than for airspaced coaxial line of equal cross-sectional area.

Obviously in the case of waveguides there is no 'flow and return' circuit and line voltages and line currents cannot be uniquely defined; however, as we shall see, nominal voltages and currents can be defined so that transmission line theory can be utilized.

[†] Physically speaking the wave is reflected at the air–metal interface and hence is trapped within the waveguide. Another type of trapping effect due to a sharp change in refractive index (eg in going from glass to air) is utilized in light pipes which are finding applications as the pressure for increased informed handling capacity pushes telecommunications into the optical region of the electromagnetic spectrum.

Transmission lines and waveguides

The modes of propagation of electromagnetic waves in waveguides are governed by Maxwell's equations and the boundary conditions on the alternating electric and magnetic fields at the inner surfaces of the waveguide walls. A detailed analytical description of the physical situation lies outside the province of this text‡ and so only the general features will be outlined:

(a) There can be a component of electric field E or magnetic field H in the direction of propagation of the wave along the waveguide; this is in contrast to the situation for electromagnetic waves in an unbounded medium where **E** and **H** are transverse to the direction of propagation.

A mode of propagation is denoted as an E_{mn} or an H_{mn} mode if there is a component of electric or magnetic field, respectively, in the direction of propagation (an alternative convention denotes these modes as TM_{mn} and TE_{mn}, respectively, standing for 'transverse magnetic' and 'transverse electric').

(b) The wavelength λ_g of a mode in a waveguide is related to the free space wavelength λ_0 through

$$\frac{1}{\lambda_g^2} = \frac{1}{\lambda_0^2} - \frac{1}{\lambda_c^2} \qquad (8.38)$$

λ_c is related to the transverse dimensions of the waveguide and you can see that $\lambda_g > \lambda_0$. The physical corollary of the fact that λ_g tends to infinity as λ_0 increases and tends towards λ_c is that the waveguide acts as a high-pass filter and only waves of frequency greater than c/λ_0 are propagated.

The attenuation factor as a function of free space wavelength for a number of modes in rectangular waveguide is shown schematically in Fig. 8.13; similar curves exist for circular waveguide. We can see that if $\lambda_{c_{20}} < \lambda_0 < \lambda_{c_{10}}$ then only the H_{10} mode can be propagated. This is very convenient since it means that if we operate in the appropriate frequency region we are certain of the mode of propagation (the H_{10} mode is called the dominant mode) and hence of the pattern of electric and magnetic fields in the waveguide. N.B. modes other than the dominant one can be set up in a waveguide, for instance at a discontinuity, but they are very highly attenuated and so have significant amplitude only in the vicinity of the discontinuity. Such modes are termed evanescent modes.

For the dominant mode in a rectangular waveguide of cross-sectional dimensions $a \times b$ $(a > b)$ then $\lambda_c = 2a$. For waveguide of circular

---

‡For more detailed treatments see, for example, Robinson, F.N.H. *Electromagnetism*, Oxford Physics Series, Oxford (1973) and Bleaney, B.I. *Electricity and magnetism* (Third edition) Oxford University Press, Oxford (1976).

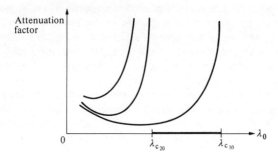

FIG. 8.13. A sketch of the attenuation factor as a function of free space wavelength for three modes in rectangular waveguide including the dominant H_{10} mode.

section of diameter d the dominant mode (the H_{11} mode) has a critical wavelength $\lambda_c = 1.71d$. A waveguide of circular section has many uses but a disadvantage is that the planes of polarization of the electromagnetic waves are not fixed as in a rectangular waveguide.

(c) The pattern of electric and magnetic fields for the dominant mode in rectangular waveguides is shown in Fig. 8.14; the pattern propagates along the waveguide at the phase velocity of the wave. It is very important to notice that eddy currents are induced in the walls of the waveguide (within the skin depth of course) and that these form an integral part of the process of wave propagation. If the current pattern is distorted by a hole or slot in the waveguide wall then the pattern of electric and magnetic fields in the waveguide is disturbed also.

At the beginning of this section it was hinted that transmission line theory could be applied to waveguides; in order to do so we must define a line voltage, a line current and, by implication, a characteristic impedance. Before doing that let us first consider the question of power flow along a transmission line. We saw in Section 3.5 that if $V = \hat{V} \cos \omega t$ and $I = \hat{I} \cos(\omega t + \phi)$ then the average power P_{av} is given by

$$P_{av} = \tfrac{1}{2} \hat{V}\hat{I} \cos \phi.$$

It should be noticed that if V, I are written as $V = \hat{V} e^{j\omega t}$ and $I = \hat{I} e^{j(\omega t + \phi)}$, then

$$P_{av} = \tfrac{1}{2} \operatorname{Re}(VI^*) \qquad (8.39)$$

where VI^* is called the *complex power*.

Now the average power per unit area incident on a plane surface

Transmission lines and waveguides

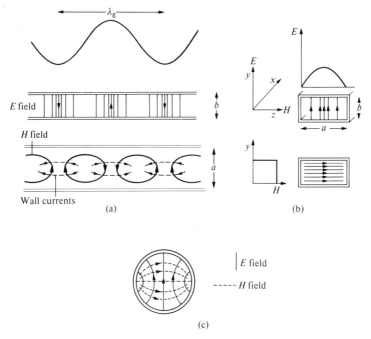

FIG. 8.14. (a), (b) Electric and magnetic field patterns for the dominant H_{10} mode in rectangular waveguide. (c) Electric and magnetic field patterns for the dominant H_{11} mode in circular waveguide.

normal to the direction of propagation of an electromagnetic wave is given by[†]

$$P_{av} = \tfrac{1}{2} \hat{E} \hat{H} \text{ W m}^{-2} \qquad (8.40)$$

where $\hat{E}$, $\hat{H}$ are the amplitudes of the electric and magnetic field vectors in the wave.

We are now going to derive an expression for the characteristic impedance of rectangular waveguide by making use of a defined line voltage and a defined line current. Using these definitions we will derive P_{av} according to eqn (8.39) and equate it to P_{av} as given by eqn (8.40). Hence we will be able to eliminate one of the two unknown constants employed in the definitions of line voltage and line current but the other constant will be chosen arbitrarily to be unity in accordance with widespread practice. This is not a great impediment since most measurements on waveguide systems are relative measurements e.g. ratios of voltages as in the measurement of the vswr.

[†] See the previously cited texts (p. 257) on electromagnetism, for instance.

260 Transmission lines and waveguides

On a transmission line the voltage (or, more strictly, potential difference) between the two conductors is equal to the line integral of the electric field along any path which joins the two conductors. An obvious problem arises in waveguides, as demonstrated by the case of rectangular waveguide (see Fig. 8.14), in that the integral of the electric field E_y over a path between the lower and upper waveguide walls, parallel to the narrow wall, say, depends on the value of the coordinate z at which the integral is evaluated; so a unique value cannot be allocated to the line voltage. We will define the line voltage $V(x)$ through

$$V(x) \equiv A \int_{y=0}^{b} [E_y(x)]_{z=a/2}\, dy \qquad (8.41)$$

where A is a constant and the integral is evaluated at $z = a/2$.

The nominal line current is defined in terms of the magnetic field **H** of the waveguide mode but there is an analogous problem to that involved in the definition of the line voltage and so we will define the line current $I(x)$ in terms of the magnitude of the magnetic field adjacent to the broad waveguide wall and at a value of z corresponding to the centre of the waveguide viz:

$$I(x) \equiv D[H_z(x)]_{\substack{y=b \\ z=a/2}}. \qquad (8.42)$$

In order to obtain an expression for the characteristic impedance Z_0 of the waveguide (which we will assume to be lossless) we will obtain expressions relating $dV(x)/dx$ to $I(x)$ and $dI(x)/dx$ to $V(x)$ and, by comparison with the Telegrapher's equations (see eqn (8.4)) we will be able to obtain an expression for Z_0. Now

$$E_y(x) = \hat{E} \sin\frac{\pi z}{a}\, e^{j(\omega t - \beta x)} \qquad (8.43)$$

or, at $z = a/2$,

$$E_y(x) = \hat{E} e^{j(\omega t - \beta x)}$$

and

$$V(x) = bA\hat{E} e^{j(\omega t - \beta x)}. \qquad (8.44)$$

The magnitudes and directions of the electric and magnetic fields in an electromagnetic wave are interrelated and it can be shown, using Maxwell's equations (see Appendix 3), that

$$\frac{\partial E_y(x)}{\partial x} = -\mu_0 \frac{\partial H_z(x)}{\partial t} \qquad (8.45)$$

Transmission lines and waveguides 261

and, since all of the components of **E** and **H** have an $e^{j\omega t}$ time dependence, that

$$H_z(x) = \frac{\beta E_y(x)}{\omega \mu_0} \qquad (8.46)$$

and so

$$I(x) = \frac{D\beta E_y(x)}{\omega \mu_0}. \qquad (8.47)$$

Now in the waveguide

$$E_y(x) = \hat{E} \sin \frac{\pi z}{a} \cdot e^{j(\omega t - \beta x)}$$

or

$$E_y(x) = \frac{V(x)}{bA} \cdot \sin \frac{\pi z}{a} \qquad (8.48a)$$

$$H_z(x) = \frac{I(x)}{D} \cdot \sin \frac{\pi z}{a}. \qquad (8.48b)$$

Also

$$H_x(x) = -\frac{j}{\omega \mu_0} \frac{\partial E_y(x)}{\partial z}$$

(using Maxwell's equations; see Appendix x)

or

$$H_x(x) = -\frac{j\pi \hat{E}}{a\omega \mu_0} \cdot \cos \frac{\pi z}{a} \cdot e^{j(\omega t - \beta x)}. \qquad (8.48c)$$

Using eqns (8.45), (8.48a), and (8.48b) we have that

$$\frac{\partial V(x)}{\partial x} = -j\omega \mu_0 \cdot \frac{bA}{D} \cdot I(x). \qquad (8.49)$$

Again from Maxwell's equations (as shown in Appendix 3) we have

$$\frac{\partial H_x}{\partial z} - \frac{\partial H_z}{\partial x} = j\omega \epsilon_0 E_y$$

and using eqn (8.48a), (8.48b), and (8.48c) we find

$$\frac{\partial I(x)}{\partial x} = \frac{jD}{\omega \mu_0 bA} \left[\left(\frac{\pi}{a}\right)^2 - \omega^2 \mu_0 \epsilon_0 \right] V(x). \qquad (8.50)$$

So, by comparison with the Telegrapher's equation we have

$$Z_0 = \sqrt{\left\{\frac{\frac{\mu_0 bA}{D}}{\left(\frac{D}{\omega^2 \mu_0 bA}\right)\left(\omega^2 \mu_0 \epsilon_0 - \frac{\pi^2}{a^2}\right)}\right\}}.$$

By using $\mu_0 \epsilon_0 = 1/c^2$; $c = \omega\lambda/2\pi$; $\lambda_c = 2a$ and eqn (8.38) it follows that

$$Z_0 = \frac{bA}{D} \cdot \sqrt{\left(\frac{\mu_0}{\epsilon_0}\right)} \cdot \frac{\lambda_g}{\lambda_0}. \qquad (8.51)$$

The constant A/D remains to be defined and to accomplish this definition we now equate the power in the wave to the power given by the nominal line voltage and current. For the electromagnetic wave there is an expression for the complex power which is analogous to that given in eqn (8.39) namely

$$\text{Average power in wave} = \frac{1}{2}\int_0^b \int_0^{a_0} \text{Re}\,(E_y H_z^*)\,\mathrm{d}y\,\mathrm{d}z.$$

If we use eqn (8.46) to substitute for H_z in terms of E_y and then use eqn (8.43) we find

$$\text{Average power in wave} = \frac{ab\beta \hat{E}^2}{4\mu_0 \omega}. \qquad (8.52)$$

Turning now to eqn (8.39) and substituting for $V(x)$ from eqn (8.44) and $I(x)$ from eqn (8.47) (using eqn (8.43) for E_y) we find

$$\tfrac{1}{2}\,\text{Re}\,[V(x)I(x)^*] = \frac{b\beta AD\hat{E}^2}{2\mu_0 \omega}. \qquad (8.53)$$

On equating the right-hand sides of eqns (8.52) and (8.53) we obtain the requirement that $AD = a/2$; for convenience we choose $A = 1$ so that $D = a/2$.

Thus with these definitions

$$Z_0 = \frac{2b}{a} \cdot \sqrt{\left(\frac{\mu_0}{\epsilon_0}\right)} \cdot \frac{\lambda_g}{\lambda_0}.$$

For the H_{10} mode in a rectangular waveguide $a = 2b$ usually and also $(\mu_0/\epsilon_0)^{1/2} = 120\pi\,\Omega$ (the so-called wave impedance of free space) and so

Transmission lines and waveguides

$$Z_0 = 120\pi\lambda_g/\lambda_0. \qquad (8.54)$$

In normal operating conditions $1 \leqslant \lambda_g/\lambda_0 \leqslant 2$. It is important to note that $Z_0 \propto b$, the narrow dimension of the waveguide — analogous arguments prevail for other modes of propagation and for other types of waveguide. So we have attained our objective of defining a characteristic impedance for a waveguide and thenceforward can use standard transmission line theory to describe and analyse microwave circuits.

An alternative, useful, approach to the description of microwave systems uses 'scattering' coefficients' to describe the properties of the elements of the system but we will not enter this field here.

It is of interest to make an estimate of the maximum power handling capacity of air-filled waveguide; if for an 'X'-band waveguide we assume $a \approx 2.5$ cm, $b \approx 1$ cm, $\lambda_g \approx 1.5\,\lambda_0$, and $c = 3 \times 10^8$ m s^{-1}, then, since $\beta = 2\pi/\lambda_g$, so that $\beta/\omega = \lambda/(c\lambda_g)$, we have

Average power ≈ 1.5 MW

where it has been assumed that the maximum value of $\hat{E}$ is that corresponding to the breakdown of air, namely 3×10^6 V m^{-1} approximately. If a safe working value of the maximum electric field is taken to be half of this value then the maximum average power becomes about 0.4 MW.

So far our discussion has been concerned almost entirely with uniform transmission lines. Any form of discontinuity in an otherwise uniform line whether it be an obstruction, an aperture in a containing conductor (e.g. the outer sheath of a coaxial line or the wall of a waveguide), a change in physical dimensions, a bend, a twist, or a junction, will cause a partial reflection, at least, of an incident wave. As far as bends, twists, and changes in physical dimensions are concerned the golden rule is to make the 'change' as gradual as possible so that the vswr created by the reflection is tolerably small to avoid the necessity of having a matching unit in association with each such element. A wide variety of 'obstacles' act as circuit elements in waveguides: obstacles such as metal posts or diaphragms containing apertures (see Fig. 8.15) can be capacitive, inductive, or resonant in nature. A prediction of the effects of such obstacles requires a detailed calculation of the pattern of the electromagnetic field around the obstacle in both the propagating mode and the evanescent modes. Such calculations are very complex in nature and even then may not yield useful results because of the approximations which have to be made; so resort is made to tabulated data[†] which have been assembled from experimental measurements in the main.

It is worth pointing out that a circuit element behaves capacitively

---

[†] See, for instance, *Microwave engineer's handbook and buyer's guide.* Horizon House, Dedham, Mass. (published annually).

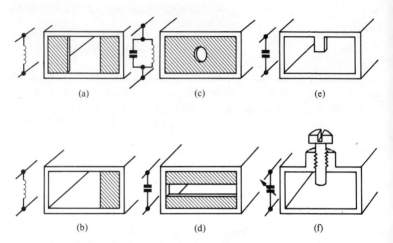

FIG. 8.15. Waveguide circuit elements: (a), (b), (d) diaphragms; (c) a resonant iris; (e) a fixed post; (f) a 'tuning' screw (capacitive for small penetrations).

if the stored energy in the electromagnetic field pattern which is set up around it is associated principally with the electric field component (c.p. the energy stored in the electric field in a charged capacitor; see Problem 1.2). On the other hand if the stored energy is associated principally with the magnetic component of the field then the element behaves inductively (c.p. the energy stored in the magnetic field of an inductor). If the maximum stored energy in the electric component of the field is equal to the maximum stored energy in the magnetic component then the element is resonant. For example a metal post whose penetration into a waveguide can be varied acts capacitively for small penetrations, resonantly for intermediate penetrations and inductively for large penetrations. Capacitive, resonant, and inductive apertures (or irises) are illustrated in Fig. 8.15; design data are tabulated extensively and/or given in graphical form.†

If an obstacle extends for a small distance only along the waveguide then the electric field will have the same value, substantially, on each side of it. Hence in thinking of the waveguide as a transmission line the obstacle acts as a *shunt* element (see Fig. 8.15). It follows from this that the shunt stub matching procedure which was described in Example 6 of Section 8.4 can be used; a unit operating on this principle is called a 'sliding-screw' matching unit in that the position of a variable penetration screw can be altered over a distance of a few guide wavelengths. Other commonly-used forms of matching unit consist of two or three fixed screws spaced a distance $3\lambda_g/8$ apart. A shunt stub can also

† See footnote on p. 263.

be realized by connecting a variable-length, short-circuited waveguide to the main line via a shunt-T junction.

Another important type of microwave circuit element is the directional coupler; this is a four-port element in which power fed into port 1, from a matched generator, divides in a certain ratio (the 'coupling ratio') between ports 2 and 4 with no power emerging from port 3, providing that ports 2, 3, and 4 are terminated by matched terminations. Common values for the coupling ratio are 3 db, 6 db, and 10 db. In a special type of directional coupler, which is usually called a hybrid junction, the power divides equally between ports 2 and 4; two particular realizations are the 'hybrid ring' and the 'magic-T'. Hybrid junctions have some very important applications of which one in particular is the microwave bridge circuit: the junction is fed from a matched generator connected to port 1 and port 2 is terminated by a matched load. If the termination of port 4 is not matched then power is reflected and some enters port 3; however, if the termination to port 3 is matched then no power enters port 3.

In simple terms attenuators and phase-shifters consist of resistive and dielectric vanes, respectively, which can be made to project into the waveguide. Matched loads for low power levels (less than 1 W, roughly) consist of wedge-shaped blocks of lossy materials such as a mixture of finely-ground carbon and a binder, or of 'polyiron'; the wedge shape assists in minimizing the reflection of waves from the load. At high power levels forced cooling of the load may be necessary.

Microwave generators range from low power klystrons, and a variety of solid-state sources, (up to 1 W roughly) to high power magnetrons (~ 1 MW peak pulse power). There is a wide variety of detectors also but the most commonly-used form is a solid-state p–n junction rectifier which is connected to a metal probe which projects into the waveguide. Voltage standing wave ratios are determined by means of movable probes which project through a narrow longitudinal slot in the broad wall of the waveguide.

Microwave cavities are of very great importance in three respects:

(i) If their length is equal to $n\lambda_g/2$, they behave like resonant circuits and since their Q-values can be in the range $10^3 - 10^4$ the electric and magnetic fields inside them are very much higher than in non-resonant lines; this feature is made use of in the study of the microwave spectra of materials.

(ii) Variable length cylindrical cavities can be used as wavelength meters since they can be machined very precisely and because of the above-mentioned high Q-values.

(iii) Because of the properties outlined in (ii) they can be used as the basis of microwave filters.

266 Transmission lines and waveguides

The fact that a cavity behaves like an L–C–R circuit when it is near resonance is shown by the following analysis: Consider a length of waveguide which is short-circuited at both ends (see Fig. 8.16(a)). Using eqn (8.13) the impedance Z, looking from one end, is given by

$$\frac{Z}{Z_0} = \tanh(\alpha + j\beta)l$$

or

$$\frac{Z}{Z_0} = \frac{\tanh \alpha l + \tanh j\beta l}{(1 + \tanh \alpha l \tanh j\beta l)}.$$

If we assume that the losses are small then (αl) is small and $\tanh \alpha l \approx \alpha l$; furthermore $\tanh j\beta l = j\tan \beta l$. Now near resonance $l \approx n\lambda_g/2$ so that $\beta l \approx n\pi$ and $|\tan \beta l| \ll 1$. So we have

$$\frac{Z}{Z_0} \approx \frac{\alpha l + j\tan \beta l}{1 + j\alpha l \tan \beta l} \quad \text{or} \quad \frac{Z}{Z_0} \approx \alpha l + j\tan \beta l.$$

For an L–C–R circuit

$$Z = R + j\left(\omega L - \frac{1}{\omega C}\right)$$

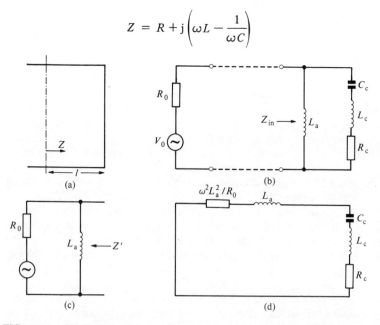

FIG. 8.16. Equivalent lumped circuits for microwave cavities: (a) a short-circuited length of line; (b) a network representing a generator coupled to a cavity; (c) the effective impedance 'seen' from the cavity; (d) the equivalent network for the loaded cavity.

and since $\omega_0^2 \equiv (LC)^{-1}$ this can be rewritten as

$$Z = R + jL(\omega^2 - \omega_0^2)/\omega = R + jL(\omega + \omega_0)(\omega - \omega_0)/\omega.$$

Near resonance $\omega \approx \omega_0$ so that $(\omega + \omega_0)/\omega \approx 2$ and so we have

$$Z = R + j2L\delta\omega$$

where $\delta\omega \equiv \omega - \omega_0$.

For the microwave cavity $\beta l = 2\pi l/\lambda_g$ so that

$$\delta(\beta l) = \frac{\partial(\beta l)}{\partial \lambda_g} \cdot \delta\lambda_g = \frac{\partial(\beta l)}{\partial \lambda_g} \cdot \frac{\partial \lambda_g}{\partial \omega} \cdot \delta\omega.$$

Now we can show from eqn (8.38) that $\partial\lambda_g/\partial\omega = -\lambda_g^3/\omega\lambda_0^2$ where λ_0 is the free space wavelength, so that

$$\delta(\beta l) = \frac{2\pi l \lambda_g}{\lambda_0^2} \frac{\delta\omega}{\omega}.$$

At resonance $l = n\lambda_g/2$; $\beta l = n\pi$ and so for the near-resonance condition we can write

$$\beta l = \pi + \delta(\beta l)$$

$$= \pi + \frac{\pi \lambda_g^2}{\lambda_0^2} \cdot \frac{\delta\omega}{\omega_0}.$$

Now $\tan \beta l = \tan\left(\pi + \frac{\pi\lambda_g^2}{\lambda_0^2} \cdot \frac{\delta\omega}{\omega_0}\right) \approx \frac{\pi\lambda_g^2}{\lambda_0^2} \cdot \frac{\delta\omega}{\omega_0}$ since $\tan(\pi + \theta) = \tan\theta \approx \theta$ for θ small, and so

$$\frac{Z}{Z_0} \approx \alpha l + j\frac{\pi\lambda_g^2}{\lambda_0^2}\frac{\delta\omega}{\omega_0}.$$

Hence, near resonance, the cavity behaves like a resonant circuit with

$$R_c = Z_0 \alpha l$$

$$L_c = \frac{Z_0}{2} \cdot \frac{\lambda_g^2}{\lambda_0^2} \cdot \frac{\pi}{\omega_0}$$

$$C_c = \frac{2}{Z_0} \cdot \frac{\lambda_0^2}{\lambda_g^2} \cdot \frac{1}{\pi\omega_0}.$$

The quality factor

$$Q_u = \frac{\omega_0 L_c}{R_c} = \frac{\pi}{2\alpha l} \cdot \frac{\lambda_g^2}{\lambda_0^2}$$

Typical values for R_c, L_c, and C_c for a cavity having a resonant frequency of 10 GHz are of the order of 10^{-1} Ω, 10^{-2} μH, and 10^{-9} μF respectively and $Q_u \approx 6000$.

This Q-factor is an 'unloaded' Q-factor since the cavity is isolated from the outside world; this is not of much use since it means that we cannot couple power into or out of the cavity. Imagine that one of the end walls of the cavity consists of a thin diaphragm having a small aperture, or iris, in it. For a small enough iris diameter a 'thin' iris acts as a shunt susceptance $-\mathrm{j}/\omega L_a$ and the microwave network can be represented as shown in Fig. 8.16(b). Looking 'from' the cavity (Fig. 8.16(c)) the effective impedance Z' is given by the parallel combination of the inductance L_a and the generator resistance R_0 (which is assumed to be equal to the characteristic impedance of the connecting line),

$$Z' = \frac{\mathrm{j}\omega L_a R_0}{(R_0 + \mathrm{j}\omega L_a)} = \frac{\omega^2 L_a^2 R_0 + \mathrm{j}\omega L_a R_0^2}{(R_0^2 + \omega^2 L_a^2)}.$$

For a small enough iris $\omega L_a \cdot < R_0$ and

$$Z' \approx \frac{\omega^2 L_a^2}{R_0} + \mathrm{j}\omega L_a.$$

Hence the resultant equivalent circuit is as shown in Fig. 8.16(d) and the 'loaded' Q-factor Q_L is given by

$$Q_L = \frac{\omega_0(L_c + L_a)}{R_c + \dfrac{\omega_0^2 L_a^2}{R_0}}$$

or, assuming $L_a \ll L_c$,

i.e.

$$\frac{1}{Q_L} = \frac{R_c}{\omega_0 L_c} + \frac{\omega_0^2 L_a^2}{\omega_0 L_c R_0},$$

$$\frac{1}{Q_L} = \frac{1}{Q_u} + \frac{1}{Q_u}\left(\frac{\omega_0^2 L_a^2}{R_c R_0}\right).$$

The coupling factor β for the iris is defined through $\beta \equiv \omega_0^2 L_a^2 / R_c R_0$ so that

$$\frac{1}{Q_L} = \frac{1}{Q_u}(1 + \beta).$$

Looking from the generator (see Fig. 8.16(b))

$$Z_{\text{in}} = \frac{j\omega L_a \left(R_c + j\omega L_c - \dfrac{j}{\omega C_c}\right)}{j\omega L_a + R + j\omega L_c - \dfrac{j}{\omega C_c}}. \tag{8.55}$$

Providing that the coupling iris is not too large the resonance frequency is close to ω_0 and so at, or near, resonance

$$Z_{\text{in}} \approx \frac{\omega_0^2 L_a^2 R_c + j\omega_0 L_a R_c^2}{\left(R_c^2 + \omega_0^2 L_c^2 + \dfrac{1}{\omega_0^2 C_c^2} - \dfrac{2L_c}{C_c}\right)}$$

where we have neglected L_a with respect to L_c in the denominator of eqn (8.55). But

$$Q_u^2 = \omega_0^2 L_c^2 / R_c^2 = (\omega_0^2 C_c^2 R_c^2)^{-1} = L_c / C_c R_c^2$$

and so

$$Z_{\text{in}} \approx \beta R_0 + j\omega_0 L_a.$$

If $L_a \sim L_c/10^2 \approx 10^{-10}$ H so that $\omega_0 L_a \sim 2\pi\,\Omega$ and $R_0 = Z_0 \approx 500\,\Omega$ then, to the approximation of neglecting $\omega_0 L_a$ compared to βR_0 we have

$$Z_{\text{in}} \approx \beta R_0.$$

The case of critical coupling for the cavity is when $\beta = 1$ and $Z_{\text{in}} = R_0$ and the cavity is matched to the line. In this case $Q_L = Q_u/2$.

A cavity of the type analysed above is called a 'reflection' cavity since it has one coupling iris only; 'transmission' cavities have an iris at each end.

The loaded Q-factor of a cavity depends on the size of the coupling iris but it also depends on the material from which the cavity is constructed (the better the electrical conductivity of the walls the higher the Q-factor), the precision with which the cavity is machined, and the 'finish' of the cavity walls; highly polished, clean surfaces are necessary in order to obtain the highest Q-factors (Q-factors of 10^4 can be obtained if sufficient care is taken).

To conclude this chapter we will give brief consideration to the important features of microwave striplines. The demand for decreased size and weight and increased ruggedness has stimulated the development of striplines and integrated circuitry for use at microwave frequencies; the geometry of microstrip in particular lends itself to the integrated circuit technology which had been developed already for circuits designed to operate at much lower frequencies.

The most common forms of striplines and 'microstrip' are shown in Fig. 8.17; the conducting strips are separated by a low-loss solid material such as alumina or sapphire. The calculations of the electromagnetic field patterns for both uniform lines and for circuit elements is a more formidable problem even than for waveguides and designs are based to a large extent on empirical data. Nevertheless the fundamental ideas of transmission line theory remain useful although it should be re-emphasized that, if a circuit element's linear dimensions are less than about 1/40 th of a wavelength, then it can be described in terms of lumped element parameters: Inductors of value ~ 10 nH have linear dimensions of about 2 mm and a capacitor ~ 10 pF might be about 1 mm square.

PROBLEMS

8.1. A wave of frequency $10^2/2\pi$ MHz propagates along a uniform lossless transmission line for which $L = 0.5\,\mu\text{H m}^{-1}$; $C = 200\,\text{pF m}^{-1}$; Calculate (a) the characteristic impedance of the line; (b) the phase change coefficient; (c) the wavelength of the wave on the line; and (d) the input impedance of a quarter-wavelength section which is terminated by an impedance $-\text{j}\,50\,\Omega$.

8.2. A transmission line has the following characteristics: $L = 0.2\,\mu\text{H m}^{-1}$; $C = 100\,\text{pF m}^{-1}$; $R = 2 \times 10^{-2}\,\Omega\,\text{m}^{-1}$; $G = 10^{-8}\,\text{S m}^{-1}$. What are the value of Z_0, α, β, λ, v_{ph} at a frequency of $100/2\pi$ MHz?

8.3. Find the attenuation per wavelength for a wave of frequency $50/2\pi$ kHz travelling on a transmission line having the following distributed characteristics:

$$R = 0.05\,\Omega\,\text{m}^{-1}; L = 10^{-6}\,\text{H m}^{-1}; C = 50\,\text{pF m}^{-1}.$$

(The shunt conductance can be assumed to be negligible).

8.4. A transmission line of characteristic impedance $Z_0 = (75 + \text{j}0)\,\Omega$ is terminated by a load impedance $(50 - \text{j}50)\,\Omega$. What is the voltage reflection coefficient at the termination of the line?

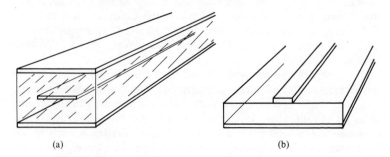

FIG. 8.17. (a) Microwave 'stripline'; (b) 'microstrip'.

8.5. A lossless transmission line has a distributed capacitance of 30 pF m^{-1} and a distributed inductance of 0.2 μH m^{-1}. If the line is operated at 1000/2π MHz what is the shortest length of short circuited line that will have a capacitive input susceptance of magnitude 0.213 S?

8.6. The measured value of the vswr on a lossless transmission line is 3.5 and the distance between successive minima of the vswr pattern is 50 cm. If the first vswr minimum from the termination is distant 35 cm what is the normalized impedance of the terminating load?

8.7. The voltage minima in the standing wave pattern on a lossless transmission line shift towards the termination by a distance 0.45 λ when the load is replaced by a short circuit. If the original vswr on the line was 3.0 find the normalized impedance of the load.

8.8. There is a vswr of 3.0 on a lossless transmission line and the first minimum occurs at a distance 0.35 λ from the terminating load. If $\lambda = 1$ m, where on the line should a variable shunt stub be placed in order to remove the standing wave in the line on the source side of the stub. (Hint: At which points on the line is $R(d)/Z_0 = 1.0$?)

8.9. Show that for the case of a lossy line eqn (8.26) becomes

$$|v(x, t)| \propto \{\sinh^2(a + \alpha d) + \cos^2(b + \beta d)\}^{1/2}.$$

9. Transducers

THE class of transducers which we are concerned with here give an electrical signal in response to a change in a physical variable (and vice versa) and it hardly needs stating that transducers lie at the heart of all measuring and control systems in all areas of scientific and technical research, industry, transportation systems, and communications. In an industrial plant it is necessary to measure and control such physical variables as temperatures, pressure, flow rate; in the development and testing of new types of aircraft it will be necessary to measure strain and the amplitude and frequency of vibrations in structural members, for example; in sound recording and reproduction systems sound vibrations are converted into electrical signals and then back into sound vibrations. A comprehensive list of further examples of the use of transducers would be extremely lengthy.

A block diagram illustrating the general features of a transducer and its associated systems is given in Fig. 9.1. Some examples of the physical variables/processes etc. which could be involved in such a system are listed in Table 9.1. Measurands are sometimes described as 'direct' (e.g. speed) or as 'indirect' (e.g. acceleration $a = dv/dt$).

The point in the over-all process at which the various forms of signal processing are enacted depends on the special features of the situation such as, for example:

(a) The length, and type, of the transmission path from the input transducer to the display (e.g. the path may be via a cable or a radio-link).

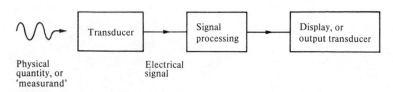

FIG. 9.1. A block diagram showing the main features of a transducer system.

Table 9.1

Measurands	Signal processing	Display or output transducers
Temperature	Amplification	Meter (analogue or digital)
Position/displacement (linear)	Filtering	Chart recorder
Position/displacement (angular)	Modulation	Loudspeaker
Speed	Analogue/digital Conversion	
Acceleration	Demodulation	
(force = mass × acceleration)		
Torque	Digital/analogue	
(couple = moment of inertia × angular acceleration)	Conversion	
Pressure		
Magnetic field		
Light intensity		

(b) The signal-to-noise ratio and the level of interference (see Section 10.5.1) at the output of the input transducer. For instance it may be necessary to filter out large interference spikes at an early stage since they would saturate a subsequent amplifier stage.

9.1. General features of transducers

The transfer function of a transducer is the functional relationship between the input variable Q_i and the output variable Q_0;

$$Q_0 = f(Q_i).$$

The sensitivity, or scale factor, S of the transducers is defined by

$$S \equiv dQ_0/dQ_i$$

and if the transfer function is linear so that

$$Q_0 = mQ_i + C$$

then $S = m$ is a constant. No transducer is exactly linear but it may be so to a good enough approximation, particularly over a limited range of the measurand.

As an example of an input transducer consider a copper–constantan thermocouple; this consists of a pair of junctions between copper wire and constantan wire, one junction being at temperature T greater than the other (see Fig. 9.2(a)). The measured relationship between E (the e.m.f. developed between the two junctions) and T (with the 'cold' junction at $0°C$ say) is non-linear and can be described by a power series

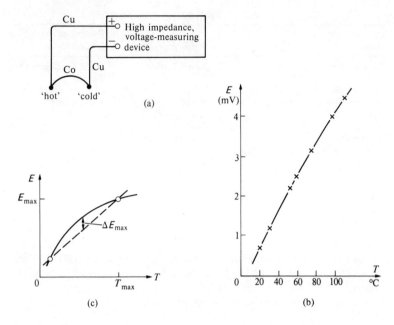

Fig. 9.2. (a) An arrangement for measuring the e.m.f. of a thermocouple (a copper–constantan thermocouple, say). (b) Thermoelectric e.m.f. E vs. temperature for a copper–constantan thermocouple. (c) A schematic representation of the E vs. relationship for a thermocouple, with the non-linearity exaggerated.

$$E = A + BT + CT^2 + DT^3 + \ldots \quad (9.1)$$

Obviously the coefficient A is zero in this case since $E = 0$ for $T = 0$ and also $B, C, D \ldots$ must be small since there is only a small deviation from linearity over the temperature range 20–100°C. By measuring E at two known temperatures for a particular thermocouple the coefficients B and C were found to be

$$B = 4.9 \times 10^{-2} \text{ mV}/°C$$
$$C = -8.0 \times 10^{-5} \text{ mV}/(°C)^2 \quad (9.2)$$

So in this case the sensitivity (or 'thermoelectric power') dE/dT is given by

$$S = dE/dT = B + 2CT + \ldots$$

and is approximately constant at a value of 0.04 mV/°C, approximately, over the specified temperature range.

The accuracy of a transducer system is a descriptive term which relates to the closeness of the indicated value of a physical quantity (e.g. temperature, pressure, etc.) to the actual value of the quantity. In a 'display' the accuracy is quoted, usually, as a percentage of full scale deflection (fsd).

The resolution of a system is determined by the fineness of the graduations on the scale of the display unit. For example consider a straight scale which has graduations at 1-mm intervals: If we can estimate reliably a position of the pointer to the nearest graduation then the reading we quote can be in error, on this account alone, by not more than ± 1/2 mm. This is called a resolution error, usually, and we must be able to distinguish between the 'accuracy' (of a scale) and the 'resolution'; the accuracy of a scale depends on the accuracy of its calibration.

In some situations for a given value of the measurand the output of the transducer depends on whether the value of the measurand is approached 'from below' or 'from above'. A system which behaves in this way is said to exhibit hysteresis. We have met this already in connection with the properties of ferromagnetic materials (see Section 5.5.3) but it does occur in other physical situations. 'Stickiness' and 'backlash' in mechanical movements may produce hysteresis effects also.

The repeatability (reproducibility) of a system is the closeness of agreement among a number of consecutive indications of a particular value of the measurand during wide range traverses of the value of the measurand. This property may be more important, practically, than absolute accuracy.

A transducer/display system has a zero error, or zero shift, if the whole scale is shifted up or down by the amount of the zero shift (see Fig. 9.3).

Very often a scale is interpolated between, and extrapolated beyond, two calibration points, with the assumption that the transfer function is linear. Consider the e.m.f. versus temperature difference relationship for a thermocouple (see Fig. 9.2(c)); If a linear calibration is assumed, based on two calibration points, then there will be a certain maximum calibration error (ΔE_{max}) as shown in the figure. The linearity is specified, commonly, as the percentage ($\Delta E_{max}/E_{max}$) × 100 or ($\Delta Q_{0\,max}/Q_{0\,max}$) × 100 in general terms. For the thermocouple calibration shown in Fig. 9.2.(b), and for the E versus T relationship specified by eqns (9.1) and (9.2), we have

$$\Delta E = BT + CT^2 - (B + CT_{max})T$$

or

$$\Delta E = C(T^2 - T_{max}T).$$

To find the maximum value of ΔE we find $d(\Delta E)/dT$ and set the resulting expression equal to zero.

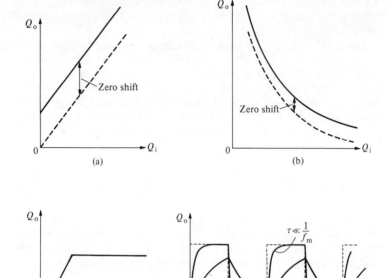

FIG. 9.3. Some general features of transducers: (a), (b) zero shifts; (c) limiting; (d) the effect of the response time

Now

$$d(\Delta E)/dT = 2CT - CT_{max}$$

and so for $d(\Delta E)dT = 0$ we have $T = T_{max}/2$. So

$$\Delta E_{max} = -CT_{max}^2/4$$

and since $E_{max} = BT_{max} + CT_{max}^2$ we have

$$\text{Linearity} = \frac{-\frac{CT_{max}^2}{4}}{(BT_{max} + CT_{max}^2)} \times 100.$$

Using $C \approx -10^{-4}$ mV/(°C)^2; $B \approx 3 \times 10^{-2}$ mV/°C; and $T_{max} = 100°$C we find

$$\text{Linearity} \approx 12 \text{ per cent.}$$

Similar analyses can be applied in the case of other transducers, e.g.

resistance thermometer $\quad R = R_0(1 + \alpha T + \beta T^2 + \ldots)$

strain guage $\quad R = R_0(1 + \epsilon S + \ldots)$

where S is the mechanical strain.

Any transducer system saturates for large enough values of the measurand and if the onset of this manifestation of non-linearity is relatively sudden then it is usually called limiting (see Fig. 9.3(c)). Limiting sets an upper limit to the useful range of a transducer whereas noise, interference, and drift set a lower limit.

For a first-order transducer the response Q_0 is related to Q_i through a linear first-order differential equation, e.g.

$$dQ_0/dt = A - Q_0/\tau.$$

Such an equation was solved in Section 3.1 and the result is

$$Q_0 = (Q_0)_\infty \{1 - e^{-t/\tau}\}$$

where $(Q_0)_\infty$ is the value of Q_0 as $t \to \infty$. The response time τ characterizes the speed of response of the transducer and if the measurand is periodic, with frequency f_m, then only if $f_m < 1/\tau$ will the output of the transducer be a reasonable facsimile of the input (see Fig. 9.3(d) and Section 6.6).

The operation of a second-order transducer is described by a linear second-order differential equation and the results of the description of $L-C-R$ networks can be taken over to describe the response of such transducers.

Some examples of first-order transducers are thermocouples, thermistors, and photodetectors and examples of second-order transducers are accelerometers (e.g. a mass on a spring) and piezo-electric crystals.

9.2. Resistive transducers

In this section, and the succeeding ones, a representative selection of transducers is discussed and to begin with we will consider some resistive transducers.

9.2.2. Potentiometers

Under this heading we are referring to a resistor (wire wound or carbon film) which is circular in form and which has a sliding contact (see Fig. 9.4). If $r = k\theta$, where θ is the angular position of the rotatable shaft then such a potentiometer (or 'pot' as they are referred to usually) can be used as an angular position transducer. For wire wound pots

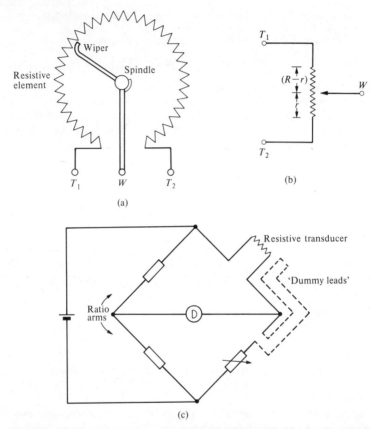

FIG. 9.4. (a) A schematic diagram of a potentiometer ('pot'). (b) The circuital representation of a 'pot'. (c) A resistive transducer incorporated into a Wheatstone bridge circuit, showing the use of 'dummy', compensating, leads.

the resolution is limited by the spacing of the turns of wire. In carbon film resistors the resolution is limited by the granularity of the film and is better than for wire wound versions. A source of noise which may be of practical significance arises from the irregular contact between the wiper and the wire, or carbon film, as the wiper moves; the use of a suitable lubricant should reduce this problem.

9.2.2. Resistance thermometers

The temperature dependence of the resistance of a sample of metal, in the form of a wire, say, can be formulated through

$$R = R_0 \{1 + \alpha(T - T_0) + \beta(T - T_0)^2 + \ldots\}$$

where R, R_0 are the resistances at temperatures T, T_0 respectively; α is the temperature coefficient of resistance.

Platinum is used commonly since it is available in very pure form, is stable under a wide spectrum of environmental conditions, has a relatively simple R versus T characteristic with good linearity ($\alpha = 3.92 \times 10^{-3}$ K^{-1}; β negligible for many purposes; $R_0 \sim 1\,\Omega$), and can be used over a wide temperature range (-260 to $550°$C). Another metal which is used occasionally is nickel.

For metals the temperature coefficient of resistance is positive whereas for semiconductors it is negative and the variation of R with temperature is of the form

$$R = A e^{-T/B}. \qquad \text{(see Fig. 9.3.b.)}$$

Thermistors have a large temperature coefficient of resistance but are extremely non-linear and have a narrow operating range (-60 to $250°$C).

A resistance thermometer is usually incorporated into a standard Wheatstone bridge circuit except that it is usual to have a set of 'dummy' leads which are arranged as shown in Fig. 9.4(c). In this way thermal effects in the leads to the element of the resistance thermometer, or to the thermistor, are compensated by similar effects in the dummy leads.

In a platinum resistance thermometer a typical construction used to be in the form of a non-inductively wound wire element on a mica frame, the whole being enclosed in a borosilicate glass tube; one disadvantage of this form of construction is the relatively large thermal capacity of the unit which gives a long response time. Today 'thick film' techniques are used to deposit a film of platinum on a suitable substrate; since this process can be automated production costs are reduced. Typical characteristics for this type of element are

$$R_0(T = 0°\text{C}) \approx 100\,\Omega$$

Temperature range: -70 to $500°$C
Self-heating $< 0.02°$C mW^{-1} when immersed in well-stirred water at $0°$C.
Thermal response time < 0.15 s to 63 per cent level.
Such a response time is about three times better than for the wire-wound type of element.

9.2.3. Resistive strain gauges

For a wire of uniform area of cross-section A, length l, and resistivity ρ, the resistance R is given by

$$R = \frac{\rho l}{A}$$

i.e. $R = R(\rho, l, A)$. If such a wire is strained in a longitudinal direction then if l increases by δl say then the mechanical strain ϵ is defined by $\epsilon = \delta l/l$. Poisson's ratio μ is defined in terms of the concomitant reduction δr in the lateral dimension of the wire (radius r say), i.e.

$$\mu \equiv \frac{\delta r/r}{\delta l/l}.$$

The expression for the total change δR in the resistance R is

$$\delta R = \left[\frac{\partial R}{\partial l}\right]_{\rho,A} \delta l + \left[\frac{\partial R}{\partial A}\right]_{\rho,l} \delta A + \left[\frac{\partial R}{\partial \rho}\right]_{A,l} \delta \rho$$

and so

$$\delta R = \frac{\rho}{A} \delta l - \frac{\rho l}{A^2} \delta A + \frac{l}{A} \delta \rho$$

or

$$\frac{\delta R}{R} = \frac{\delta l}{l} - \frac{\delta A}{A} + \frac{\delta \rho}{\rho}$$

where the last term represents the fact that the physical properties of the material of the wire change under the effect of a stress.

If the radius of the wire is r then $A = \pi r^2$ and, since $\delta A = (\partial A/\partial r)\delta r$, we have $\delta A/A = 2\delta r/r$ or $\delta A/A = 2\mu \delta l/l$. Strictly speaking $\mu = -(\delta r/r)/(\delta l/l)$ since r decreases as l increases and so we have $\delta A/A = -2\mu\epsilon$. Finally then

$$\frac{\delta R}{R} = \epsilon(1 + 2\mu) + \frac{\delta \rho}{\rho}.$$

A resistive strain gauge consists essentially of a fine wire, of diameter 0.025 mm, which is bonded to the member which is to be put under strain. The sensitivity factor $S \equiv (\delta R/R)/\epsilon$ or

$$S = (1 + 2\mu) + \left(\frac{\delta \rho}{\rho}\right)\bigg/\epsilon.$$

For elastic deformation of a wire $\mu \approx 0.3$ so that $S \approx 1.6 + \left(\frac{\delta \rho}{\rho}\right)\bigg/\epsilon$; the second term is usually not significant compared to 1.6 and so $S \sim 2$ for the metals used most commonly, such as constantan and nichrome.

Typical values of R and $(\delta R/R)_{max}$ are $10\,\Omega$ and 1 per cent respectively. Again R is measured by means of a Wheatstone bridge circuit and dummy leads are used for the same general reason as with a resistance thermometer.

From eqns (2.27) the out-of-balance current in the detector of a Wheatstone bridge is

$$i = \frac{E(R_3R_6 - R_2R_5)}{R_2R_5R_6 + R_3R_5R_6 + R_2R_3R_5 + R_2R_3R_6 + R_4(R_5 + R_6)(R_2 + R_3)}.$$

Very often four strain gauges are connected as the elements R_2, R_3, R_5, R_6. We will assume that for no strain $R_2 = R_3 = R_5 = R_6 = R$; under strain we have $R = R + \delta R_2$, $R = R + \delta R_3$ etc. Hence,

$$i \approx \frac{E(\delta R_3 + \delta R_6 - \delta R_2 - \delta R_5)}{4R(R + R_4)}$$

where we have neglected δR with respect to R and $(\delta R)^2$ with respect to δR. The gauges are usually arranged to work in push–pull so that δR_2 and δR_5 are of opposite signs to δR_3 and δR_6. If we assume, for simplicity, that $|\delta R_2| = |\delta R_3| = |\delta R_5| = |\delta R_6| = \delta R$ then, since $\delta R_2 = -\delta R_3$ and $\delta R_5 = -\delta R_6$, we have

$$i = \frac{E}{(R + R_4)} \cdot \frac{\delta R}{R}.$$

If only R_3 say, is an active strain gauge, then we can show easily that

$$i = \frac{E}{4(R + R_4)} \cdot \frac{\delta R}{R},$$

i.e. the sensitivity has been increased by a factor of four by using four active strain gauges in push–pull.

It is useful to obtain an estimate of the out-of-balance current. Suppose that the power dissipation in the transducer must be limited to W_T then, since the source 'sees' a bridge resistance R at balance, we have

$$E = 2I_T R$$

where

$$I_T^2 R = W_T.$$

So

$$E = 2\sqrt{(W_T R)}$$

and the out-of-balance current is given by

$$i = 2\sqrt{\frac{W_T}{R}} \cdot \frac{\delta R}{(R + R_4)}.$$

For $W_T = 10^{-1}$ W then $i \approx 10\mu A$ for $\delta R/R = 0.01$ per cent.

If a high impedance voltage amplifier is used as detector (perhaps if the excitation for the bridge is an a.c. generator) then for $\delta R/R = 0.01$ per cent we have

$$V_{\text{detector}} = iR_4 = \frac{ER_4}{(R+R_4)} \cdot \frac{\delta R}{R}.$$

However, since $R_4 \gg R$, we can write

$$V_{\text{detector}} \approx E \frac{\delta R}{R}$$

and if $E \sim 10$ V then $V_{\text{detector}} \approx 1$ mV.

Arguments analogous to those above could have been applied to other resistive transducers such as resistance thermometers.

9.3. Capacitive transducers

The capacitance of a parallel plate capacitor $C = \epsilon\epsilon_0 A/s$ where A is the area of the plates and s is their separation. So $\delta C = (\partial C/\partial s)\delta s = -(\epsilon\epsilon_0 A)\delta s/s^2$ and the variation of C with s is very non-linear. Hence capacitive transducers, which have the very important feature that there is no mechanical contact between the stationary and moving parts, are usually used in a 'differential' mode. Referring to Fig. 9.5 we have

$$C_1 = \frac{\epsilon\epsilon_0 A}{(s+x)} \qquad C_2 = \frac{\epsilon\epsilon_0 A}{(s-x)}$$

where x is the displacement of the movable plate from the median position. If the angular frequency of the source is ω then

$$V_1 = \frac{1/j\omega C_1}{1/j\omega C_1 + 1/j\omega C_2} \cdot E \qquad V_2 = \frac{1/j\omega C_2}{1/j\omega C_1 + 1/j\omega C_2} \cdot E$$

whence the output $V_1 - V_2$ to a differential amplifier is $(V_1 - V_2) = xE/s$, i.e. the output voltage is a linear function of displacement x. The sensitivity $d(V_1 - V_2)/dx = E/s$ V m^{-1}, e.g. for $E = 10$ V, $s = 1$ mm; then the sensitivity is 10 mV μm^{-1}.

For an air-spaced capacitor having a plate area of 50 cm^2 and a plate spacing of 1 mm the capacitance is 50 pF approximately. Also since C_1 and C_2 are in series the effective capacitance of the unit is 25 pF. Now coaxial cable has a capacitance per unit length in the region of $20-50$ pF m^{-1} and so the first stage of the detection/amplification system (i.e. the pre-amplifier) must be placed very close to the transducer unit. Also the reactance of 25 pF at 1 kHz say is ≈ 6 MΩ and so

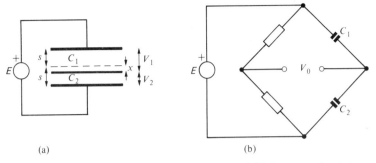

FIG. 9.5. (a) A differential capacitive transducer and (b) its connection into an a.c. bridge circuit.

the pre-amplifier must have an input impedance greater than this which points to the use of a field effect transistor (F.E.T.).

A differential capacitor can be connected into a bridge circuit as shown in Fig. 9.5(b), where $V_0 = Ex/2s$.

Concentric cylinder capacitors can also be used as displacement transducers; in this case $C = 2\pi\epsilon_0 l/\ln(b/a)$ and $\delta C = 2\pi\epsilon_0 \delta l/\ln(b/a)$, i.e. $\delta C \propto \delta l$. If $b = 10$ mm and $a = 2$ mm then $\delta C \approx 3.5 \times 10^{-2}$ pF mm^{-1}.

A good example of the use of a differential capacitor transducer is in the role of a fluid pressure indicator; a thin, flexible, metal diaphragm is positioned between two fixed metal plates and deforms in response to a pressure difference between the fluid on its two sides.

The deformation of a thin metal diaphragm in response to pressure changes in a fluid is exploited also in the condenser microphone (see Fig. 9.6). If we assume that the time constant CR is so long that the charge q on the capacitor remains constant, to a good enough degree of approximation, during a period of a sound wave of interest then

$$E - V_C - V_R = 0 \quad \text{or} \quad V_R = E - V_C.$$

Now $V_C = q/C$ and $q = C_0 E$, say, where C_0 is the equilibrium value of the capacitance C. So, if $C = C_0(1 + (\hat{s}/s_0) \sin \omega t)$, then

$$V_R = E - \frac{C_0 E}{C_0 \left(1 + \dfrac{\hat{s}}{s_0} \sin \omega t\right)}.$$

Assuming that $\hat{s}/s_0 \ll 1$ then $V_R \approx E - E(1 - (\hat{s}/s_0) \sin \omega t)$ or

$$V_R \approx \frac{E\hat{s}}{s_0} \sin \omega t.$$

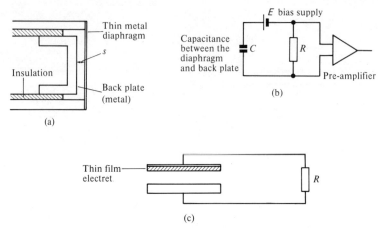

FIG. 9.6. (a) A schematic cross section of a condenser microphone. (b) The coupling of a condenser microphone to a load R and pre-amplifier. (c) A schematic cross-section of an electret microphone.

For $s_0 = 1$ mm, $E = 100$ V, and $\hat{V}_R = 10\,\mu$V then $\hat{s} = (\hat{V}_R/E)s_0 = 10^{-10}$ m.

If we let f_{min} be the frequency of the lowest frequency sound signal which we wish to detect then our assumption above means that we require $C_0 R > 1/f_{min}$ or $R > 1/(f_{min} \times C_0)$. Supposing $C_0 = 100$ pF and $f_{min} = 30$ Hz then this condition requires $R > 300$ MΩ which means that a pre-amplifier of very high input impedance must be used.

Electret microphones are similar, in essence, to condenser microphones the principal difference being that the bias field is provided by the 'frozen-in' electric field in a so-called electret. The electrets which are employed as microphone diaphragms are usually made from thin 'mylar' or polycarbonate film having a thickness $\sim 10\,\mu$m. The film is polarized in a high electric field while at a relatively high temperature; the film is then cooled while still under the influence of the field and the polarization is 'frozen-in' and remains after the polarizing field has been removed.† A schematic cross-section of an electret microphone is shown in Fig. 9.6(c); one side of the film has a metal coating and there is a small gap which has a width of the same order of magnitude as the thickness of the film. The frozen-in polarization provides an electric field in the capacitor and vibrations of the film in response to incident sound waves cause variations in the voltage across R; sensitivities of a few mVμbar^{-1} can be obtained which is of the order of ten times more sensitive than condenser microphones.

†This is an extremely simplified description of the process of creating a long-lasting polarization but it will suffice for our purposes; the relevant point is that a long-lasting polarization can be created.

9.4. Piezo-electric transducers

In certain naturally occurring crystalline materials, of which quartz is the best known example, an electric field is set up inside a specimen of the material when a mechanical strain is induced in it i.e. the material becomes electrically polarized (see Fig. 9.7). Conversely, dimensional changes occur in a specimen when it is subjected to an electric field. The first named phenomenon is exploited in strain gauges, accelerometers, and acoustic and ultrasonic detectors and the second in ultrasonic generators.

For instance an accelerometer can be made by sandwiching a piezo-electric transducer which is in the form of a disc between a suitable base and a relatively massive disc; vibrations of the base cause strains in the transducer because of the inertia of the massive disc.

Polarized ferroelectric ceramics such as barium titanate and lead zirconate (PZT) which display piezo-electricity also, have many practical and commercial advantages over naturally occurring materials, particularly the fact that they can be moulded to desired shapes, and so they are in widespread use. The value of the piezo-electric strain constant is $\sim 10^{-12}$ CN^{-1} in quartz and up to 600×10^{-12} CN^{-1} in certain grades of PZT.

Many modes of deformation other than the thickness expansion/contraction illustrated in Fig. 9.7 are possible and by making suitable electrical connections transducers can be fabricated which are sensitive to bending and twisting, for example.

It would be out of place to discuss here the detailed theory of piezo-electric transducers and the wide field of their application but there are two important features which should be mentioned.† First the acoustic impedance of a medium is related to the ratio force/particle velocity (c.p. electrical impedance where the p.d. across a

FIG. 9.7. A schematic (and highly exaggerated) indication of the polarization of a deformed sample of a piezo-electric material.

---

†For fuller descriptions of piezo-electric transducers see, for example, Blitz, J. *Fundamentals of ultrasonics.* Butterworths, London (1967) and Mason, W.P. *Piezo-electric crystals and their application to ultrasonics.* van Nostrand, New Jersey (1950).

uniform element is proportional to the electric field and hence to the force acting on a charge and the expression for the current contains the drift velocity of the charge). It will be appreciated from the form of this ratio that the acoustic impedance of a solid is much greater than that of gas. The compressibility of liquids is of the same order of magnitude as that of solids and so piezo-electric transducers operate quite efficiently in liquids whereas they are very inefficient in gases unless some mechanical transformation is arranged to provide impedance matching e.g. a lever system. A lever system is employed in piezo-electrical record player pick-ups in order to match the force/velocity ratio of the stylus to that of the piezo-electric material; in piezo-electric ('ceramic') microphones the mechanical linkage is via a diaphragm and rod.

Secondly, since the constituent material is elastic a transducer will exhibit a mechanical resonance which is exploited in generators. However wide-band receiving transducers are operated below their resonant frequency, the resonance being damped deliberately in any case.

The mechanical Q-factors of quartz crystals can be as high as 10^6 when measured under specially arranged conditions but in more general use losses in mounting cements and acoustic radiation result in working Q-factors of up to 10^3 at most in piezo-electric ceramics.

9.5. Inductive transducers

We will give here a description of the basic principles of operation of a selection from the wide variety of transducers which exploit the phenomenon of electromagnetic induction. It can be pointed out at this stage that inductive transducers have much smaller impedances than the capacitive transducers which were described in Section 9.3.

A 'magnetic cartridge' pick-up for a record player is an example of what is perhaps the simplest type of inductive transducer. A small permanent magnet, which is coupled mechanically to the stylus of the pick-up, moves inside a coil in response to the vibrations of the stylus. A voltage is induced in the coil whose instantaneous value is proportional to the instaneous speed of the magnet and hence of the stylus. For a given amplitude of vibration of the stylus the speed increases in proportion to frequency (i.e. at 6 db per octave) and this effect has to be 'equalized' in the subsequent processing of the electrical signal.

The important details of a ribbon microphone are sketched in Fig. 9.8.; the ribbon-like conductor moves in response to an incident sound wave and the area A of the circuit linked with the magnetic field B changes

$$V = \frac{d\phi}{dt} = \frac{d}{dt}(BA) = B\frac{dA}{dt} = Bl\frac{dx}{dt}.$$

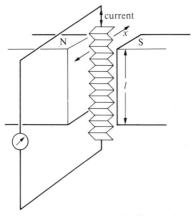

FIG. 9.8. A schematic diagram of a ribbon microphone.

We see again that the induced voltage is related to the speed of the displacement and, for a given sound amplitude, the output voltage is proportional to frequency.

The principle of a linear variable differential transformer (LVDT) is illustrated in Fig. 9.9.; an alternating exciting current in a primary coil induces equal voltages in two balanced coaxial secondary coils if the movable ferromagnetic core is situated at a position which is symmetrical with respect to the secondary coils. The secondary coils are connected in push–pull so that the net output voltage is zero if the core is positioned symmetrically. If the core is displaced, however, then there is an output voltage the phase of which depends on the sense of the displacement. A typical value of the sensitivity, for 50 Hz excitation, could be $1 \text{ mV} \mu\text{m}^{-1}$.

Tacho-generators exploit the e.m.f. induced in an armature which is rotating in a constant magnetic field. Depending on the type of commutator used the output can be either d.c. or a.c. with a sensitivity of

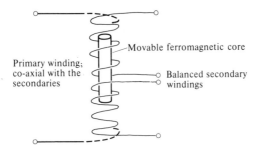

FIG. 9.9. The basic features of a linear variable differential transformer.

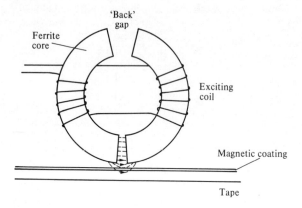

FIG. 9.10. A schematic illustration of a tape recorder 'head'.

~ 10 V per r.p.m. typically; providing that a high resistance load is used then the output voltage is proportional to the speed of rotation of the armature.

A tape recorder head is a good example of a magnetic circuit: In the 'record' mode the alternating current in the exciting coils generates an alternating magnetic field in the ferrite core and the fringing field at the 'active' gap (see Fig. 9.10) magnetizes the magnetic coating of the plastic tape which is being pulled past the head.

From eqn (5.47) the magnetic induction B in the magnetic circuit is given by

$$B = \frac{\phi}{A} = NI \bigg/ \left\{ \frac{1}{\mu_0} \left(l_g + \frac{l_m}{\mu(\phi)} \right) \right\}$$

where l_m is the length of path in ferrite and l_g is the combined length of the gaps; note that the fact that the magnetic permeability μ is a function of the value of ϕ (see Section 5.5.3) has been included. If $l_g \gg l_m/\mu(\phi)$, then, to this approximation, B is independent of $\mu(\phi)$ and the transducer behaves in close to linear fashion.

Now the frequency of a signal to be recorded must not be higher than v/l, where l is the width of the 'active' gap, and v is the speed of the tape, otherwise the alternating magnetic field will execute more than one cycle whilst any section of tape is passing the gap and the net magnetization would be zero. Hence the width of the active gap is kept small enough to accomodate the required maximum signal frequency and the total gap width required to give an acceptable approximation to linearity is made up by the 'back' gap (at some cost to the magnitude of the magnetic induction B, of course).

In the play-back mode the spatially variable magnetization of the tape sets up a time-dependent magnetic field in the ferrite core which in turn induces a voltage in the coils.

A loudspeaker is an example of an output transducer: In the most common form the exciting current flows in a movable coil which sits over a permanent magnet. The coil is attached also to the cone of the loudspeaker and the interaction between the exciting current and the permanent magnet causes the coil to vibrate and these vibrations are coupled to the cone.

A popular type of metal detector has a coil which acts as both transmitter and detector. The magnetic field generated by the exciting current induces eddy currents within nearby metal objects; these eddy currents themselves interact with the coil via the magnetic field which they generate and a voltage is induced. Thus the alternating magnetic field in the coil is reduced in magnitude and its phase is altered, in general. Hence there is a change in the effective impedance of the coil and this can be detected if it is large enough.

9.6. Photodetectors

There are many different types of detector of electromagnetic radiation which vary widely in their mode of response, sensitivity, and also with regard to the spectral region in which they operate. Broadly speaking there are four types of detector:

(i) *Bulk detectors.* These rely on photoconductivity in materials such as CdS, Ge, Si, InSb. The incident electromagnetic radiation creates mobile charge carriers so that the resistance of the sample is reduced. For instance in a typical commercially made CdS photocell which operates in the visible region of the spectrum the resistance R depends on the illumination I (in lux) according to $R = 80 I^{-0.9}$ kΩ to a reasonable degree of accuracy. For 10 V bias and an illumination of 50 lux the sensitivity is of the order of 0.2 mA lux^{-1}. The response time will be of the order of 100 ms typically.

For radiation in the infra-red region of the spectrum InSb and PbS bulk detectors are commonly used.

(ii) *Photovoltaic cells.* If a p–n junction, which is open-circuited is exposed to electromagnetic radiation having a frequency above a threshold value then, as a result of the creation of mobile charge carriers and their subsequent diffusion across the junction, a p.d. is developed across the junction in the 'forward' sense. The exploitation of this principle is more familiar, probably, in 'solar' batteries than in photo-detectors.

(iii) *Photo-emissive cells.* These cells contain a photo-sensitive cathode in a gas-filled, or evacuated, glass envelope and, for a cathode voltage ~ 100 V a sensitivity of the order of $0.2\,\mu\text{A lux}^{-1}$ can be obtained, with a response time as low as 15 ns. A photomultiplier consists, in essence, of a photocathode and a number of 'dynodes', a p.d. of about 100 V being maintained between successive electrodes. Electrons released from the cathode by the photoelectric effect are accelerated to the first dynode which is coated with a material which has a good secondary electron emission coefficient e.g. say 6 secondary electrons are emitted for every incident electron. This multiplication process is repeated at the succeeding dynodes and in a 9-stage multiplier electron gains of from $10^7 - 10^8$ can be obtained.

(iv) *Radiation detectors.* In this context by radiation we mean ionizing radiations such as X-, γ-, α-, and β-rays. In Geiger–Mueller tubes the incidence of a particle or high energy photon initiates the electrical breakdown of the gas between the cathode and anode which are maintained at p.d. of between 500 and 1500 V. The current pulse caused by the flow of electrical charge during the breakdown is detected by measuring the p.d. across an external resistor connected between the cathode and anode. The response time (the so-called 'dead-time') of a Geiger–Mueller tube itself can be as low as $15\,\mu\text{s}$.

Solid state radiation detectors made from germanium and silicon are used widely also.

9.7. Magnetic fields

There are many methods of measuring magnetic field strengths but we will discuss two, only, which are, perhaps, the most widely used for one or more of the following reasons; simplicity, linearity, insensitivity to temperature changes.

(i) *Search coil.* An e.m.f. is induced in a coil if the magnetic flux linked with the coil changes;

$$E = d(n\phi)/dt$$
$$= nA\,dB/dt \qquad (9.3)$$

where B is the magnetic flux density (the magnetic field is assumed to be homogeneous over the region of space occupied by the search coil), A is the area of cross-section of the coil, and n is the number of turns on the coil.

If B is time-independent then it can be measured by connecting the coil to an integrator and then removing the search coil from the magnetic field (or by switching-off the field, temporarily):

$$\int E \, dt = nA \int_{t=0, B=B_0}^{t, B=0} \frac{dB}{dt} \cdot dt$$

$$= nA \int_{B_0}^{0} dB$$

$$= nAB_0$$

$$B_0 = \frac{1}{nA} \int E \, dt.$$

Another way of obtaining a time-dependent flux linkage with the search coil is to move it at a known rate through the magnetic field; the most common method is to rotate the search coil about an axis perpendicular to the direction of the magnetic field.

If B is time-dependent then there will be an induced e.m.f. in the search coil and the amplitude of B can be found from eqn (9.3) although corrections may have to be made if the time dependence is not sinusoidal.

For accurate measurements the magnetic field must be reasonably uniform over the cross-section of a search coil. This being so search coils have good linearity, are simple to use, and can measure fields as low as 10^{-9} T (rotating coil) or 10^{-7} T (static coil).

(ii) *Hall effect.* If a plate of a conducting material is placed at right angles to a magnetic field with electrical connections made to it as shown in Fig. 9.11, then a so-called Hall voltage V_H is developed,

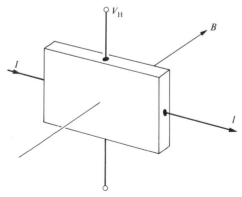

FIG. 9.11. The relative directions of the bias current I, the magnetic field B to be measured, and the Hall voltage V_H, in a Hall effect magnetometer.

292 Transducers

as shown, where

$$V_H = \frac{R_H I B}{t}.$$

Here the Hall coefficient R_H is closely equal to $(ne)^{-1}$ where n is the concentration of mobile charges in the material of the plate. For practical purposes semiconductors are the most useful materials, particularly the compound semiconductor InSb. The useful range of Hall effect devices is from about 10^{-4} to 2 T with a linearity of between 2 and 10 per cent; the Hall voltage is of the order of 1 mV per mA per T, typically.

PROBLEMS

9.1. A 10-V d.c. supply, a circular wire-wound 'pot', and a digital voltmeter (whose input resistance may be assumed to be infinite) are used as a transducer to indicate the angular position of a shaft:
  (a) Calculate the average sensitivity in mV degree^{-1} (the 'pot' has a maximum angular displacement of 300°).
  (b) If the 'pot' has 600 turns of wire and the digital voltmeter displays voltages to 5 significant figures on its 10-V range is the resolution, in degrees, of the system limited by the voltmeter or by the 'pot'?
  (c) Calculate the angular resolution in degrees.

9.2. A 'pot' of total resistance R is used as an angular position transducer and is connected into a circuit as shown in Fig. 9.12 so that $r = r(\theta)$ and $V_0 = V_0(\theta)$. Show that

$$V_0 = \frac{Er}{R\left(1 + \dfrac{r}{R_0} - \dfrac{r^2}{R_0 R}\right)}.$$

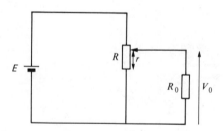

FIG. 9.12.

9.3. For the circuit of Problem 9.2 (Fig. 9.12) show that, to terms in second order in $(r/R_0)^2$, V_0 can be expressed as

$$V_0 = \frac{Er}{R}\left\{1 - \frac{r}{R_0} + \frac{r^2}{R_0^2}\left(1 + \frac{R_0}{R}\right)\right\}.$$

9.4. Show that the linearity of the angular position transducer described in Problems 9.2 and 9.3 is given by

$$\text{Linearity} = \frac{400R}{27R_0\left(1 + \dfrac{R}{R_0}\right)^2} \text{ per cent.}$$

What is the linearity if $R = 1\,\text{k}\Omega$; $R_0 = 10\,\text{k}\Omega$?

9.5. A photomultiplier, with an average electron gain of 2×10^5 and a probability of 10^{-2} that an incident photon will produce an anode pulse, has an average output current of 10^{-8} A. What is the radiant flux, in photons per second, at the photocathode?

10. Instrumentation

TODAY the electronic instrument field is a rapidly changing one due largely to the advent of integrated circuits and the increasing use of digital techniques. However, although these changes are not superficial, there are certain basic principles underlying the operation of instruments and it is with these that we are concerned in this chapter.

Information is transferred between a measurand, arising in the process of interest, and the instrument via an input transducer, as we have seen in Chapter 9, and between the instrument and the user via a display device (such as a meter, a digital display, a C.R.O., or a chart recorder). In analogue instruments the amplitude of the electrical signal is proportional, ideally at least, to the amplitude of the measurand; hence the use of the term analogue. On the other hand, in digital instruments the electrical signal (initially in analogue form) is converted into digital form, usually in binary code. The design of analogue instruments imposes stringent requirements on the linearity and stability of the performance and tolerance of the constituent components; in contrast the requirements on the components of digital instruments are less severe. However other relevant factors are that most transducers produce electrical signals in analogue form and that the presentation of data in analogue form for human use is often much more satisfactory than in digital form; so analogue-to-digital converters are an integral part of many digital instruments. In the end the choice of a digital or analogue form of instrument will be made on the grounds of cost balanced against convenience.

In the case of d.c. circuits, voltages, currents, and resistances can be measured by using instruments, such as moving-coil ammeters, whose calibrations rest ultimately on the precisely defined international standards of e.m.f. and current. Such high precision is not required generally with a.c. instruments since the e.m.f.s of a.c. generators are less stable than those of standard d.c. cells. The calibrations of a.c. instruments rely on the availability of transfer instruments which can be calibrated on d.c. currents and whose a.c. behaviour can be accurately predicted from a knowledge of their constructional details. Measurements of capacitance, inductance, phase, power, and frequency are also required, in general, in the a.c. situation.

10.1. Voltage and current measurements

10.1.1. Meters

The majority of instruments that are called 'meters' have a moving system which rotates about a fixed axis and carries a pointer which moves over a graduated scale. The deflecting torque acting on this system is due to the current flowing through the coil in the meter and is opposed by a controlling torque usually produced by a hair-spring. Since the period of the mechanical system is, in general, much longer than that of the a.c. currents to be measured, the rotating system takes up an equilibrium position in which the average torque due to the current is balanced by the controlling torque.

The relation between the deflecting torque Γ and the current I determines the law of response of the meter: for example, if $\Gamma = KI^2$ then $\Gamma_{av} = T^{-1} \times \left(K \int_0^T I^2 \, dt \right)$ where T is the period of the cyclical, but not necessarily sinusoidal, current. If Γ_c is the controlling torque per unit angular deflection then for an equilibrium deflection of θ radians $\Gamma_{av} = \Gamma_c \theta$. Now since $T^{-1} \int_0^T I^2 \, dt = (I_{rms})^2$ it follows that $\theta = K(I_{rms})^2/\Gamma_c$ and the meter is said to have a 'square-law' response.

If $\Gamma = KI$, such as for a permanent magnet moving-coil meter, then $\Gamma_{av} = T^{-1} \left\{ K \int_0^T I \, dt \right\} = KI_{av}$; if I is sinusoidal $I_{av} = 0$ but if a sinusoidal current which is to be measured is first 'full-wave' rectified (see Fig. 10.1) then $I_{av} = 2T^{-1} \left\{ \int_0^{T/2} \hat{I} \sin \omega t \cdot dt \right\}$: in this case $\theta = KI_{av}/\Gamma_c$ and the meter is said to have a linear response.

It is important to note that the scale of a meter may not follow the basic response law either because of intrinsic features of the instrument, such as K being a function of θ for instance or because the instrument has been designed specifically to produce an expanded scale over certain portions of its over-all range of deflection.

'Moving-iron' meters consist basically of a vane of material of high permeability which is attracted towards a coil through which the current to be measured I is flowing. If the inductance of the coil-plus-vane is L then the potential energy E_p of the system is given by $E_p = LI^2/2$ and the magnitude of the torque is given by $\Gamma = dE_p/d\theta = (I^2 \, dL/d\theta)/2$. Hence the deflection is proportional to $\propto I^2$ and such a moving-iron meter has a 'square-law' response.

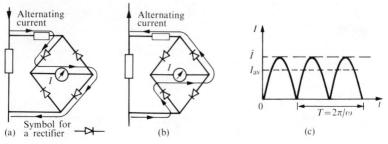

FIG. 10.1. A moving-coil d.c. meter connected into a full-wave rectifier circuit to enable alternating currents to be measured: (a) and (b) current paths for successive half-cycles of an alternating current; (c) wave-form of the current through the meter.

The operating principle of a thermocouple meter is that a fine thermocouple is in thermal contact with an element which is heated by the current which is to be measured; the thermocouple is connected to a sensitive moving-coil d.c. meter which indicates the thermo-e.m.f. Since the heating effect is proportional to the square of the current, and the thermo-e.m.f. is approximately proportional to the change in temperature of the element, a thermocouple meter also has a square-law response, to a close approximation.

A rectifier instrument employing a moving-coil meter as indicator (see Fig. 10.1) responds to a sinusoidal input current $\hat{I} \sin \omega t$ with a deflection proportional to the average value of the rectified current which is $2\hat{I}/\pi = 0.637\hat{I}$. Hence a reading on a moving-coil meter which has been calibrated for d.c. currents must be multiplied by $\sqrt{2} \times (0.637) = 1.11$ to give the rms value.

As a result of the relatively large current drawn, moving-iron meters can only be used as voltmeters across load impedances of relatively low value.

Moving-coil meters are characterized by a relatively high sensitivity and a much lower power consumption than moving-iron meters (less than 1 mW for a 50 μA meter). Rectifier instruments incorporating sensitive moving-coil meters draw relatively low currents and hence can be used as high resistance voltmeters; for this reason and because of their low power consumption they find widespread use for measuring voltages at audio-frequencies where the available power is usually rather small. The capacitance of the rectifiers, which is effectively in parallel with the bridge, allows a fraction of the alternating current to bypass the bridge thus making the calibration of the instrument frequency-sensitive. Since this effect increases with frequency, rectifier meters are, broadly speaking, restricted in their uses to frequencies below 10 kHz.

Thermocouple meters can be used up to radio-frequencies (100 MHz)

Instrumentation

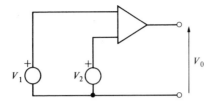

FIG. 10.2. A general representation of a differential amplifier; $V_0 \propto (V_1 - V_2)$, ideally.

their main disadvantages being their relatively small overload capability compared to rectifier and moving-iron instruments (several hundred per cent of full-scale deflection).

Digital voltmeters are now in common use; a converter produces trains of pulses within sampling intervals, the number of pulses within an interval being proportional to the magnitude of the d.c. voltage. The display will show six digits typically.

In sensitive voltmeters an amplifier is interposed between the input terminals and the display; by using a field effect transistor (F.E.T.)† in the first stage of the amplifier an extremely high input impedance can be obtained. The input resistance of the F.E.T. itself may be 10^9 Ω and the input impedance of the voltmeter as a whole will be limited by the bias circuit of the F.E.T. and/or the degree of insulation it is possible to achieve between the input terminals, in the input leads, and in the physical system to which the leads are attached: an input impedance of 100 MΩ is common place. Very high impedance voltmeters are sometimes referred to an 'electrometers' since they can be used to measure charge by measuring the voltage across a capacitor of known value. Also the amplifier could be a 'differential' amplifier for which the output voltage to the display is proportional to the difference of the potentials applied to the input terminals (see Fig. 10.2). An important feature of differential amplifiers is the 'common mode rejection ratio'. This ratio describes the effectiveness of the amplifier as regards the rejection of signals which appear simultaneously at the two input terminals and is defined as

$$\frac{\text{Gain for differential mode input}}{\text{Gain for common mode input}}.$$

Typical values are 140 db for d.c. signals and 100 db for signals at 50 Hz.

†Field effect transistor is a generic term; an individual device may be a junction device JFET or a metal oxide-semiconductor device MOSFET (see a modern text on electronic devices).

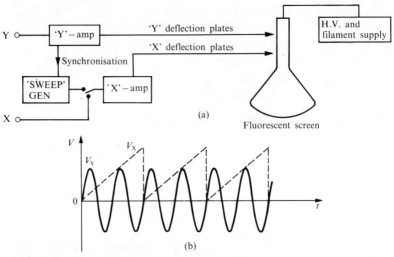

FIG. 10.3. The cathode ray oscilloscope. (a) block diagram; (b) synchronization of the time-base voltage V_X and V_Y.

10.1.2. Cathode ray oscilloscope

A block diagram of a typical C.R.O. is shown in Fig. 10.3: an electron gun projects a stream of electrons to produce a small bright spot on the fluorescent face of a cathode ray tube. The electric fields produced by voltages applied between two sets of mutually perpendicular deflector plates can cause deflections of the electron stream and hence of the spot: thus the tube can be used as a device to plot one voltage against another. In many applications the 'X'-deflection is produced by a sawtooth voltage which is generated in a 'sweep generator': this drives the spot across the fact of the screen from left to right in the X-direction at a known constant speed and then returns it, virtually instantaneously, to its starting point. Hence the X-displacement of the spot is proportional to the time within a cycle of the sawtooth waveform. The persistence time of the fluorescence of the screen is long enough so that, for all but very low sweep speeds, the moving spot is perceived as a bright line on the screen.

If the sweep generator is synchronized with a periodic Y-voltage then a stationary pattern is displayed on the C.R.O. screen which represents the Y-voltage as a function of time. The synchronization is achieved electronically and ensures that successive cycles of the sawtooth X-voltage always begin at equivalent points in the cycles of the Y-voltage (see Fig. 10.3(b)). The sawtooth X-voltage is said to provide the 'time-base' for the C.R.O. screen.

The 'Y-amplifier' (or Y_1 and Y_2 amplifiers in a double-beam C.R.O.)

is designed to have a high input impedance (at least 1 MΩ) and so the C.R.O. can be used essentially as a voltmeter if the deflection sensitivity, in $V\,cm^{-1}$, in the 'Y-direction' is known. (The maximum deflection sensitivity is about $10\,mV\,cm^{-1}$ in the oscilloscopes in most widespread use). A C.R.O. is not a very precise voltmeter but in the testing of electronic circuits great precision is not usually required. As well as having a rough idea of the magnitudes of a signal at various points in an amplifier, say, it is equally useful to be able to examine the waveform of the signal in order to look for various types of distortion or for excessive electrical noise.

The inertia of electrons is so small that a limit to the speed at which the spot can be swept across the screen is set by the frequency response of the sweep generator: the shortest sweep periods available are between 10^{-8} and 10^{-9} s. The unavoidable capacitance in parallel with the input resistance of the 'Y-amplifier' due to connecting leads, and the inherent input capacitance of the amplifier, also limit the gain of the 'Y-amplifier' at high frequencies; for a general purpose C.R.O. the input impedance of the 'Y-amplifier' might well be due to a capacitance of 10 pF in parallel with a resistance of 1 MΩ.

10.2. Power

In Section 3.5 it was shown that if there is a difference in phase ϕ between the current through, and the voltage across, a circuit element then the power P being dissipated is given by $P = (\hat{I}\hat{V}\cos\phi)/2$. Hence an a.c. power meter must respond to the current, the voltage, and the phase difference between them. A dynamometer wattmeter is illustrated schematically in Fig. 10.4: the current passes through Coils 1 and 2 and Coil 3 is connected in parallel with the circuit element. Providing that the impedance of Coil $3 \gg |Z|$ then the torque acting on it, which is proportional to the product of the current through Coils 1 and 2 and that through Coil 3, is given by $\Gamma \propto IV\sin\theta$ or $\Gamma \propto \hat{I}\hat{V}\sin\theta\,\sin\omega t\,\sin(\omega t + \phi)$. The value of Γ averaged over a complete cycle is proportional to $(\hat{V}\hat{I}\cos\theta)/2$ and if the controlling torque is $K\theta$ then $\theta \propto P\sin\theta$: hence such an instrument responds to the true power although not with a linear scale.

Dynamometer wattmeters have a relatively large internal power consumption and so can be used only in situations where the power to be measured is relatively large. This restricts their use to those low frequencies at which large amounts of power are transmitted, notably mains frequencies. At audio- and radio-frequencies, where the available power is comparatively small, $\hat{V}$, $\hat{I}$, and $\cos\phi$ have to be measured separately although sometimes a dummy load which is purely resistive can be substituted for a real load and the power dissipation then measured.

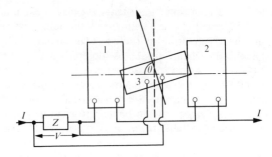

FIG. 10.4. A dynamometer wattmeter to measure the power dissipated in a load of impedance Z.

10.3. Impedance measurements

10.3.1. Bridges

a.c. bridges can be used up to frequencies of 100 MHz although resonance methods are often used at frequencies of 30 kHz and higher. The balance conditions for a basic a.c. bridge were described in Section 3.9 and many permutations of $Z_1 \ldots Z_4$ are possible in principle: however only some of the more common bridge circuits will be described here. A general feature is that inductors are not employed as calibrated, variable impedances since they inevitably have a significant resistance as well as reactance and since they are not readily available in a convenient variable form.

A general practical problem is the existence of stray capacitances between the detector and earth and between the generator and earth (see Fig. 10.5(a)): the existence of such capacitances can cause the measured value of Z_1 to be in error and/or can make the balancing operations tedious and imprecise if their magnitudes are not constant. For instance, if the detector is a set of earphones then a.c. currents can flow between them and the operator's head (presumed to be at earth potential).

In a Wagner earthing system impedances Z_5 and Z_6 are connected in series across the generator and their junction, E, is earthed. The bridge is first balanced with the detector connected to E and then with it connected to M; this procedure is repeated until no further change in either Z_5 or Z_6 or Z_3 or Z_4 is required in order to achieve the balanced condition. Thus M and N are also brought to earth potential and the reactances of the stray capacitances between M and E and N and E are effectively infinite since no a.c. current flows through them. It follows that C_3 and C_4 shunt Z_5 and Z_6 and do not affect the balance condition of the bridge $Z_1 Z_2 Z_3 Z_4$.

Many inductors have relatively large losses (low Q-factor) and a

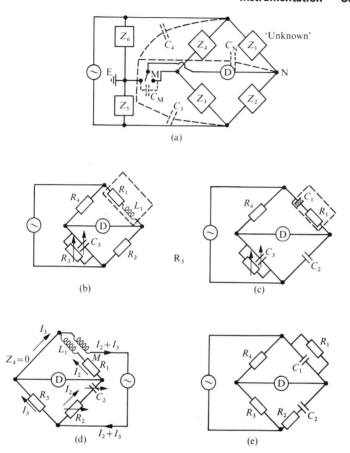

FIG. 10.5. a.c. bridges: (a) a general bridge circuit showing stray capacitances; (b) Maxwell bridge; (c) Schering bridge; (d) Carey–Foster bridge; (e) Wien bridge.

suitable bridge for measuring their inductance and resistance is Maxwell's bridge (Fig. 10.5(b)): from eqn (3.60) the balance conditions are $L_1 = C_3 R_2 R_4$ and $R_1 = R_2 R_4 / R_3$ and the Q-factor is equal to $\omega L_1 / R_1 = \omega R_3 C_3$. Since $R_3 = Q/(\omega C_3)$ its value may be required to be inconveniently large if Q is large. This problem is overcome in the Hay bridge in which R_3 and C_3 are in series and substitution in eqn (3.60) yields $L_1 = R_2 R_4 C_3 / (1 + \omega^2 R_3^2 C_3^2)$. This expression for L_1 is frequency dependent but since $\omega R_3 C_3 \ (= R_1/(\omega_1 L_1)) = 1/Q$ is much less than unity this feature is not usually of great practical significance.

The Schering bridge (Fig. 10.5(c)) finds widespread use both for the precise measurement of capacitance and dielectric loss at low voltages and for the study of insulating structures at high voltages: $C_1 = C_2 R_3/R_4$, $R_1 = C_3 R_4/C_2$. The loss-angle δ of the dielectric of the capacitor is given by $\tan\delta = \omega C_1 R_1 = \omega R_3 C_3$ (see Section 3.6.1).

A Carey–Foster bridge (Fig. 10.5(d)) is designed for the measurement of a coefficient of mutual inductance M in terms of a capacitance. Assuming $Z_4 = 0$ then it follows that, at balance

$$\{I_2 R_2 - j/(\omega C_2)\} = I_3 R_3 \text{ and } I_2(R_1 + j\omega L_1) - (I_2 + I_3)j\omega M = 0.$$

Hence, it follows that $M = R_1 R_3 C_2$ and $L_1 = M(1 + R_2/R_3) = C_2 R_1 (R_2 + R_3)$.

A variety of devices may be used as detector, e.g. vibration galvanometers, earphones, a.c. amplifiers, and indicating meters, C.R.O., 'magic-eye'. The detector must obviously be sensitive enough for the measurement in hand but the use of an oversensitive detector may not only be economically unsound but also may increase the difficulty, and tediousness, of the balancing adjustments.

Vibration galvanometers, which are tuned to resonance at the operating frequency, may be almost as sensitive as moving-coil d.c. galvanometers; their use is confined to frequencies of a few hundred hertz and below. In fact they are usually used at the mains frequencies of 50 or 60 Hz.

Earphones may be used at all audio-frequencies although maximum sensitivity is usually obtained between 1 and 2 kHz.

Amplifiers, which are often tuned to the operating frequency, can be used to amplify the out-of-balance signal: the amplified signal can be displayed on a meter, C.R.O., or 'magic eye' balance indicator. Amplifiers can be designed to have non-linear (for example logarithmic) responses so that sensitivity is increased as the balanced condition is approached.

10.3.2. The 'Q-meter'

The problems associated with stray capacitances and extraneous interference becomes greater with increasing frequency in the case of a.c. bridges; in resonance methods of measuring impedance stray capacitances merely alter the resonant frequency by a small amount and extraneous interference is not a significant problem. The basic circuit of a simple 'Q-meter' which could be used up to 20 MHz say is shown in Fig. 10.6: a known a.c. voltage at a known frequency is provided in a series resonant circuit by passing a standard current I through a small resistance r. The circuit is brought to resonance by adjusting a calibrated capacitor C and the voltage across C (equal to QrI; see Section 4.3) is measured. Thence, since $Q = \omega_0 L/R$ and $\omega_0 = (LC)^{-1/2}$, and C is known, L and R can be calculated. Alternatively an 'unknown' capacitance can be measured if C is adjusted

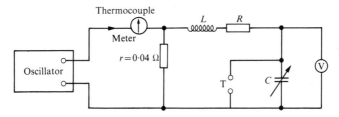

FIG. 10.6. A Q-meter circuit.

to give resonance at some frequency before and after connecting the unknown capacitor between the terminals T. The difference in the two values of C is equal to the capacitance of the unknown capacitor.

10.4. Frequency measurements

Frequency measurements can be made by:
  (i) Comparing one signal with another of known frequency;
  (ii) by balancing a frequency-sensitive bridge; or
  (iii) by counting cycles over a known time interval.

If a sinusoidal voltage of unknown frequency f is applied to the Y-plates of a C.R.O. and another sinusoidal voltage of known variable frequency f_0 to the X-plates (the time-base generator being disconnected), then the spot describes a pattern which is a Lissajous's figure if f_0 is adjusted so that $f = nf_0$ or f_0/n where n is an integer. Lissajous's figures are fascinating in themselves, and constitute a way of comparing frequencies but in practice methods (ii) and (iii) are much more likely to be used, particularly method (iii).

For the Wien bridge (see Fig. 10.5(e)) the balance condition is frequency sensitive: $\omega^2 = 1/(R_1 R_2 C_1 C_2)$ and $C_1/C_2 = R_3/R_4 - R_2/R_1$. Thus the frequency of a signal can be measured in terms of the components of the bridge.

Today frequency measurements up to a few hundred megahertz are most likely to be made by using a commerical 'Timer/Counter'. In basic terms such instruments operate by counting the number of cycles of a signal in a specified time interval (or 'gate time'), the resulting number being displayed on a digital display system; this process is repeated at regular, short, intervals. A typical general purpose instrument might have a frequency range from 2 Hz to 50 MHz with 'gate times' ranging from 1 ms to 10 s, and with a six-digit display panel. A feature of such instruments which can be a disadvantage in certain situations is that the minimum amplitude of signal on which they will operate is relatively high ($\sim$ 10 mV perhaps).

A C.R.O. may be used to make relatively crude, but nevertheless useful, measurements of the frequency of small-amplitude signals, or indeed or large amplitude signals, by counting the number of cycles of the 'Y-voltage' in one period of the (calibrated) time-base generator.

10.5. Interference suppression, noise reduction, and signal enhancement

Interference ('hum' and 'pick-up') and noise are ever-present features of amplifiers, as was mentioned in Section 8.5, and we will describe below, in outline, some of the techniques for the alleviation of these problems. Interference arises largely from man-made sources, particularly mains supply lines, electrical machinery, and radio-frequency radiation from communications systems, but there are also natural sources such as lightning discharges.

10.5.1. Interference suppression

The tracing and elimination of sources of interference in an amplifying system of only moderate complexity can be a very frustrating experience and so the aim here is to discuss some of the more obvious and easily avoidable pitfalls.

Consider a commonly encountered situation where two amplifiers, I and II, are connected in cascade; for instance amplifier I could be a high-gain pre-amplifier, with a transducer connected at its input port, and amplifier II could be a C.R.O. (see Fig. 10.7(a)). The impedance of the signal source will be assumed to be resistive for present purposes and is denoted by R_g, a d.c. supply is used to supply amplifier I (a C.R.O. has its own supply, of course), and the cases of both the C.R.O. and the d.c. supply are connected to earth via the earth wires of their mains leads.

There are three ways, principally, by which interference can occur, namely:

(i) capacitive pick-up;
(ii) inductive pick-up;
(iii) through the non-zero resistance of the earth rail.

As regards case (i) any point of the system which has a p.d. relative to the input line to the amplifier I will be coupled to A_1 via stray capacitance which is denoted by C_s; the problem of the capacitance between wires and between a wire and a conducting plane was discussed in Section 1.1.2. The fraction of the p.d. which appears across the input terminals $A_1 C_1$ of the amplifier I is equal to $R/(R - j/\omega C_s)$ where R is the parallel combination of R_g and the input resistance pf the amplifier I; if the p.d. varies at the mains frequency of 50 Hz and C_s is 10 pF say then, if $R = 10 \text{ k}\Omega$, this fraction $\approx 3 \times 10^{-5}$. Hence, if the amplitude of the p.d. is ~ 1 V then an interference signal of amplitude

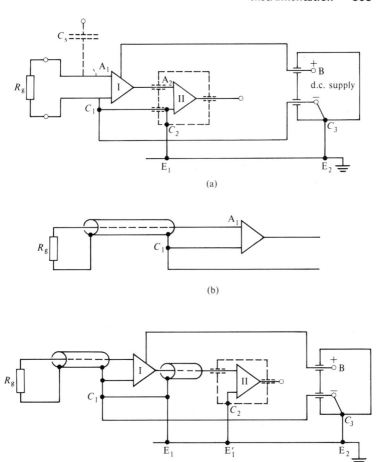

FIG. 10.7. (a) A two-stage amplifier showing the power supply lines, earth line, and a source of capacitive pick-up. (b) Capacitive pick-up at the input can be eliminated largely by using a coaxial cable connection. (c) The elimination of the ground loop $C_3 C_1 C_2 E_1 E_2 C_3$ of system (a).

$\sim 30\,\mu\text{V}$ appears across the input terminals of the amplifier I. Obviously the situation becomes worse the larger are the source and amplifier impedance and the higher is the frequency of the interfering signal. If the input lead is shielded by a metal screen, by using coaxial cable for instance, as shown in Fig. 10.7(b), then this source of interference is eliminated; it may be necessary to screen the signal source/transducer also.

Magnetic pick-up (case (ii)) can occur wherever there is a closed conducting loop in the system; an alternating current flowing in one part of the system generates an alternating magnetic field which induces a current in any closed path with which it is linked. Since the effect is related to the flow of current it is of more significance in low impedance paths in contrast to capacitive pick-up which is more significant in relation to high impedance paths. An obvious path in the system of Fig. 10.7(a) is $C_3 C_1 C_2 E_1 E_2 C_3$. The principal, but not the sole, source of magnetic pick-up is likely to be currents at the mains frequency flowing in power supply lines and power supply transformers for instance. So it is a wise procedure to give as much physical separation as possible between power lines and sensitive parts of the system, particularly the transducer and amplifier I. Pick-up of this type may be reduced also by arranging for suitable relative orientations of the relevant parts of the system, particularly a mains transformer, to reduce magnetic flux linkage, e.g. the coefficient of mutual inductance is zero between two loops whose planes are perpendicular. So-called ground loops, of which $C_3 C_1 C_2 E_1 E_2 C_3$ is an example, can be eliminated by 'breaking' the loop as shown in Fig. 10.7(c) where the screen of the coaxial cable linking amplifier I to amplifier II is connected to earth at one end only. Screens made from high permeability materials, such as μ-metal, can be used to provide magnetic shielding at low and audio frequencies through the agency of the skin-effect (see Section 5.4.1) and at higher frequencies, above 10 kHz, effective screening is provided by sheets or meshes of ordinary good metallic conductors such as copper, brass, and aluminium.

A different source of interference (case (iii)) associated with ground loops arises from the fact that there is resistance (and inductance and capacitance) associated with the ground and common lines; for instance for 14 S.W.G. copper wire (2 mm diameter approx.) the resistance is about $10^{-2} \, \Omega \, m^{-1}$ and about $1 \, \Omega \, m^{-1}$ for a typical copper track on a printed circuit board. If a 50 Hz current is flowing in the earth line then there will be a p.d. between E_1 and E_2 and a fraction of this p.d. will appear between C_1 and C_2 and hence will be amplified; the remedy for this problem is to bring points E_1 and E_2 (and E_1') together to the same point of the earth rail.

10.5.2. Noise suppression and signal enhancement

For 'white' noise (thermal noise, shot noise) the mean noise power is proportional to the bandwidth of the amplifying system. Hence, other things being equal, in order to reduce the noise level at the output of an amplifier the bandwidth must be reduced although it should be remembered that as the bandwidth is reduced the response time of the amplifier increases. Although a comprehensive discussion of the design of filters does not lie within the aims of this text we have given an outline of the basic features of filters in Chapter 7. In passive networks the Q-factors of $L-C-R$ networks are ~ 100 at best which gives a passband/stop-band width of 1 MHz at a centre frequency of 100 MHz and at microwave frequencies the Q-factors of resonant cavities can be $\sim 10^3$ which gives a bandwidth of 10 MHz at a centre frequency of 10 GHz. For frequencies below about 100 kHz active filters can have better characteristics than passive networks.

In a situation where the signal-to-noise ratio is poor, but where the signal is coherent with a periodic reference signal (i.e. where there is a fixed phase relationship between the signal of interest and the periodic reference signal) then the signal-to-noise ratio may be enhanced by means of a synchronous rectifier.† A common example of the type of situation where a synchronous rectifier could be employed to advantage is where the measurand is the intensity of a weak light beam. The light beam incident on the photodetector is modulated by a rotating blade, or light 'chopper', which intercepts the beam periodically at a frequency f_0 say (see Fig. 10.8). The reference signal, which is also modulated at frequency f_0, operates an electronic switch‡ which connects the electrical signal to points T_1, T_2 alternatively. The form of the voltage at P depends on the phase of the reference relative to the signal. For illustrative purposes let us suppose that the electrical signal is sinusoidal and in phase with the reference signal (see Fig. 10.9). During a positive half-cycle of the reference the switch connects the signal to T_1 and to T_2 during a negative half-cycle; however because of the action of the inverter the waveform of the signal at P as shown in Fig. 10.9(c). A low-pass filter, bandwidth f_b say, produces a smoothed d.c. output signal of magnitude proportional to the input signal (see Fig. 10.9(d)). Any signals, and particularly incoherent noise and interference signals, which have frequencies greater than f_b average to zero, effectively, over a period $\sim 1/f_b$ and herein lies the most important feature of the system since f_b can be made very small in relative terms.

In typical situations a value for f_b/f_0 of 10^{-3} can be arranged comfortably and if $f_0 = 1$ kHz say, then the averaged magnitudes of noise and interference components having frequencies greater than 1 Hz will be insignificant.

An additional factor which affects the chosen value for f_b, however, is the desired response time for the system which is $\sim 1/f_b$; in practice a compromise has to be made between the noise bandwidth and the response time.

If there is a phase difference ϕ between the reference and the signal,

---

† Also referred to, commonly, as a 'phase sensitive detector' or 'lock-in' amplifier.
‡ There are many different ways, in practice, in which switching and inversion can be realized.

308 Instrumentation

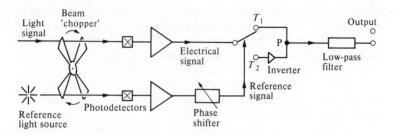

FIG. 10.8. A block diagram of a synchronous rectifier system for detecting the variations in intensity of a beam of light.

then (see Fig. 10.9(e)), for a sinusoidal signal

$$V_{P_{average}} = \frac{1}{(1/f_0)} \int_{t=\phi/2\pi f_0}^{[(1/2f_0)+(\phi/2\pi f_0)]} \hat{A} \sin 2\pi f_0 t \, dt$$

$$= \frac{\hat{A}}{2\pi} \int_{\phi}^{\phi+\pi} \sin 2\pi f_0 t \, d(2\pi f_0 t)$$

or

$$V_{P_{average}} = \hat{A} \frac{\cos \phi}{\pi}.$$

For a signal in the form of a square wave of amplitude $\hat{A}$ having a phase difference of ϕ with respect to the reference then the average value of the voltage at P is

$$V_{P_{average}} = \hat{A}\left(1 - \frac{2\phi}{\pi}\right).$$

The results of these two cases illustrate the point that the magnitude of the d.c. output signal in zero if there is a phase difference of $\pi/2$ between the signal and reference and that the sign of the output is reversed if $\phi = \pi$.

It is obvious that noise and interference components which are contained in the signal before it is modulated will not be averaged out. In order that large amplitude interference signals, from the mains for example, should not saturate the amplifier it is common practice to insert a band-pass filter, tuned to the frequency f_0, between the transducer and the amplifier. So, to summarize, in situations where this technique can be applied a very narrow bandwidth can be obtained but, as always, there is a price to be paid; in this case it is that changes in

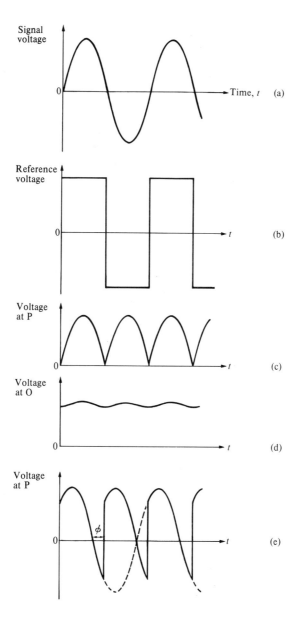

FIG. 10.9. The signal and reference waveforms at various points in the synchronous rectifier system of Fig. 10.8.

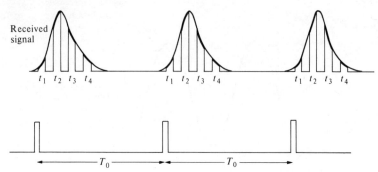

FIG. 10.10. The basic principle of signal averaging; a repetitive signal, of period T_0, is sampled a number of times (here shown as only four) in each interval T_0.

signal level must occur at frequencies lower than f_b otherwise they will be filtered out.

Another technique of enhancement for signals buried in noise, which has become widely available with the advent of cheap digital circuits, is that of signal averaging; this technique can be used when the time of occurrence of the signal is known. The stimulus (a radar pulse directed to a planet say) is repeated at regular intervals, with period T_0 say, and the receiver samples the received signal at a number of times in each interval T_0. In Fig. 10.10 the number of sampling channels per interval is shown as four but in a digital averager the number could be 1000; note that in the figure the signal is shown without the obscuring noise, for clarity. For each channel the successive sampled values of the signal plus noise are added and the cumulative result is stored. As a result of these successive additions the noise components tend to average out whilst the signal builds up. After an appropriate time the contents of the store associated with each channel can be read to give a display of the signal as a function of time. Again the better the signal-to-noise ratio that is wanted the longer it will take to achieve; this parallels the bandwidth/response time relationship of the synchronous rectifier system.

Appendix 1. Colour codes for the values of resistors and capacitors

Resistors

THE first two digits, and the number of following zeros, in the numerical value of a resistor are indicated by coloured bands on the body of the resistor:

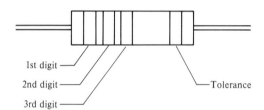

FIG. A1.1.

Colour	Number
Black	0
Brown	1
Red	2
Orange	3
Yellow	4
Green	5
Blue	6
Violet	7
Grey	8
White	9

e.g. a 3900 Ω resistor would be described by the colours

1st coloured band	Orange
2nd coloured band	White
3rd coloured band	Red (number of zeros)

In addition the manufacturer's tolerance on the value of the resistor is indicated by a further coloured band:

Colour codes

Colour	Tolerance (per cent) B
Brown	1
Red	2
Gold/yellow	5
Silver	10
No additional colour	20

The power rating and maximum working value of a capacitor are given on the body of some resistors but in general it will be necessary to consult the manufacturer's specifications in order to obtain this information.

Capacitors

Many capacitors are large enough for the value of capacitance, working voltage, and tolerances to be given on the body of the capacitor. In other cases a system of coloured bands, or 'spots', is used with the same colour code as with resistors. However there is less uniformity than with resistors in the manner of coding between types and makes, and manufacturer's information sheets may have to be consulted.

It should be pointed out that it is becoming increasingly common practice to indicate the values of resistors and capacitors as follows:

Resistors	*Capacitors*
4k7 (4700 Ω)	68 p (68 pF)
5M6 (5.6 MΩ)	4n7 (4700 nF)
	6μ8 (6.8 μF)

Appendix 2. Partial fractions—general rules

$$P(x) = \frac{N(x)}{D(x)}.$$

Cases

(i) Degree of $N(x)$ < degree of $D(x)$.
(ii) Degree of $N(x)$ ⩾ degree of $D(x)$.

Case (i)

(a) To every *linear* factor $(a_1 x + b_1)$ of $D(x)$ there will be a corresponding partial fraction $A_1/(a_1 x + b_1)$.

(b) To every quadratic factor $(a_2 x^2 + b_2 x + c_2)$ of $D(x)$ there will be a corresponding partial fraction $(A_2 x + B_2)/(a_2 x^2 + b_2 x + c_2)$.

(c) To every repeated linear factor $(a_3 x + b_3)^2$ of $D(x)$ there will be two corresponding partial fractions

$$\frac{A_3}{(a_3 x + b_3)} + \frac{B_3}{(a_3 x + b_3)^2}$$

(d) To every repeated quadratic factor $(a_4 x^2 + b_4 x + c_4)^2$ of $D(x)$ there will be two corresponding partial fractions

$$\frac{(A_4 x + B_4)}{(a_4 x^2 + b_4 x + c_4)} + \frac{(C_4 x + E_4)}{(a_4 x^2 + b_4 x + c_4)^2}$$

(e) To every thrice-repeated linear factor $(a_5 x + b_5)^3$ of $D(x)$ there will be three corresponding partial fractions

$$\frac{A_5}{(a_5 x + b_5)} + \frac{B_5}{(a_5 x + b_5)^2} + \frac{C_5}{(a_5 x + b_5)^3}.$$

(f) To every cubic factor $(a_6 x^3 + b_6 x^2 + c_6 x + d_6)$ of $D(x)$ there will be a corresponding partial fraction

Partial fractions

$$\frac{(A_6x^2 + B_6x + C_6)}{(a_6x^3 + b_6x^2 + c_6x + d_6)}$$

and so on.

Case (ii)

(a) If $N(x)$ be of the same degree as $D(x)$, then A will be added to the partial fractions as given in case (i).

(b) If $N(x)$ be one degree higher than $D(x)$, then $(Ax + B)$ will be added to the partial fractions as given by case (i).

(c) If $N(x)$ be two degrees higher than $D(x)$, then $(Ax^2 + Bx + C)$ will be added to the partial fractions as given by case (i); and so on.

Appendix 3. Maxwell's equations

$$\nabla \cdot \mathbf{D} = \rho \qquad \nabla \cdot \mathbf{B} = 0$$

$$\nabla \times \mathbf{E} + \frac{d\mathbf{B}}{dt} = 0 \qquad \nabla \times \mathbf{H} - \frac{d\mathbf{D}}{dt} = \mathbf{J}.$$

To find the relationships between the components of the electric and magnetic fields in an electromagnetic wave propagating in the H_{10} mode in a rectangular waveguide we first write Maxwell's equations in their 'free space' form. Using $\mathbf{B} = \mu_0 \mathbf{H}$ we have

$$\nabla \cdot \mathbf{E} = \frac{\rho}{\epsilon_0} \quad \text{(i)} \qquad \nabla \cdot \mathbf{H} = 0 \quad \text{(ii)}$$

$$\nabla \times \mathbf{E} + \mu_0 \frac{d\mathbf{H}}{dt} = 0 \quad \text{(iii)} \qquad \nabla \times \mathbf{H} - \epsilon_0 \frac{d\mathbf{E}}{dt} = 0 \quad \text{(iv)}$$

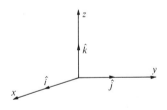

FIG. A3.1.

Eqn (iii) can be written in terms of cartesian components as

$$\hat{\mathbf{i}}\left(\frac{dE_z}{dy} - \frac{dE_y}{dz}\right) + \hat{\mathbf{j}}\left(\frac{dE_x}{dz} - \frac{dE_z}{dx}\right) + \hat{\mathbf{k}}\left(\frac{dE_y}{dx} - \frac{dE_x}{dy}\right)$$

$$= -\mu_0 \frac{d}{dt}(\hat{\mathbf{i}}H_x + \hat{\mathbf{j}}H_y + \hat{\mathbf{k}}H_z) \qquad \text{(v)}$$

316 Maxwell's equations

where $\hat{i}, \hat{j}, \hat{k}$ are vectors of unit magnitude in the directions of the x-, y-, and z- axes respectively.

Since the x-, y-, and z-components of each side of eqn (v) must be equal, separately, we have, in particular, that for the z-components,

$$\frac{\partial E_y}{\partial x} - \frac{\partial E_x}{\partial y} = -\mu_0 \frac{\partial H_z}{\partial t}.$$

For the particular choice of cartesian axes which was made in Chapter 8 to define the polarization, and direction of propagation, of the electromagnetic wave, $E_x = 0$, and so we have

$$\frac{\partial E_y}{\partial x} = -\mu_0 \frac{\partial H_z}{\partial t}.$$

Also, for the x-components in eqn (v)

$$\mu_0 \frac{\partial H_x}{\partial t} = \frac{\partial E_y}{\partial z}$$

since $E_z = 0$ for the H_{10} mode. So on taking into account the time dependence of H_x we have

$$H_x = -\frac{j}{\omega \mu_0} \frac{\partial E_y}{\partial z}.$$

From eqn (iv)

$$\hat{i}\left(\frac{\partial H_z}{\partial y} - \frac{\partial H_y}{\partial z}\right) + \hat{j}\left(\frac{\partial H_x}{\partial z} - \frac{\partial H_z}{\partial x}\right)$$

$$+ \hat{k}\left(\frac{\partial H_y}{\partial x} - \frac{\partial H_x}{\partial y}\right) = \epsilon_0 \frac{\partial}{\partial t}(\hat{i}E_x + \hat{j}E_y + \hat{k}E_z)$$

and for the y-components

$$\frac{\partial H_x}{\partial z} - \frac{\partial H_z}{\partial x} = \epsilon_0 \frac{\partial E_y}{\partial t}$$

or

$$\frac{\partial H_x}{\partial z} - \frac{\partial H_z}{\partial x} = j\omega\epsilon_0 E_y.$$

Answers to problems

1.1. 4.3×10^5 J; 29 kV
1.3. $0.17 \, \Omega$; 71 C
1.6. $M = (\mu_0/4\pi)2a \ln\{1 + (b/d)\}$
2.1. 1 kV; 100 Ω
2.2. $R_3 \geqslant 240$ kΩ
2.5. 'I_0' = 8/19 A; 'R_0' = 19 Ω
2.6. (a) 'E_0' = 12 V; 'R_0' = 4 Ω. (b) 'I_0' = 3 A; 'R_0' = 4 Ω
2.7. (a) $(E_1 R_2 + E_2 R_1)/(R_1 + R_2)$; $R_1 R_2/(R_1 + R_2)$ in series.
  (b) $\{(E_1/R_1) + (E_2/R_2)\}$; $R_1 R_2/(R_1 + R_2)$ in parallel
2.8. 0.21
2.9. Thévenin: 1.46 V; 3.1 kΩ. Norton: 0.48 mA; 3.1 kΩ
2.10. 6-V battery: 3 W. 2-V battery: 1 W

3.2. 1 mA
3.3. (a) CR_2. (b) $C(R_1 + R_2)$
3.4. $CRR_g/(R + R_g)$
3.6. (a) E. (b) E/R_2. (c) $L/(R_1 + R_2)$
3.7. 5 A; $\tan^{-1}(0.75)$
3.8. $(\phi_1 + \phi_2)$
3.9. $2 \cos(\omega t - 60°)$
3.10. $i = \hat{V}\{(1/R^2) + (1/\omega^2 L^2)\}^{1/2} \sin(\omega t - \phi)$
  where $\phi = \tan^{-1}(R/\omega L)$.
3.11 334 V; $-72.6°$
3.12 15.7 mA; $-81°$
3.13 Power factor = 0.84; power = 41 W

Answers to problems

3.14. (a) $100\,\Omega$. (b) $55\,\mu F$

3.15. $10^{-1}\sin(1000t - 2\pi/3)$; $0.112\sin(1000t - 63.4°)$

3.16. $G = (2R^2 + \omega^2 L^2)/\{R(R^2 + \omega^2 L^2)\}$;
$B = (\omega L)/(R^2 + \omega^2 L^2)$

3.17. $\hat{V} = \hat{I}/\{(1/R^2) + \omega^2 C^2\}^{1/2}$; $\tan\phi = -(\omega RC)$

3.18. $G = R(R^2 + \omega^2 L^2)^{-1}$;
$B = \omega C\{R^2 - L/C + \omega^2 L^2\}(R^2 + \omega^2 L^2)^{-1}$

3.19. 0.954; 262 W

3.20. 3.2×10^{-3}; 29 mW

3.22. $L_{eff} = 1.11\,\text{mH}$; $R_{eff} = 278\,\Omega$

3.24. $(1 + \omega^2 C^2 R^2)^{-1/2}$; $\tan^{-1}(-\omega CR)$

3.25. $Y = 1/R$

3.27. $2.70\,\underline{/-59.1°}$ A; $8.1\,\underline{/-57.2°}$ A

3.28.

$C' = L$; $G' = 1/R$; $L' = C$

$L' = C$; $G' = 1/R$; $C' = L$

3.29. $0.16\,\underline{/-61.3°}$

3.30. $0.72\,\underline{/-23.7°}$ A; $(2.54 + j1.67)\,\Omega$

3.31. $4.47\,\underline{/63.4°}$ V; $(6 + j2)\,\Omega$

3.32. $1.59\,\underline{/5.7°}$ A; 15.8 W

4.2. $Y = R/(R^2 + \omega^2 L^2) + j\omega[C - \{L/(R^2 + \omega^2 L^2)\}]$

4.3. 2.5 nF; 0.25 pF; $2\pi\,\text{k}\,\Omega$; $200\pi\,\text{k}\,\Omega$

4.4. $X = (1 - \omega^2 LC)/\{\omega C(\omega^2 LC - 2)\}$;
$\omega_0^2 = 1/(LC)\ (X = 0)$; $\omega_0^2 = 2/LC\ (B = 0)$

4.5. $C = 6.3\,\text{nF}$; $R = 4.3\,\Omega$; $2.5\,\text{k}\,\Omega$

5.2. $\tau = 100\,\mu s$; $\epsilon' = 4.96\ (160\,\text{Hz})$; $1.10\ (160\,\text{kHz})$

5.3. 8×10^{-4}; 8×10^{-3}; 6.4×10^{-1}; 3.5×10^{-2}; 3.5×10^{-2}

5.5. 9.4 mm; 0.94 mm; 94 μm; 9.4 μm; 0.94 μm

6.4. $\hat{V}/2$; $\hat{V}/2$; $\hat{V}/2\sqrt{3}$

6.6. $v(t) = (AL/R)(1 - e^{-\frac{Rt}{L}})$

6.7.

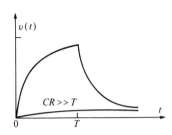

7.3. $h_{11} = j2\,\Omega; h_{12} = 2; h_{21} = -2; h_{22} = j\dfrac{3}{2}\,\text{S}$

7.7 Series elements $2R_0/3 = 33\,\Omega$
Shunt elements $5R_0/12 = 21\,\Omega$

7.8 $y_{11} = y_{22} = (7+j6)/10\,\Omega$
$y_{21} = y_{12} = -(3+j4)/10\,\Omega$

7.9
$$\begin{bmatrix} \left(Z_1 + R_1 - \dfrac{j}{\omega C}\right) & 0 & \dfrac{j}{\omega C} & -R_1 \\ 0 & \left(Z_2 + j\omega L_1 - \dfrac{j}{\omega C}\right) & \dfrac{j}{\omega C} & -j\omega L_1 \\ \dfrac{j}{\omega C} & \dfrac{j}{\omega C} & \left(R_2 - j\dfrac{2}{\omega C}\right) & 0 \\ -R_1 & -j\omega L_1 & 0 & (R_1 + j\omega L_1) \end{bmatrix} \begin{bmatrix} i_1 \\ i_2 \\ i_3 \\ i_4 \end{bmatrix} = \begin{bmatrix} V \\ 0 \\ 0 \\ 0 \end{bmatrix}$$

'Δ'

$$i_2 = \dfrac{\begin{bmatrix} \left(Z_1 + R_1 - \dfrac{j}{\omega C}\right) & V & \dfrac{j}{\omega C} & -R_1 \\ 0 & 0 & \dfrac{j}{\omega C} & -j\omega L_1 \\ \dfrac{j}{\omega C} & 0 & \left(R_2 - \dfrac{j2}{\omega C}\right) & 0 \\ -R_1 & 0 & 0 & (R_1 + j\omega L_1) \end{bmatrix}}{'\Delta'}$$

7.10.
$$\begin{bmatrix} \left(\dfrac{1}{Z_1} + \dfrac{1}{R_2} + j\omega C_2\right) & -j\omega C_2 & -\dfrac{1}{R_2} & 0 \\ -j\omega C_2 & \left(\dfrac{1}{R_1} + j\omega C_2 + j\omega C_3\right) & 0 & -j\omega C_3 \\ -\dfrac{1}{R_2} & 0 & \left(\dfrac{1}{R_2} + \dfrac{1}{R_3} + j\omega C_1\right) & -\dfrac{1}{R_3} \\ 0 & -j\omega C_3 & 0 & \left(\dfrac{1}{Z_2} + \dfrac{1}{R_3} + j\omega C_3\right) \end{bmatrix} \begin{bmatrix} V_1 \\ V_2 \\ V_3 \\ V_4 \end{bmatrix} = \begin{bmatrix} \dfrac{V_1}{Z_1} \\ 0 \\ 0 \\ 0 \end{bmatrix}$$

320 Answers to problems

7.11. 41 μV

7.12. $\begin{bmatrix} 5/3 & 2 \\ 2 & 3 \end{bmatrix}$

7.18. $s/\{L(s^2 + 1/LC)\}$

7.19. Internal zero $\omega = 1$; internal pole $\omega = \sqrt{2}$; External zero $\omega = \infty$; external pole $\omega = 0$

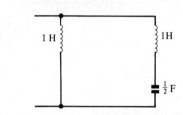

7.21.

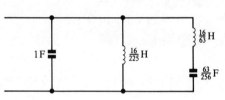

7.22.

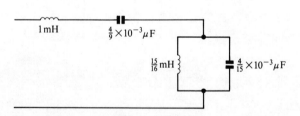

7.23. External pole at zero; external pole at ∞; internal pole at $\omega = \sqrt{2}$; internal zeros at $\omega = 1; \sqrt{3}$

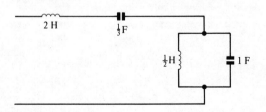

Answers to problems 321

7.24. External zeros at zero and ∞; internal zero at $\omega = 3$; internal poles at $\omega = 2; 4$

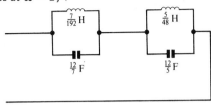

7.25.

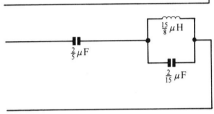

7.26. Zeros: $s = -(CR_C)^{-1}; -(R_L/L)$;
poles: $s = -(R_C + R_L)/2L \pm j\,[(1/LC) - \{(R_C + R_L)^2/4L^2\}]^{1/2}$; non-oscillatory for $(LC)^{-1} \leqslant (R_C + R_L)^2/4L^2$; limiting condition: $R_C \geqslant 1990\,\Omega$

7.27.

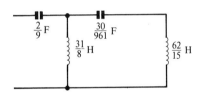

7.28.

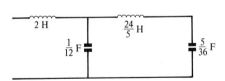

7.29. External poles at zero and ∞; internal pole at $\pm\sqrt{2}$; internal zeros at $\pm 1; \pm\sqrt{3}$

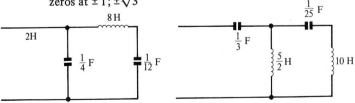

Answers to problems

8.1. (a) 50 Ω. (b) 1 rad m^{-1}. (c) 2π m. (d) j50 Ω

8.2. $Z_0 = 45\,\Omega; \alpha = 2.2 \times 10^{-4}; \beta = 0.45\,\text{rad m}^{-1};$
$\lambda = 14.1\,\text{m}; v_{ph} = 2.2 \times 10^8\,\text{m s}^{-1}$

8.3. 29 db

8.4. $0.42 \underline{/-95°}$

8.5. 1.1 m

8.6. $(0.72 + j1.09)$

8.7. $(0.45 - j0.12)$

8.8. 18.3 cm; 55.8 cm; 93.3 cm; ...

Index

active devices, equivalent circuits, 189
admittance, 56
 driving point, 200
 parameters, 170
ampere, the, 16
amplifier, bandwidth of, 195
 differential, 297
 linear 193
 noise of, 195
angular frequency, 47
Argand diagram, 48
attenuator, matched, 185

Bode plot, 211
boundary conditions, for electric field, 108
bridge (a.c.), 76, 300
 product arm, 78
 ratio arm, 78
bridge, Wheatstone, 31, 36, 39, 279, 280

capacitance,
 definition, 4
 self, 59
 stray, 12, 59, 164
capacitor, 5, 22
 concentric cylinder, 5, 283
 electrolytic, 7, 114
 parallel-plate, 5
Cauer network, 221
characteristic impedance,
 of a transmission line, 234
 of a two-port network, 174
 of rectangular waveguide, 259
charge,
 conservation of, 1
 image, 11
 of electron, 1

circuital law, Ampere's, 18, 117, 122
coaxial line, 7, 107, 231
complementary function, 90
complex frequency, 147
complex number, 48
 argument of, 49
 conjugate of, 54
 modulus of, 49
 rationalization of, 54
complex plane, 48
conductance, 32, 56
constant current generator, 24
constant voltage generator, 23
Coulomb's law, 1
coupling coefficient of a transformer, 63
coupling factor of a cavity, 268
Cramer's rule, 29
critical damping, 88, 205
critically coupled resonant circuits, 96
critical wavelength in waveguide, 258
current coupling, 75

d'Arsonval meter, 37
decibel, 194
delta function, 145, 206
dielectric constant, 6, 104
dielectrics, electrical properties of, 100, 112
dipole moment,
 electric, 100
 magnetic, 120
directional coupler, 265
displacement, electric, 104
dual networks, 56

eddy currents, 19, 61, 130, 258
electrometer, 297

Index

electromotive force,
    definition, 13
    induced, 17, 42
    sources of, 23
energy density,
    in a magnetic field, 20
    in an electric field, 14
energy (potential), 2, 4
equipotential surface, 9, 11
equivalent circuits, 34, 189
evanescent mode (in waveguide), 257, 263

ferrites, 19, 61, 128, 130, 132, 187
ferroelectric materials, 102
ferromagnetic materials, 121, 128
field,
    conservative, 3
    electric, 2
    magnetic, 15
    non-electrostatic, 13
filters, 215
flux,
    electric, 3, 4
    magnetic, 17, 61
force, velocity-dependent, 15
Foster networks, 220
Fourier series, complex form, 139
Fourier's theorem, 133
Fourier transformation, 144

Gauss' law, 3
Gaussian surface, 7, 107

Hall effect, 291
hybrid junction, 265
hybrid (h) parameters, 171, 191
hysteresis, 61, 128, 275

immittance, 200
    driving point, 217
impedance, 52
    acoustic, 285
    driving point, 200
    normalized, 213
    reflected, 65
impedance (z-) parameters, 171
inductance, mutual, 17, 21
inductance, self,
    of a coaxial line, 118
    of a straight wire, 117
    of coils, 119
    of a twin line, 118
inductor, 18, 22

Kirchhoff's laws, 26

Laplace transform, 145, 151
left-hand rule, 16
Litz wire, 119
logarithmic decrement, 88
loss angle, 60, 112, 131
losses,
    'copper', 61, 67, 131
    dielectric, 59, 109, 112
    eddy current, 130
    hysteresis, 130
    'iron', 61, 67
    ohmic, 99
lumped circuit component, 170, 230

magnetic circuit, 124
magnetic field, 15
magnetic flux, 17, 61
magnetization, 120
magnetomotive force, 125
matrix parameters,
    a-, 178
    conversion table, 183
    h-, 171
    y-, 170
    z-, 171
Maxwell's equations, 257, 315
measurand, 272
microphone,
    condenser, 283
    electret, 284
    ribbon, 280
microwave,
    attenuator, 265
    cavity, 265, 269
    phase-shifter, 265
modes (in waveguide), 257
multimeter, 38

neper, 236
network,
    Cauer, 221
    dual, 56
    Foster, 220
    high-pass, 76, 155, 162, 209
    ladder, 221
    linear, 170
    low-pass, 76, 158, 162
    passive, 170, 189
    phase-shift, 70, 172
    potential divider, 30
    symmetrical, 174
    T-. 172
    twin-T, 179
    two-port, 169

Index

noise,
  in amplifiers, 195, 306
  figure, 196
non-linear circuit element, 54
non-ohmic materials, 15
normalized frequency, 213
Norton's theorem, 37

ohmmeter, 38
Ohm's law, 14
oscillations,
  damped, 86
  forced, 89
overdamped circuit, 88, 205

particular integral, 90
permeability, relative, 19, 120, 128
permittivity,
  complex, 112
  of free space, 1
  relative, 6, 100, 104
phase angle, 46
phase characteristic, 245
phasor, 46
phase-shift network, 70
polarization, 100
poles, of a system function, 200
potential difference, 2
potential divider network, 30
potentiometer, 31, 35, 277
power, 13, 24, 57, 299
  complex, 258
  factor, 58
proximity effect, 11, 119
pulsatance, 47

quadrature, 53
quality factor, 61, 88
  loaded, 268
  mechanical, 286
  unloaded, 268
quarter-wavelength transformer, 248

reactance, 53
reciprocity theorem, 34
relative permittivity, 6, 100, 104
  complex, 112
reluctance, 125
resistance, 14
  dynamic, 14
resistance thermometer, 278
resistor, 15, 22
  non-inductive, 22
resonance, 91
root mean square value, 57

saturable reactor, 128
scalar product, 3

search coil, 290
shifting theorem, 150
shunt, 37
  universal, 37
signal averaging, 310
skin depth, 132, 258
skin effect, 61, 115
sliding screw matching unit, 264
Smith chart, 248
solenoid, 18, 20
step function, 145
strain gauge, 279
stripline, 231, 269
stub line, 255
superposition theorem, 34
susceptance, 56
susceptibility,
  electric, 103
  magnetic, 124
synchronous rectifier, 307
system function, 169, 198

Telegrapher's equations, 233, 261
thermistor, 279
thermocouple, 13, 273, 296
Thévenin's theorem, 24
time constant, 44
  of a dielectric, 100
transformer, 19, 61
  'dot' convention for, 65
  linear variable differential, 289
  magnetization current in, 66
  quarter wavelength, 248
transistor, 189
transmission lines, properties of, 246
transmission (a-) parameters, 178
transmittance, 200
twin line, 8, 246
twin-T network, 179

unit impulse, 145, 206
universal resonance curve, 210

Van de Graaff generator, 12
voltage coupling, ideal, 74
voltage standing wave ratio, 241, 265, 271

Wagner earth, 300
wattmeter, 299
waveguide, 231, 256
  circuit elements, 264
Wheatstone bridge, 31, 36, 39, 279, 280

zeros of a system function, 207